# The Fabric of Space

# The Fabric of Space

Water, Modernity, and the Urban Imagination

Matthew Gandy

The MIT Press
Cambridge, Massachusetts
London, England

© 2014 Massachusetts Institute of Technology

All rights reserved. No part of this book may be reproduced in any form by any electronic or mechanical means (including photocopying, recording, or information storage and retrieval) without permission in writing from the publisher.

MIT Press books may be purchased at special quantity discounts for business or sales promotional use. For information, please email special_sales@mitpress.mit.

This book was set in Bembo by the MIT Press. Printed and bound in the United States of America.

Library of Congress Cataloging-in-Publication Data

Gandy, Matthew.
 The fabric of space : water, modernity, and the urban imagination / Matthew Gandy.
    pages cm
 Includes bibliographical references and index.
 ISBN 978-0-262-02825-7 (hardcover : alk. paper)
1. Water resources development. 2. Sewerage. 3. Water use—Social aspects. 4. Municipal water supply—Economic aspects. 5. City and town life. 6. Urban ecology (Sociology). I. Title.
 TC405.G36 2014
 333.91'15—dc23

2014013240

10  9  8  7  6  5  4  3  2  1

# Contents

Preface  vii

Introduction  1

1  The Paris Sewers and the Rationalization of Urban Space  27

2  Borrowed Light: Journeys through Weimar Berlin  55

3  Mosquitoes, Modernity, and Postcolonial Lagos  81

4  Water, Poverty, and Urban Fragmentation in Mumbai  109

5  Tracing the Los Angeles River  145

6  Fears, Fantasies, and Floods: The Inundation of London  185

Epilogue  217

Notes  225

Selected Bibliography  303

Index  339

Preface

While doing research in New York City for my earlier book *Concrete and Clay*, I visited an exhibition of work by the nineteenth-century French photographer Félix Nadar held at the Metropolitan Museum of Art. I already knew some of his photographs, including his own striking self-portrait dating from 1855, but there was one room especially that caught my attention, with an unfamiliar set of images taken mostly beneath the streets of Paris. The photographs of the newly completed sewer system from Haussmann-era Paris fascinated me in particular because they confounded my expectations of what the underground city might have looked like. These pristine spaces appeared to mimic aspects of "scientific urbanism," yet the photographs were taken in the early 1860s, before the epidemiological aspects to public health were fully understood let alone widely accepted. This underground world was the setting for an intriguing encounter between the new optical preoccupations of modernity and the latest advances in engineering science.

Over subsequent years my fascination with the intersections between water and the technological spaces of modernity has evolved to encompass a range of developments such as changing attitudes toward the body, the politics of water and sanitation, and the symbolic significance of water for modern cities. If we trace the flow of water through urban space, the metaphorical and topographic dimensions to cities take on new and sometimes unexpected dimensions. We encounter not only entanglements between human agency and the material reconstruction of cities but also unpredictable aspects to nonhuman agency, such as the epidemiological dimensions to the different hydrological terrains of modernity or the unforeseen properties of construction materials. More recently, the question of climate change, and its impact on cities, has also raised issues that cut across disciplinary boundaries and posed dilemmas that enter the realm of science fiction.

My study of water has extended to six cities in particular—Paris, Berlin, Lagos, Mumbai, Los Angeles, and London—and in each case I have sought to explore a different facet of the relationship between water, modernity, and the urban imagination. The book combines different kinds of research materials, including archives, interviews, and ethnographic observations. During the course of my research I visited the Paris sewers and found a well-organized municipal spectacle not so unlike the encounters of early visitors in the 1860s; I swam in Berlin's lakes and felt the warmth of the sun and the cool shade of pine trees; I stepped across open drains in Lagos next to concrete surfaces festooned with faded posters; I noticed the maps and diagrams on the walls of engineering offices in Mumbai and reflected on the degree to which the cartographic imagination shapes urban consciousness; I wandered along the concrete landscapes of the Los Angeles River and noticed dramatic changes within the space of a few years; and I marveled at the afternoon sun glinting on the steel-clad gates of the Thames Barrier.

Many people have helped me over the years with my research on water and urban infrastructure. I would like to mention in particular Nikhil Anand, Bayo Anatola, Johan Andersson, Pushpa Arabindoo, Tunde Atere, Karen Bakker, Stephen Barber, Andrew Barry, Lawrence Beale Collins, Sarah Bell, Christoph Bernhardt, Nirupa Bhangar, Tim Bunnell, Helene Burningham, Ben Campkin, Federico Caprotti, Hugh Clout, Claire

# Preface

Colomb, Steven Connor, Denis Cosgrove, Olivier Coutard, Stephen Daniels, Monica Degen, Richard Dennis, Jürgen Essletzbichler, Michael Flitner, Adrian Forty, Laurent Fourchard, Susanne Frank, Jamie Gillen, David Gissen, Carlotta Giustozzi, John Goodwin, Kate Goodwin, Stephen Graham, Maren Harnack, Andrew Harris, Hartmut Häußermann, Michael Hebbert, Kristina Hill, Joseph Hillier, Alan Ingram, Sandra Jasper, Maria Kaïka, Gerry Kearns, Roger Keil, Eje Kim, Gang-Li Kim, Koku Konu, Shekhar Krishnan, Patrick Le Galès, Alex Loftus, Colin McFarlane, Savitri Medhatul, Jeremy Melvin, Gustav Milne, Leandro Minuchin, Michael Müller-Verweyen, Richard Munton, Gbenga Odele, Muyiwa Odele, Ayodeji Olukoju, Giles Omezi, Kate Orff, Ben Page, Nayan Parekh, Amala Popuri, Hugh Prince, Jennifer Robinson, Rebecca Ross, Kelly Shannon, Rachana Sheth, AbdouMaliq Simone, Stephen Smith, Ola Söderström, Rahul Srivastava, Jon Stokes, Erik Swyngedouw, Karen Till, Valerie Viéhoff, Peter Wood, Austin Zeiderman, Marie-Hélène Zérah, and Benedikte Zitouni. I would also like to thank the staff and graduate students of the UCL Urban Laboratory, since its inception in 2005, who have been an indispensable source of inspiration and encouragement.

Thanks also to staff at many archives and libraries who have helped me locate research materials, including the Akademie der Künste, Berlin; the Bauhaus-Archiv, Berlin; the Bibliothèque Historique de la Ville de Paris; the British Library, London; the Frances Loeb Library, Harvard University; the Institution of Civil Engineers, London; the Künstler-Archive, Berlinische Galerie; the Landesarchiv, Berlin; the London School of Hygiene and Tropical Medicine; the Los Angeles Public Library; the Maharashtra State Archives, Mumbai; the Moheny Memorial Library, University of Southern California; the National Archives, Ibadan; the Royal Institute of British Architects, London; the Staatsbibliothek, Berlin; and the Wellcome Library, London. Financial support for the research was provided from several sources, including the Alexander von Humboldt Foundation, the Arts and Humanities Research Council, the British Academy, and the Economic and Social Research Council. I am also grateful for a

# Preface

grant from the Graham Foundation for Advanced Studies in the Fine Arts toward the costs of illustrations for the final production of the book. The cartography was completed by Miles Irving and Susan Rowland. Thanks also to Sandra Jasper for helping me source permissions for many of the images and to the staff at Bayeux in Soho for processing my 35mm slides taken during the fieldwork. I would also like to thank Clay Morgan for his patience in seeing this long project through to completion, my editor Matthew Abbate, along with Miranda Martin, Margarita Encomienda, Yasuyo Iguchi, and the rest of the fantastic team at the MIT Press. Earlier versions of chapters 1 and 4 were published in *Transactions* and *Environment and Planning A*. The pan di ramerino from the Spence bakery in Stoke Newington kept me fueled for the completion of the manuscript, along with an eclectic mix of music ranging from Fela Kuti to Arvo Pärt. Finally, a special thank-you to Yasminah Beebeejaun, Berenice and Ben Felsenburg, Maria Gandy, and James Hewson, for their help and support in so many ways.

Introduction

> Water is truly the transitory element. It is the essential, ontological metamorphosis between fire and earth.
>
> *Gaston Bachelard*[1]

The beguiling simplicity of Gaston Bachelard's conception of water belies a tension between the material and metaphorical dimensions of nature. Bachelard draws on a series of essentialist, and often gendered, refractions of nature in order to connect with a prelinguistic and primordial realm.[2] His attempt to excavate the cultural essence of water inadvertently reveals the ideological ambiguities of nature under modernity: his search for the universal, accessed through poetry and literature, merely emphasizes the limits to a culturally distinct variant of humanism. What analytical tools, then, might enable us to excavate the specificities of human interaction with water? Can the flow of water unsettle our existing conceptions of space, technology, and landscape? The modern city, in all its cultural and material complexity, presents a distillation of these tensions, encompassing both the ambiguities and the limits of "nature" as conventionally understood.

Water lies at the intersection of landscape and infrastructure, crossing between visible and invisible domains of urban space. Water forms part of the material culture of modernity, ranging from the private spaces of the home to vast technological networks that have enabled the growth of cities, yet it is also powerfully inscribed in the realm of imagination. Although the transformation of human interaction with water has been an organizational and technological *telos* for the rationalization of urban space, the global experience remains starkly uneven.[3]

This book considers the cultural and material significance of water through studies of a series of cities, from different vantage points, in order to explore evolving relations between modernity, nature, and the urban imagination. In some instances water presents a threat, either directly or indirectly, whereas in other cases water enriches both private and public aspects of urban life. The relationship between water and modernity is not the domain of just one or even a few disciplines: established areas of research in engineering, history, and geography, for instance, can be complemented by insights drawn from fields such as anthropology, architecture, art history, comparative literature, epidemiology, hydrology, sociology, and science and technology studies. The study of water exemplifies how interdisciplinarity has itself become a contested terrain in relation to the production of knowledge.[4] Water does not constitute one object of analysis but rather an intersecting set of processes, practices, and meanings that cuts across existing disciplinary boundaries. To take account of the interdisciplinary scope of water, and its connections with urban culture, has daunting implications, since disparate concepts and methods are necessarily combined with uncertain analytical and expository implications.

Water is inextricably linked with the idea of infrastructure as a technical and organizational domain that underpins the functional dynamics of urban space. The word "infrastructure" is of French origin and was first used in relation to railway construction in the 1870s before its later extension to the coordination and control of hydrological systems, especially in an Anglo-American context. The term has gradually acquired a wider set of meanings and applications, and has proved surprisingly malleable in relation to the development of different types of technical and organizational networks.[5]

Infrastructure forms an integral part of the wider process of rationalization, state formation, and the emergence of "technoscience." In this sense the development of the modern state, and its characteristic forms of expertise and knowledge production, is to a significant degree entwined with the impetus toward greater control over water, ranging from the public health needs of cities to larger-scale interventions for agriculture, power, and flood control.[6] In Elia Kazan's film *Wild River* (1960) a young Tennessee Valley Authority official is sent to clear the last remaining landowner from an island on the Tennessee River before a new dam comes into operation. The film begins with harrowing documentary footage of the devastating floods of 1927 before cutting to a racially divided small town next to the river where to "leave nature alone" is also to leave poverty and injustice unchanged. For Kazan the TVA's program of public works, which are ironically referred to under "the general heading of progress," promises not only flood control but also electricity, roads, equal pay for black and white workers, and a sense of hope in a better future. Infrastructure is presented as modernity itself.

The related concept of landscape, of much earlier etymological provenance than infrastructure, is also closely tied with the evolution of human interactions with water. Indeed, the word enters the English language via the Dutch *landschap* in the late sixteenth century, as a "unit of human occupation" intimately tied to the drainage and regularization of land. "So it was surely not accidental," writes the historian Simon Schama, "that in the Netherlandish flood-fields, itself the site of formidable human engineering, a community developed the idea of a *landschap*, which in the colloquial English of the time became a *landskip*."[7] Although the concept of landscape has subsequently become associated with modes of visual pleasure rather than its earlier legal or functional attributes, the material connections with water remain pivotal to the transformation of "first nature" under the rationalizing impulse of modernity.

The terms "modernity" and "modernization" have been widely deployed in relation to the transformation of human interaction with water, though their etymology is closely tied with the European experience. While "modernization" is perhaps best used in relation to specific processes such as the development of technological networks, "modernity" has wider connotations including changes in the sensory realm.[8] From the late nineteenth century, the word "modernism" begins to acquire significance for specific, often experimental aspects of cultural change that find echoes between different cities. In the field of architecture, for example, the influence of Mies van der Rohe is felt in both Berlin and Lagos, though in the latter case the limitations of the architectural avant-garde are even more apparent. Terms such as "tropical modernism" reflect attempts to produce a synthesis of ostensibly contradictory elements: the term belies the limitations of an architectonic resolution to structural inequalities inherent within the European colonial project.[9] Furthermore, the self-conscious articulation of modernism as a clearly defined movement, not unlike postmodernism in the 1980s, relied on a series of ex post facto theoretical maneuvers on the part of its defenders.[10] Water, in its different relationships with urban space, touches on all three words, ranging from the "modernization" of water supply, especially from the middle decades of the nineteenth century, to the "modern" experience of bathing, hygiene, and public health, and finally to the advent of "modernist" approaches to design, which still persist in aesthetic if not ideological terms. It has become more useful to conceive of multiple and alternate forms of modernity within which earlier circulations of ideas are in a continuous process of hybridization and recombination.[11] The diversity and sophistication of many urban water systems, in use well before a clearly articulated sense of European modernity, also unsettles teleological or neatly demarcated conceptions of sociotechnical change.[12]

## I  Axes of Visibility

The history of water infrastructure, and the articulation of a putative public interest in relation to the structure and organization of urban space, have been checkered and unpredictable. The health-threatening effects of inadequate infrastructure systems did not necessarily translate easily or quickly into better technological solutions or new forms of governmental intervention. Even when expert opinion had settled on a practical solution, there was often insufficient capital to undertake the necessary public works. In other cases only repeated disease outbreaks could galvanize urban administrations into any kind of coordinated action. Time and again the intersecting flows of water and capital reveal the wider political dynamics of the urban arena.[13]

The study of water has played a key role in the emergence of "urban political ecology" during the 1990s as a theoretical synthesis between political economy and the "hydro-social cycle." A series of studies have emphasized how the production of nature, including water infrastructure, has facilitated the expansionary impetus of capitalist urbanization.[14] Yet gaps remain within this influential body of work in relation to the embodied experience of technological networks, different conceptions of agency, the microcomponents of urban space, and the significance of more recent advances within the biophysical sciences.[15]

One area that remains relatively unexplored is the intersection between ecology, epidemiology, and urban infrastructure. The biopolitical dynamics of infrastructure networks involve not only the rationalization of urban space but also more recent geographies of failure and neglect. An unforeseen consequence of the recent spate of home foreclosures across North America has been the increasing incidence of encephalitis linked to abandoned swimming pools that have become breeding grounds for insect vectors of disease.[16] In the wake of the East Asian economic crisis of 1998, similarly, standing water in abandoned building sites facilitated the spread of dengue fever in Bangkok, Jakarta, and elsewhere.[17] The worldwide spread of dengue fever since the late 1950s is related to changing patterns of urbanization and the proliferation of new habitats for insect vectors, as well as the impact of greater human mobility, warmer temperatures, and other factors.[18] A recent study of Kolkata (Calcutta) reveals an abundance of urban breeding sites for the *Aedes* genus of mosquitoes that carries dengue fever in uncollected household waste items such as broken utensils, coconut shells, bottle tops, and other container-like habitats that mimic small bodies of water in leaves or tree holes.[19] In many cities of the global South the majority of the urban population must contend with public health threats created by inadequate infrastructure networks or hazardous environmental conditions, in a scenario that is driven both by the political dynamics of urbanization and the evolutionary pathways of pathogens themselves.[20]

Water infrastructure, as a largely unnoticed skein of technological systems, forms part of an avowedly European, North American, or more recently East Asian experience of modernity.[21] The distinction between visible and invisible domains becomes especially apparent in the "horizontal cities" of the global South, where in Arjun Appadurai's words everything is "fully available to the gaze."[22] The relative absence of water infrastructure is paradoxically reflected in a jumbled landscape of pipes, open sewers, tankers (often controlled by organized crime to extract rent from the poor), water vendors (selling sachets of "pure water"), and buckets (to store available water) (figure 0.1). These exploitative and often health-threatening arrangements in the "city of buckets" can be compared with wealthy enclaves in the South or the connected cities of the global North, where the metabolic relations between the body and the city remain largely unseen except in instances of failure, interruption, or contamination; or through the mundane geographies of inconvenience engendered by the maintenance of technological networks whose unpredictability appears to evince a form of vitalism that transcends their mere status as material objects.[23] In the "plumbed city," private and public realms have been regularized and separated, apart from their intimate interface in the standardized modern bathroom, an innovation that gradually spread from the late nineteenth century to become a core element of the "auto-house-electrical-appliance complex."[24]

The layering of infrastructure networks also forms an integral element of "vertical urbanism," yet we know relatively little about the vertical axis in comparison with the more familiar horizontal dynamics of urbanization. In the nineteenth-century city the poor tended to live in the higher stories of buildings furthest from the convenience of the street and available

0.1  Ajegunle, Lagos (2003). Photo: Matthew Gandy.

sources of water.[25] In the post-elevator age these vertical distinctions have changed but not disappeared (especially in relation to low-income housing projects). The vertical axis can reveal geopolitical excavations and subterranean incursions, yet the militarized dimensions to verticality represent only one facet of contemporary developments, as the geographer Andrew Harris contends in his expanded conception of vertical urbanism across different historical and geographical contexts.[26]

II  Periodicities and Metabolisms

The identification of distinct periodicities in the history of urban infrastructure has tended to coalesce around the emergence of "ideal types," yet these heuristic constructs hide many inconsistencies; there is a frequent overemphasis on the significance of recent change in relation to the smoothed-over complexities of the *longue durée*. The fundamental break signaled by nineteenth-century modernization and the rise of industrial cities did not translate into a coherent sociotechnological formation for many decades. Indeed, there was a remarkable diversity in technological responses to the growing sanitation crisis facing nineteenth-century cities. The coevolutionary dynamics of technology and society were highly convoluted and unpredictable before the shift toward more standardized conceptions of the modern city.

For Sigfried Giedion, the history of bathing "yields searching insights into the inner nature of the period" and the degree to which "individual well-being is regarded as an indispensable part of community life."[27] But what is the "inner nature" of a historical epoch in relation to the use of water? Are there distinct periodicities that we can observe, or simply a tendency toward increasing scale and complexity? The regularization of infrastructure provision forms part of what Stephen Graham and Simon Marvin term the "modern infrastructural ideal" comprising specific ideologies of science, the modern discipline of urban planning, emergent patterns of consumption, and territorial dimensions to the modern state.[28] Yet this ostensibly universal model that diffused outward from the nineteenth-century industrial city is itself under strain with the shift to more socially and spatially differentiated systems.[29] The last quarter of the twentieth century has witnessed the development of more variegated patterns of urban infrastructure, as statist models have been partially displaced by the neoliberal impetus toward new modes of service provision. Similarly, the articulation of a "bacteriological city," emerging in the late nineteenth century, in which advances in epidemiology, civil engineering, and modern forms of public administration predominate, has proved highly restricted in its geographic scope.[30]

Although the acme of this push toward spatial rationalization, and what the historian Mark Mazower terms the "technocratic assurance" of "imperial modernism," is reached in the immediate postwar era of the late 1940s and 1950s, we should not forget that the geographic reach of modernization was highly uneven even within the global North.[31] In the late 1950s, for example, housing surveys showed that in Morley, outside Leeds, half of the homes lacked a bath and almost a third shared a lavatory; while in the Gorbals district of Glasgow only one house in thirty had a bath.[32] In parts of the Berlin district of Kreuzberg, as recently as the mid-1970s, over a third of apartments shared a lavatory and only a quarter had their own bathrooms.[33] And in the global South, patterns of urbanization militated against any straightforward replication of engineering templates derived from the cities of Europe or North America. In colonial Nigeria, for instance, emerging discourses of "scientific racism," emboldened by the strictures of fiscal conservatism, produced intensified forms of residential segregation (see chapter 3). In the postcolonial era these fractured landscapes have become subsumed within virulent forms of bourgeois environmentalism rooted in middle-class disdain for the urban poor. In contemporary India, for example, a miasmic anxiety with cultural and material contamination pervades a political discourse in relation to urbanization that ignores underlying tensions and iniquities.[34]

In much of the global South the rural water crisis and accelerated patterns of urban growth form part of a nexus of developments. The construction of hotels, leisure complexes, and the latest luxury housing projects, especially at the urban fringe, is occurring in the context of rising land values, the impoverishment of rural communities, and the intensified depletion or degradation of available water sources. In the case of Mumbai water shortages are intimately connected with parallel social, political, and environmental transformations that transcend the existing institutional and spatial parameters of the metropolitan region (see chapter 4).

Though we can find metabolic continuities with the past, in terms of basic human needs the scale and complexity of contemporary urbanization involve a different order of dynamics and interdependencies.[35] The nineteenth-century conception of the city as an assemblage of identifiable organs placed particular emphasis on the circulatory dynamics of urban space. In the twentieth century these corporeal analogies were extended to conceive of urban space as an organic machine amenable to increasingly sophisticated types of technical control. The remarkable Weimar-era depictions of the human body as machine by the physician and graphic artist Fritz Kahn, for example, show the body as an industrial production system. In Kahn's *Man as Industrial Palace* (1926), different organs are represented as factories or industrial installations, with veins and arteries replaced by metal pipes. These distinctive hierarchical representations of the body as machine owed much to the popular didacticism of the international hygiene exhibitions from 1911 onward that presented the body politic under modernity as an organic totality and technological utopia.[36]

The zenith of "scientific politics" was reached during the Cold War era, when technology appeared to hold the answer to all pressing social and environmental problems. The Promethean impulse behind water engineering is reflected in a new generation of dams and other technological structures far larger and more elaborate than those of the past. The 1950s alone saw the completion of immense new dams such as the Kakhovka (Ukraine) and the Garrison (North Dakota), followed by a generation of even larger structures in the 1960s and 1970s.[37] Though imbued with an aura of technical expertise, these structures have often revealed underlying tensions between modernity, nationalism, and the rationalization of space. For Timothy Mitchell, in his study of Egypt's Aswan Dam, completed in 1970, the unintended consequences of the project crystallized the limits to a particular form of modernity and its associated technopolitical apparatus. The power of nature did not necessarily translate into the kind of material outcomes that the main protagonists had envisaged, since "there were always certain effects that went beyond the calculations, certain forces that exceeded human intention."[38]

The technopolitical field of modernity also extends to various forms of "scientific urbanism" in which the city is conceived as a system defined by measurable parameters. The recent shift of emphasis from the organicist city of functional flows and organs to a more diffuse neo-organicist preoccupation with the digital realm or "thinking space" of the city has not dispelled the limitations of earlier corporeal conceptions of urban space. The reading of the city as a "cyborg" highlights the tension between a focus on the technological augmentation of the individual human body and the wider sets of technological networks that sustain modernity; the emphasis is not so much on augmented technological capacities as on the vulnerability of technological networks to disruption. The conceptual scope of cyborg urbanization transcends the individual human subject to consider urbanization as a larger set of sociotechnical interdependencies.[39]

At an analytical level, the metabolic dimensions to contemporary urbanization remain hampered by an uncertainty about what constitutes the city as an identifiable spatial formation.[40] The reprise of Henri Lefebvre's distinction between cities and urbanization, elaborated through emerging interest in "planetary urbanization," emphasizes the complex entanglements between "the city," as a bounded or reified object of study, and a vast expanse of "ecological frontiers," "operational landscapes," or other spatial articulations between nature and urbanization.[41]

The related idea of "technological zones," as developed by Andrew Barry, combines the material and institutional aspects of the interface between space, society, and nature. For Barry there is a periodicity to the establishment of technological zones marked by the impact of standardization, technical compatibilities, and regulatory competencies, so that the evolution of these spaces can be simultaneously considered in terms of their biophysical and institutional characteristics.[42] In this sense the urbanization of nature defies any straightforward spatial categorization, so that successive territorial recompositions of urban space blur existing boundaries between metropolitan centers, periurban spaces, and the state formations within which they are physically located.

The urbanization of nature involves far more than the physical modification of space, entailing the emergence of new sets of global relationships among production, consumption, and urban culture. The expansionary impetus of global capitalism, with its "limitless horizon of growth," has been predicated on the availability of oil, but access to water has also played a distinctive role.[43] The search for ever more elaborate technological fixes, such as nuclear-powered desalination plants, forms part of an intensified phase of capitalist urbanization predicated on distinctive sets of geopolitical, financial, and technological alliances.[44] The quest for "unconventional water," in a parallel technological and geopolitical maneuver to the extraction of "unconventional hydrocarbons," marks an escalation in the hydropolitics of global capitalism.[45]

## III   Tactics, Tubs, and Technologies

The changing relationship between water and the human body encompasses different understandings of health, morality, and pleasure. The interaction of water with the corporeal realm illuminates aspects of everyday life, yet there remain analytical tensions between finer details—the texture of space—and more generalized social structures or biophysical phenomena. The emphasis of Norbert Elias, for example, on the "process of civilization"—perhaps better expressed as a cultural history of embarrassment—has been influential in tracing the changing codes of corporeal conduct in a European context.[46] If we move from manners to networks, however, a series of interconnections between social and technological change can be discerned. The sociologist Elisabeth Heidenreich introduces the concept of *Fließräume* (flow spaces) to capture the way in which technological networks comprise a mix of material and cultural elements. For Heidenreich, echoing Alain Corbin's study of the olfactory realm, the development of water networks forms part of a multifaceted shift toward a more individualized experience of modernity.[47]

Albrecht Dürer's engraving *Das Frauenbad* (The women's bath) (1493), which includes a surreptitious glance from the artist himself, marks an ambiguous moment of corporeal conviviality before the emergence of different disciplinary codes toward nudity, sexuality, and the human body had become established (figure 0.2).[48] In fact, the bathhouses of mediaeval Europe were already in decline at the time of Dürer's travels, not least because the spread of the bubonic plague in the fourteenth century had instilled a sense of fear or suspicion of communal spaces. During the seventeenth and eighteenth centuries the use of water for washing actually declined in Europe, and contact was actively discouraged because of its "weakening" effects on the skin, so that the emphasis on cleanliness shifted to clothes, linen, and only the visible parts of the body. As living conditions in towns and cities across Europe steadily deteriorated, however, the question of water was revisited in the context of better physiological knowledge of the human body, new attitudes toward dirt, and the impact of cholera epidemics on circulatory conceptions of urban space.[49] Although "the bath" began to reappear more frequently from the 1860s onward, it was only with the arrival of more readily available hot water toward the end of the nineteenth century that the experience of indoor bathing regained something of its earlier associations with pleasure.[50] The movement from collective worlds of communal bathing in the early modern era to the private realm of the modern bathroom or the controlled space of public baths involved several processes, including the gradual standardization of technological networks, the spread of new tastes in architecture and design, the rise of greater corporeal self-consciousness, and the emerging biopolitical emphasis on health and sanitation.

The gathering momentum of the "water revolution" encompassed a multifaceted transformation of urban life. The coevolutionary dynamics of the body, space, and technology reflect a myriad of different sources of power and social compulsion to produce new manifestations of sociospatial order. Technological influences on bodily conduct have served as a rich terrain for Foucauldian readings of the "tactics" of governmentality in the modern city.[51] The Gramscian notion of hegemony, in the sense of a pervasive presence of power that is obscured by a narrow focus on formal types of political contestation, can be contrasted with the Foucauldian emphasis on the micropolitical spaces of modernity.[52] A dispersed conception of power decenters our understanding of capitalist urbanization to encompass a wider range of developments and possible points of comparison. It also problematizes assumptions about the role of the state and formal legislative interventions as the focal point for the analysis of sociotechnical aspects of historical change.

As the emphasis shifted from facade to function, the ornate hydrological paraphernalia of the early industrial city were superseded by a new technological calculus. The emergence of an increasingly sophisticated relationship between water and cities signals a paradoxical set of developments: as modernity became defined against nature, or at least through greater control over nature, the place of nature within cities began to change so that symbolic elements diverged from the increasingly hidden functional realm. "The modern town was based on a concealed infrastructure," writes the historian Georges Vigarello; "everything turned on the calculation of levels, the speed of flow through pipes, and the flexibility of the network; attention was concentrated on the diameters, gradient, and numerous intersections of pipes, in fact, on matters of engineering technique."[53] The technocratic approach to water management that spans both statist and more

0.2 Albrecht Dürer, *Das Frauenbad* (1493). Kunsthalle Bremen—Der Kunstverein in Bremen. Kupferstichkabinett. Photo: Karen Blindow.

recent neoliberal variants holds wide-ranging implications. The intersections between water, democratic deliberation, and the public realm have been extensively occluded, except in those circumstances where conflicts or crises are openly debated. This partly stems from the relatively arcane nature of civil engineering discourse, along with the constraints engendered by path dependencies from previous waves of development, but it also reflects more secretive intersections between technological networks and the strategic needs of the modern state.

The recent wave of straightforward neoliberal dogma and its associated "politics of inevitability" has receded since the 1990s.[54] While private companies have retreated from municipal water provision in some regions, notably sub-Saharan Africa, the picture in East Asia is far more variegated.[55] A more polarized landscape is emerging in which some cities have successfully won control back from underperforming private-sector providers and even developed new models of public participation in technological politics. In the case of Berlin the imposition of a privatization program for the indebted city's water supply, introduced after reunification in 1990, has now been reversed after a public campaign led by the Berlin Water Roundtable (*Berliner Wassertisch*). This antiprivatization social movement, created in 2006, has exposed the veil of secrecy over both state and corporate maladministration, thereby shattering teleological models of the inevitable demunicipalization of public services that have been enabled by the restricted production, dissemination, and discussion of knowledge.[56] Grassroots insurgencies in the field of technological politics offer the possibility for civil society to "take back the city" and are now being extended to other areas of contestation such as energy and waste.[57] The extension of democratic deliberation to technical aspects of public policy shows there are "many paths of rationalization," allowing the relationship between technology and society to be challenged or even recast.[58]

The shift away from technocratic politics is also reflected in attempts to rethink the flow of water through urban space at different spatial scales, ranging from individual households, with the differentiation and reuse of water, to more complex community negotiations over wastewater treatment technologies, floodplain restoration, and even urban agriculture.[59] At a watershed scale we can observe attempts to combine different types of knowledge and expertise. In the case of New York there has been a negotiated socioecological dimension to watershed protection since the 1990s to obviate the need for more expensive reliance on water filtration technologies.[60] In other cases, however, attempts at decentralization and higher levels of local participation have been "subverted from within," to use Amita Baviskar's phrase, where the short-term need for showcase projects or other kinds of visible "success" can obscure the more intractable dimensions to social inequality.[61]

The tensions between "adaptation" and "resilience" in the face of climatic uncertainty are being played out through a range of cultural, ecological, and technical discourses about future patterns of urbanization. The growing threat of flooding for many coastal or low-lying cities is leading to divergent strategies ranging from floodplain restoration to a new generation of technomanagerial solutions (see chapter 6). Contemporary interest in the "re-wilding" of cities through the creation of swale-type landscapes and forms of controlled flooding encompasses strands such as ecological restoration, concerns with biodiversity, and changes in urban design. We encounter a spectrum of interventions ranging from cultural and scientific interest in the spontaneous dynamics of urban nature and "cosmopolitan ecologies" to the production of elaborate simulacra of "wild urban nature."[62]

An emerging focus on the revitalization of postindustrial waterfronts now forms part of the "ecological gentrification" of cities. In some cases the reconstruction of urban rivers has served as a wider symbol for urban boosterism and the creation of new forms of public space. The uncovering of the Cheonggyecheon river in Seoul, for example, was a large-scale engineering project that also underpinned a strategic political maneuver geared toward the national stage.[63] Water is being enlisted in the "rebranding" of urban space as never before.

INTRODUCTION

## IV  Representations

Cultural representations of cities can enrich our understanding of urban environmental change. The nineteenth-century French artist Gustave Caillebotte, for instance, best known for his vivid depictions of the newly modernized Parisian boulevards of Second Empire Paris, also painted the more disordered landscapes toward the edge of the city. In Caillebotte's *A Small Branch of the Seine at Argenteuil* (1884) we see an ostensibly "natural scene" on the outskirts of Paris that belies the impending impact of urbanization. In this instance water plays an ambivalent role in the face of impending industrialization, as a vestige of nature for metropolitan contemplation. In a similar vein, the Berlin-based artist Adolph Menzel's intense eye for detail tells us much about the hidden cityscapes behind the architectural facades of the Wilhelmine city. In Menzel's *Rear Courtyard and House* (1844) the focus is a solitary water pump surrounded by trampled ground, suggesting the presence of an impoverished collective city amid the new landscapes of industrialization appearing on the horizon (figure 0.3). The work of Caillebotte, Menzel, and other nineteenth-century artists provides insights into the interstitial spaces of the modern city that parallel the nascent development of documentary photography.

Literature has also revealed intricate dimensions to urban life. Writers such as Charles Dickens, Victor Hugo, and Émile Zola transcend the scope of visual culture by incorporating observations on sound, smell, and even the subtle cadence of human conversation. In his novel *L'Assommoir* (1877), for example, Zola describes the extraordinary din of a communal laundry in nineteenth-century Paris as "a constant stream of water" accompanied by the rhythmic "pounding" and "splashing" of the beaters not unlike contemporary scenes in Mumbai's vast open-air laundry at Dhobi Ghat.[64] Castigated by conservative critics in its day, Zola's *L'Assommoir* is now considered to be one of the first modern novels to accurately depict working-class living conditions in Paris.

0.3 Adolph Menzel, *Rear Courtyard and House* (1844). Source: Bildarchiv Preußischer Kulturbesitz, Berlin.

One of the most imaginative literary engagements with water in everyday life can be found in James Joyce's *Ulysses* (1922), where the outer limits of modern experimental prose meet the technological accoutrements of modernity. The simple act of filling a kettle in early twentieth-century Dublin, and reflecting on its connection to the Roundwood Reservoir in county Wicklow, is enough to propel Leopold Bloom into an extraordinary catechism on the subject of water that ranges from "its democratic equality and constancy" to "the noxiousness of its effluvia in lacustrine marshes." Bloom's faith in the science of water and human ingenuity contrasts with his hydrophobic friend Stephen Dedalus, who has not taken a bath in many months.[65]

Water is also present in the visions of science fiction literature that explores the material and technological parameters of modernity. The novels of J. G. Ballard, for example, have investigated the "near future" involving themes such as societal collapse, infrastructure failure, and fantastical combinations of the body and technology. One of Ballard's recurring motifs is the abandoned swimming pool which serves as the exposed anatomy of modernity, variously imagined as places of violence with easily cleaned tiles, "psychic zero stations," or entire landscapes of drained concrete pits like "votive offerings."[66]

Landscapes have also served as a focus for photographic experimentation, posing specific challenges in terms of light, scale, or perspective. In the work of Bernd and Hilla Becher, for instance, landscape features such as coking ovens, grain silos, and water towers are arranged through an aesthetic taxonomy that is unrelated to their structural relationship to modernity. The repetitiveness of the image sequences, dating from the mid-1960s onward, bears an echo of Fordist mass production and the historical era that gave rise to these distinctive landscapes. The Bechers' fascination with the industrial landscapes of Europe and North America coincides with the precise moment of their dissolution or partial reconfiguration as landscape parks or other land uses. At one level the Bechers' work can be interpreted as an attempt to historicize modernism by aestheticizing industrial objects as archaeological artifacts, even if they still play a functional role within the landscape. The diversity of these structures, revealed by the intricate detail of large-format photographs, counters conceptions of industrial installations or technological networks as standardized systems within which human creativity has been extirpated.[67] An emphasis on scale is also present in recent studies of infrastructure installations by Jane and Louise Wilson. Their photographs, videos, and other works depict the utilitarian spaces of technomodernity as a mysterious form of spectacle, often comprised of multiple screens, soundscapes, and other features. In *South Corridor, Hoover Dam* (1999), for example, we observe the interior structure of the dam. The deserted internal corridor with its smooth surface and functional lighting curving into infinity evokes a sense of disorientation within the space and time of modernity (figure 0.4).

Moving from ground to air, the vertiginous waterscapes of Edward Burtynsky are redolent of the technological sublime. The emphasis is not so much on water technologies (evidenced in his earlier studies of dams) as on the imprint of water on human landscapes. Burtynsky's striking depiction from 2011 of the outskirts of Phoenix, Arizona, for example, reveals an abrupt dividing line between a tessellated suburban sprawl enabled by unseen technological networks and the Salt River Pima-Maricopa Indian Reservation. He presents a rarefied vision in which the human presence has been rendered absent or vestigial. These panoramic renditions of late modernity record the scale of human impact, even if the structural relationships between different elements of the landscape remain obscure. The highly aestheticized representations parallel photographic surveys of the "new ruins" of Detroit and other cities in which an air of mystery pervades the human landscape.[68] Indeed Burtynsky's waterscapes, along with his other landscape studies, have an ambiguous relationship with their subject matter, a tension that is shared by much documentary photography. We encounter an "excess of beauty"—to borrow Julian Stallabrass's phrase used in relation to the highly stylized work of Sebastião Salgado—that effectively muffles its elucidatory potential, rendering process into pattern for the delectation of the observer.[69]

0.4 Jane and Louise Wilson, *South Corridor, Hoover Dam* (1999). Courtesy of the 303 Gallery, New York.

But how do these imaginative engagements with urban technology relate to public culture? Different modes of representation can unearth ironic or unexpected elements that might otherwise remain hidden. Existing parameters of analysis or interpretation can be extended in new ways. A difficulty, however, is how disparate realms of individual creativity can be reconciled with public discourse or more sustained forms of social or political dialogue. Tensions between the philosophical positions of Jürgen Habermas and Richard Rorty hinge on the distinctions between the individual human subject, with capacities for "self reflection," and the search for shared meaning or "intersubjective communication."[70] Advances in knowledge involve more than metaphorical redescription, even if we reject the myth of scientific neutrality.[71] Yet Rorty's antifoundationalism may ultimately be too restrictive in relation to the role of cultural expression in political discourse or the need to differentiate between rival interpretations of material reality.[72] We encounter an implicit dematerialization of social theory whereby technical challenges remain ultimately "a matter for the engineers" rather than society as a whole.[73] With Habermas, similarly, we have a corpus of work that is marked, in Andrew Feenberg's words, by a "blissful indifference" toward technology.[74] Both Rorty and Habermas, in different ways, effectively exclude the emancipatory possibilities of technology from their analysis. Different worlds originate within the human imagination before their concrete realization, whether as a poem, an engineering solution, or a designed landscape.

The cities included in this book all have an established place within existing accounts of urbanization; indeed some of them have become archetypes for wider epochs, exemplified by Paris as "capital of the nineteenth century" or more recently by Los Angeles as the architectural citadel of "postmodern urbanism." The specific developments explored, ranging from the construction of technological networks to distinctive metropolitan cultures of nature, unsettle teleological or synecdochal assumptions, so that the idea of a singular modernity is displaced by a more polyvalent set of developments that effectively decenters existing narratives of urban change.

None of the cities included in this study can be considered in isolation: London, for example, exerted significant influence over colonial Bombay and Lagos, as well as providing capital for the modernization of nineteenth-century Berlin and other cities.[75] And more recently the attempt to both reimagine and reconstruct the Los Angeles River since the 1980s has proved highly influential for new approaches in landscape design that seek to bring more naturalistic elements of nature, including new types of flood defenses, into the heart of the city (see chapter 5).

The cities explored here reveal occasional connections or cross-cutting themes, but the emphasis is on a series of distinctive developments rather than a search for more obvious points of comparison.[76] Key protagonists in the history of engineering or urban planning can sometimes be read as "ciphers" for wider change, in part because their words or actions appear to crystallize specific developments, but tensions remain between historiographies that emphasize the role of singular interventions and the search for broader modes of explanation. The story of urban environmental change, and the politics of infrastructure networks, are marked to a significant degree by the largely unseen and unrecorded dimensions to human experience. These "historiographies of absence," as we shall see in the case of Lagos, serve to dispel those accounts of urban history that focus overwhelmingly not just on human agency but on the agency of a few individuals.

There is an implied chronological sequence to the chapters, since we start with nineteenth-century Paris and end with future uncertainties facing London. The first chapter begins with Nadar's photographs of the Paris sewers. These images capture a transformation in subterranean Paris initiated in the early 1850s as part of the comprehensive reconstruction of the city's infrastructure. The reconstruction of subterranean Paris reveals a series of cultural and political tensions that were only resolved in the post-Haussmann era in response to the combined influence of growing water usage, the persistent threat of disease, and changing conceptions of public health. Chapter 2 moves our focus to Weimar-era Berlin. We consider attempts to reconnect the city with nature that reached their zenith in attempts to protect public access to bathing lakes. The emergence of a distinctive metropolitan sensibility toward nature is explored through various sources, including the cinematic depiction of an excursion to the edge of the city. It is suggested that water, through its incorporation within a distinctively utopian vision of metropolitan nature, reveals much about the fragility of twentieth-century modernism as a coherent aesthetic and political project. Chapter 3 considers the threat of malaria in Lagos, where the first large-scale efforts to combat the disease were not undertaken until the 1940s as a result of changing geopolitical circumstances. We find that the politics of public health rested on a contradictory set of discourses, producing a fractured modernity that is exemplified by the continuing threat of waterborne disease. Chapter 4 shows how the tortuous flow of water through Mumbai presents one of the most striking indicators of persistent social inequalities within the postcolonial metropolis. The city's dysfunctional water infrastructure has its roots in the colonial era, but these incipient weaknesses have been exacerbated in recent years by rapid urban growth, authoritarian forms of political mobilization, and the dominance of middle-class interests within a fragile and denuded public realm. Chapter 5 explores the incongruous concrete landscapes of the Los Angeles River, which have served as a rich source of reflection on the fragmentary characteristics of the city and its precarious relationship with its geophysical setting. The "rediscovery" of the river since the 1980s has posed tensions between collective memory, global ecologies, and the remodeling of riparian environments. The final chapter uses the fictional scenario of a partially submerged London as the starting point for a reflection on the actual hydrological threats facing the city. Tensions between various forms of "adaptation" and "resilience" are played out in relation to different responses to flood risk, ranging from ecological conceptions of floodplain restoration to a perpetuation of the search for technomanagerial solutions.

## The Paris Sewers and the Rationalization of Urban Space

> … un enchevêtrement difforme de sentines et boyaux à défier l'imagination de Piranèse.
>
> *Félix Nadar*[1]

> Les grands égouts de Paris ont toujours préoccupé l'attention publique et ont été honorés des plus illustres visites. Il n'est pas un souverain étranger, pas un personnage important qui ait quitté Paris sans avoir visité les collecteurs.
>
> *Eugène Belgrand*[2]

The rebuilding of Paris between 1850 and 1870 is a crucial moment in urban history. The attempt to rationalize urban space by Emperor Napoléon III and his Préfet de la Seine, Baron Georges Haussmann, is one of the formative legacies in the development of urban planning. For Frederick Hiorns, the Second Empire reconstruction of Paris was a time in which "the evils of long-continued civic neglect were redeemed and Paris placed in the forefront of modern cities by imaginative reforms applied to the most onerous of human problems."[3] Edmund Bacon echoed similar sentiments in describing the new spatial structure of Paris as a "reversal in the direction of energy, from the outward explosion of avenues and palaces of the Louis Kings to the implosion of the connecting and life-giving boulevards of Haussmann."[4] Many authors have read the Haussmann era as axiomatic of modernity, but the reality is far more complex, involving an interweaving of ideas and developments spanning both modern and pre-modern conceptions of urban form.[5] In fact, the flow of water in Paris did not become modern, in the sense that we would now recognize, until after the fall of the Second Empire, with new legislative developments in the 1890s in response to rising water usage and the continuing threat of cholera. One of the least studied of these extensive public works projects is the reconstruction of the Paris sewers.[6] This chapter describes how the reorganization of subterranean Paris held implications far beyond the modernization of drainage and sanitation. Metaphors of progress and the application of scientific knowledge became entangled with wider cultural and political developments surrounding the transformation of nineteenth-century Paris.

Sewers enjoy a special place in the pantheon of urban mythology. Sewers are one of the most intricate and multilayered symbols and structures underlying the modern metropolis: they form a poignant point of reference for the complex labyrinth of connections that bind urban space into a coherent whole. Sewers have long been used as metaphors for the hidden worlds of crime, poverty, and political insurrection. There is a rich legacy of representations ranging across literature, cinema, and music.[7] In *Les Misérables*, perhaps the most famous literary evocation of the underground city, Victor Hugo depicted the Paris sewers of the 1830s as "the evil in the city's blood," a place where the poor and the outcasts of society lurked together as a formation threatening to the world above ground.[8] This chapter develops a rather different perspective from the genre of urban horror by emphasizing how sewers have also been portrayed as symbols of progress. Sewers are considered in this context as an integral element in the emergence of what the architectural historian Anthony Vidler terms the "technical ideology of metropolis."[9] Just as sewers are repeatedly associated with dirt, danger, and the unseen, they are also physical manifestations of new patterns of water usage, bodily hygiene, and the progressive application of new advances in science and technology.

The search for spatial order has been an integral element in the contradictory experience of modernity, yet hidden within the more progressive conceptions of urban transformation lie imperial and premodern conceptions of social and elemental harmony. This chapter argues that the process of "Haussmannization" was predicated on a holistic conception of the relationship between the body and the city that drew on a series of organic analogies to compare the new city with a healthy human body:

> These underground galleries would be the organs of the metropolis and function like those of the human body without ever seeing the light of day. Pure and fresh water, along with light and heat, would circulate like the diverse fluids whose movement and replenishment sustain life itself. These liquids would work unseen and maintain public health without disrupting the smooth running of the city and without spoiling its exterior beauty.[10]

The material presented here, however, suggests that the circulatory dynamics of economic exchange were to overwhelm organic conceptions of urban order and institute a new set of relationships between nature and urban society. By tracing the history of water in urban space we can begin to develop a fuller understanding of changing relations between the body and urban form under the impetus of capitalist urbanization. This interdisciplinary task involves exploring changing relationships between the body, architecture, and ideological conceptions of nature as part of a broader project to expand our understanding of modern cities and their cultural meaning.[11]

We begin with an examination of the photographs taken by Félix Nadar in the early 1860s of the newly modernized Paris sewers. These images introduce a series of ideas around progress, modernity, and the aesthetic representation of the modern city. It is suggested that Nadar's photography, and his passionate advocacy of the progressive potential of technological innovation in society, hold important implications for our understanding of the often contradictory dynamics behind capitalist urbanization. After this, the reconstruction of the subterranean city is set in its broader political and historical context in order to draw out some of the tensions inherent in the drive to modernize urban space. It is argued that these contradictory dimensions to the control of water are only satisfactorily resolved in the post-Haussmann era. The complexities are then traced to the rapid growth of Paris and the growing consumption of water for washing and bathing in private dwellings, which led to a breakdown in premodern conceptions of the organic cycle linking the body and the city. The representation of riparian urban leisure in the art of Georges Seurat is used to delineate an emerging distinctively "metropolitan" experience of nature and new patterns of everyday life. Finally, the chapter considers how the reconstruction of subterranean Paris involved the reworking of corporeal metaphors in the development of aesthetic sensibilities toward urban infrastructure. These changes are related to wider developments in French society, including the sharpening sense of self-identity under modernity in the context of widening social and economic polarities across the city. It is suggested that sewers form an enduring element of the "urban uncanny" through their integral relationship with changing conceptions of bodily abjection and urban order.

## 1.1   Photographing the Paris Sewers

Among the strangest images we have of nineteenth-century Paris are the underground photographs of Félix Nadar (1820–1910). The photographic legacy of Nadar provides a remarkable record of the complex and often contradictory interweaving of political, technological, and scientific developments underlying the rebuilding of Second Empire Paris. Nadar was born Gaspard-Félix Tournachon in Paris in 1820, just two years before Nicéphore Niépce (1765–1833) began the first experimentation with photography (using a bitumen ground technique on pewter and glass). Following a brief spell as a medical student in the late 1830s, Nadar began to devote increasing energy to literary and political pursuits. In the 1840s he worked on the republican daily *Le Commerce* and also as a cartoonist for the satirical journals *Le Corsaire-Satan* and *La Silhouette*.[12] The political turmoil of 1848, and the subsequent coup that brought Napoléon III to power, were to have a decisive impact on Nadar along with many of his contemporaries. With the utopian and revolutionary left greatly weakened, there was now a growing divide between the romantic attachment to artisan labor and newer political ideas that embraced technological change.[13] Nadar looked increasingly to a dynamic and progressive combination of science and politics as the most realistic means to transform society. In 1856, for example, he published a collection of stories, *Quand j'étais étudiant*, dedicated to the socialist writer George Sand. In these stories he brought together a number of his main concerns: the power of science and medicine to dispel death, disease, and ignorance; a sense of society in rapid and chaotic change; and the power of reason to bring about both individual and collective advancement.[14] Under the repressive regime of Napoléon III, Nadar turned increasingly to the use of satire and allegory as a vital means of expression under the tight censorship and surveillance of the time. He soon established his reputation as a novelist, journalist, and caricaturist and worked closely with the radical publisher Charles Philipon on his satirical papers *La Caricature*, *Le Charivari*, and *Le Petit Journal pour rire*.[15]

From the late 1840s onward Nadar began to take an increasing interest in photography. In 1854 he set up his own studio under the direction of Adolphe Bertsch and Camille d'Arnaud. The poor sales of his last major lithographic venture, the *Panthéon*, must also have encouraged him to focus his energies on the new medium of photography. Nadar's reputation for portrait photography developed rapidly, and he was soon able to charge some 100 francs a sitting: Charles Baudelaire, Claude Debussy, Gustave Doré, Hector Berlioz, and many other leading cultural and political figures of the time had their portraits taken at his Paris studios.[16] Nadar became convinced of photography's status as a new art form equal to painting and uniquely capable of capturing the ephemeral and fragmentary qualities of modern life. With photography he was able to convey images with unprecedented speed and accuracy, introducing an extraordinarily intense kind of realism into the aesthetic representation of Paris and its people.[17] By the late 1850s he had begun to gain international critical acclaim for his work, and in 1858 he filed a patent for the first aerial photography based on a series of urban panoramas taken from a hot air balloon.[18]

In 1861 Nadar sought to radically expand the medium of photography by transcending any reliance on natural light. In order to do this he began making a series of underground photographs using electric light. The first outcome of this experimentation was to be seventy-three views of the Paris catacombs taken in 1861–1862. Then in the winter of 1864–1865 he produced twenty-three photographs of the Paris sewers under the invitation of the city's chief water engineer, Eugène Belgrand (1810–1878), who had been appointed by Haussmann in 1853 to oversee the reconstruction of the city's sewer system. The extension of photography into the underground city not only radically extended the possibilities for the meticulous visual documentation of hitherto unknown places and spaces but also reinforced the ambiguous role of modern technologies in providing an illusion of complete control and comprehension of complex urban societies. In photographing the sewers Nadar contributed powerfully to new ways of seeing and understanding the city by challenging a series of metaphorical axes ranging across light, cleanliness, verticality, knowledge, and control. As Walter Benjamin observed, reflecting on Nadar's photographs of the new sewer system, "for the first time, the lens was deemed capable of making new discoveries."[19]

When first shown, these unfamiliar images intrigued and astounded the French public. A completely new and strange world lying beneath the streets of Paris had been revealed: not a threatening and chaotic mass of tunnels but a clean, well-lit network of structures at the leading edge of engineering science. In figure 1.1 we can observe a spacious, symmetrical and well-lit tunnel amenable to easy movement, observation, and control. In figure 1.2 a mannequin is shown seated next to an example of the new sewer technologies introduced as part of the modernization program (the lengthy eighteen-minute exposure time precluded the inclusion of a real worker). Before their improvement, the dominant imagery of the Paris sewers had been of an unexplored urban realm shrouded in darkness and mystery, a threatening maze beneath the streets of the city. Yet as curator Maria Hambourg suggests, "in the photographs of the vaulted sewers, which might have conveyed the horrors of Piranesi's prisons, one sees rational structure and channeled cleanliness."[20] The sewers were no longer to be feared but now celebrated and enjoyed as symbols of progress. Victor Hugo, for example, was quick to recognize that his imaginary representation of the 1830s sewers bore little resemblance to the new reality of the 1860s. "The sewer today has a certain official aspect," he reflected. "Words referring to it in administrative language are lofty and dignified. ... Nothing is left of the cloaca's primitive ferocity."[21]

With the introduction of electric light into the sewers, the "spectacle of enlightenment" now extended both above and below ground.[22] The new boulevards and shopping arcades had their subterranean counterpart. The transformation of Paris made urban space comprehensible and visible to the public, thereby dispelling much of the opacity and heterogeneity of the premodern city. The Paris sewers were rapidly acknowledged to be "unequalled in any other city in the world" and attracted a steady stream of international delegations of engineers and urban planners.[23] The subterranean photographs of Nadar played a key role in fostering the growing popularity of sewers and catacombs among middle-class Parisians, and from the 1867 Exposition onward the city authorities began offering public tours

1.1 Félix Nadar, *The Sewers* (1864–1865). Courtesy of the Caisse Nationale des Monuments Historiques et des Sites, Paris.

1.2 Félix Nadar, *The Sewers (Sluice System)* (1864–1865). Courtesy of the Caisse Nationale des Monuments Historiques et des Sites, Paris.

of underground Paris.[24] Yet as the historian David Pinkney wryly notes, most visitors to the Paris sewers over the years have probably been disappointed to find "not the dark and dangerous caverns through which Jean Valjean made his perilous escape in 1832 but the spacious, clean and well-lighted galleries of the Second Empire."[25]

## 1.2   From City of Mud to City of Light

In order to understand the significance of Nadar's images, we need to explore the background to the transformation of the Paris sewer system. If, as Raymond Williams has suggested, the modern city becomes "the physical embodiment of a decisive modern consciousness," then what do the Paris sewers tell us about changes in nineteenth-century French society?[26] The historical context to the reconstruction of Paris is well known. As early as 1827 an official report on the city's health had noted how "the sense of smell gives notice that you are approaching the first city in the world, before your eyes could see the tips of its monuments."[27] The population of Paris had increased from around 786,000 in 1831 to over a million by 1846. Growing congestion threatened to bring social and economic life to a standstill. The devastating cholera epidemics of 1832–1835 and 1848–1849 had spread panic in rich and poor quarters alike. And by 1848 the Paris economy was also facing a deep downturn, a major factor behind the political turmoil that was to usher in the regime of Napoléon III.[28]

In June 1853, less than a year after his successful *coup d'état*, Emperor Napoléon III appointed Baron Georges-Eugène Haussmann (1809–1891) as Préfet de la Seine with responsibility for the reconstruction of Paris. Before his appointment, Haussmann had already acquired extensive public service experience in Vienne, the Gironde, and the Var. More critically, though, he had lent vociferous political support to Napoléon III and closely shared his republican ideals rooted in a powerful role for the French state.[29] Since no accurate map of Paris existed, one of Haussmann's first tasks was to undertake a detailed survey and triangulation of the whole city.

Napoléon III envisaged that the new Paris would be an imposing "city of marble," worthy of comparison with Augustinian Rome, and a lasting symbol of French international power and imperial ambition.[30] Haussmann was charged with the responsibility of transforming a congested medieval city into a dynamic modern metropolis. The changing role of Paris within the newly integrated national economy demanded an urgent transformation in the city's physical structure. Central to its modernization lay a combination of faith in the application of scientific principles and a program of centrally directed public investment. The underused capital and labor behind the economic depression and political violence of 1848 were to be channeled into the reconstruction of the built environment through a deficit-financed economic strategy rooted in Saint-Simonian ideas.[31]

By 1870 Haussmann had carried out some 2.5 billion francs' worth of public works through innovative debt financing equivalent to some forty-four times the total city expenditure on all services in 1851.[32] At the time of peak reconstruction one in five Parisian workers were employed in construction activity, and one fifth of the Paris streets were rebuilt.[33] The program of reconstruction required not only financial innovation but also a radical reorganization in the balance of political and economic forces in the city. Second Empire Paris was to grow out of the articulation of a general interest resting on the imposition of a new form of capitalist rationality that was "alien to the privatism of traditional property owners."[34] For the historian Anthony Saalman, the reconstruction of Second Empire Paris was the most influential nineteenth-century solution to the problem of rapid urbanization. The transformation of Paris gave the nineteenth-century bourgeois revolution its most radical architectural expression in any European city. The reconstruction reflected the needs of an urban mercantile class that faced the consequences of modernity not by an escape into romantic anti-urbanism but through a celebration of the possibilities for the technological mastery of urban space and progressively greater degrees of social and spatial order.[35] Yet, as we shall see, these new discourses of order and control both reflected and constituted emerging tensions and inequalities driven by the processes of capitalist urbanization.

When Haussmann and Belgrand began their work in the early 1850s, the city was still served by a medieval network of sewers clustered around the city center (figure 1.3).[36] The preliminary investigations of Haussmann and his chief engineer Eugène Belgrand soon discovered a series of design faults in the existing sewer system. The size of the sewers had been determined by the height of a sewerman, and they were inadequate for handling large quantities of water after heavy rain. Given their layout, elevation, and gradient, the sewers were unable to prevent water periodically flooding onto the streets, and much of the growing city was not even integrated into the existing drainage system.[37]

In 1857 the sewer reconstruction program began in earnest. The first major project was the construction of the Collecteur Général d'Asnières, a new elliptical structure some fourteen feet high and eighteen feet wide. The purpose of this complex channel, far bigger than the existing Collecteur de la Rue de Rivoli, was to ensure that wastewaters would be diverted into the River Seine downstream of the city.[38] Both Haussmann and Belgrand believed that a modern sewer system should, as far as possible, be mechanically cleaned in order to eliminate the need for dangerous and degrading human labor (the sewers had hitherto been cleaned by hand using the most rudimentary of tools). Their conception of spatial rationalization thus extended to the application of new labor practices as well as the use of the latest advances in the engineering and empirical sciences. The most significant technological achievement of all, however, which effectively completed the major part of the new sewer system, was the construction of a vast siphon under the Seine in order to connect the two sections of the Collecteur de la Bièvre in 1868.[39] By 1870 the city was served by a network of 348 miles (560 kilometers) of sewers: a virtual fourfold increase from twenty years earlier (figure 1.4).

But what were these new sewers actually for? When the reconstruction of the sewers began in the 1850s, it was assumed that only limited quantities of human feces from individual homes would enter the sewer system (only a fifth of private dwellings were connected at this time) and

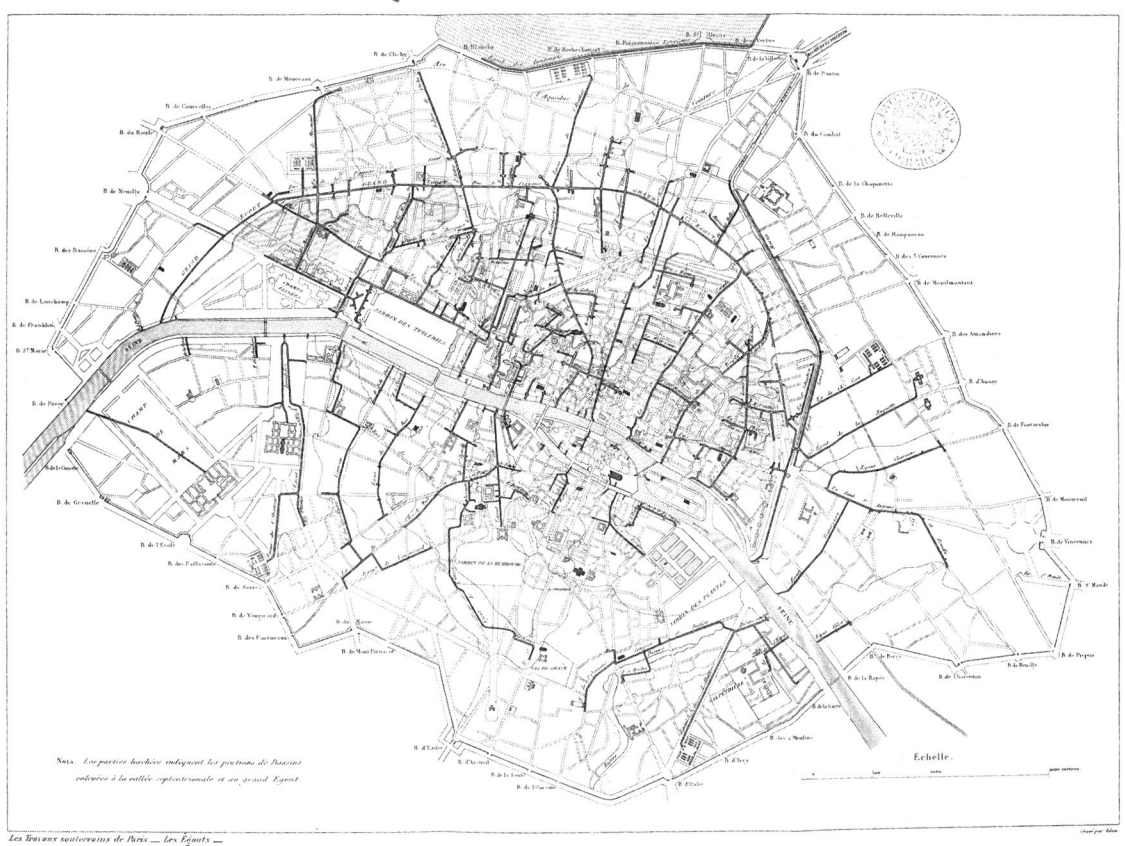

1.3  The Paris sewers in 1837. Eugène Belgrand, *Les travaux souterrains de Paris V: les égouts et les vidanges* (Paris: Dunod, 1887).

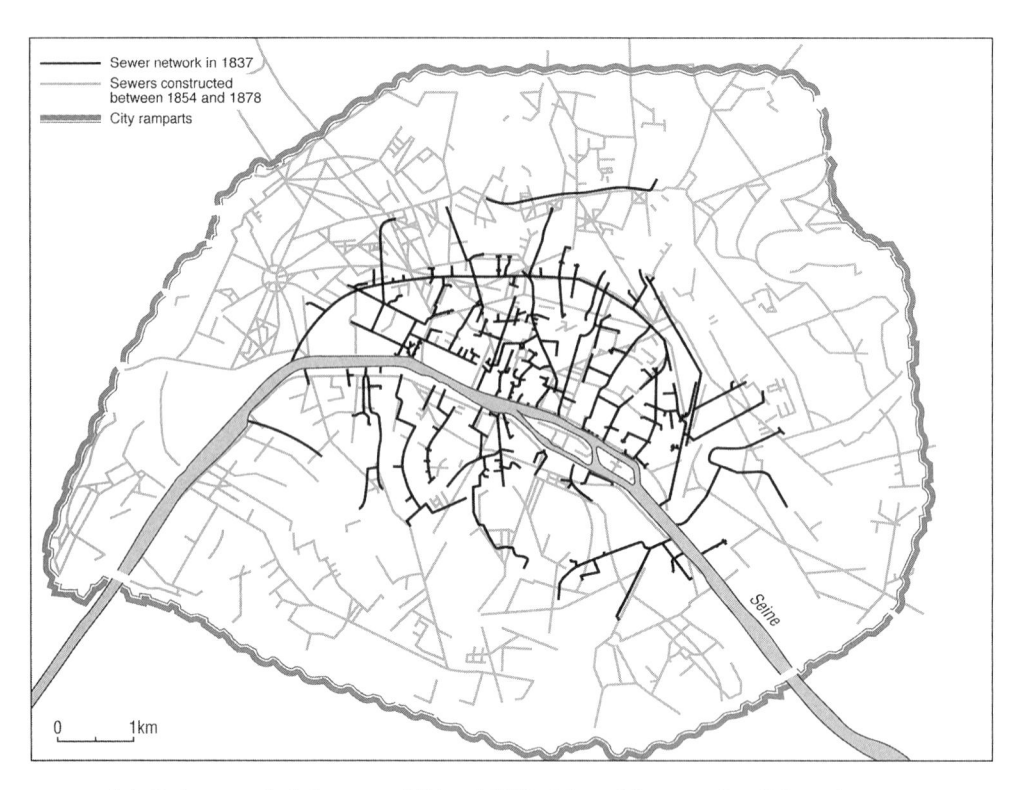

1.4 Paris sewers built between 1854 and 1878. Adapted from Eugène Belgrand, *Les travaux souterrains de Paris V: les égouts et les vidanges* (Paris: Dunod, 1887), and other sources.

that there would be a continuation in the work of night soil collectors.[40] The initial scope of the reconstruction was thus concerned primarily with the drainage of storm waters, but the steady increase in personal water consumption unsettled this conception of the public works that would be needed. Haussmann was reluctant to allow *any* human feces to enter the magnificent collecting channels of the new sewer system, and only did so under intense pressure from the city's municipal authorities.[41] The desire to separate "clean" storm water from "dirty" human waste was integral to Haussmann's conception of an orderly flow of water through urban space. His objections to human excrement entering the sewer system were not only out of his concern with the contamination of the underground city. He feared that the dilution of human wastes in water would reduce their value as fertilizers and thereby disrupt the organic economy of the city.[42] Human feces, collected as night soil, had long been profitably used in northern France as fertilizers for agriculture and in the manufacture of saltpeter for gunpowder, thereby allowing a cyclical integration of bodily functions into the regional economy.[43]

Haussmann was not alone in his desire to separate the drainage of storm water from the continuing reliance on cesspits for human waste. In the 1850s, opposition to the connection of sewers to individual homes came from various quarters. A vociferous source of hostility was the cesspool cleaning companies, who feared that they would be ruined by alternative means of treating wastewater. The city itself also made money out of night soil collection and the processing operations at Montfaucon and Boncy, which suited a continuation in existing arrangements.[44] The users of night soil in agriculture also drew attention to the declining nitrogen content caused by the greater mixing of feces with water. Consequently the lowest-value material was being collected from the richer parts of the city where the use of water closets was gaining popularity. And before the extensive application of deficit financing for public works, there was also concern over how the costs of sewer construction would be spread, since only a third of dwellings were directly supplied with running water.[45]

In the 1860s an apparent compromise was reached by allowing human feces to enter parts of the sewer system and diverting this water for agricultural use. The leading city engineer, A. A. Mille, successfully used sewer water for the irrigation and fertilization of vegetables in the fields of Clichy and Gennevilliers near the city's main sewage outlets. He advocated the use of the same sewers for the handling of storm water and human wastes based on his knowledge of sanitary improvements carried out in England, but this solution met with official resistance in Paris where most engineers continued to insist on just two options: a separate sewer system for human waste or an improvement in night soil collection.[46] In nineteenth-century Amsterdam, for example, the Lieurnur sewer system had been adopted with separate networks for storm water and human waste, but this was to be rejected for Paris on the grounds of cost.[47] Property owners fiercely resisted higher taxes for sewer construction as well as the construction of impermeable cesspools to replace the *fosses à fond perdu* that seeped their contents into the subsoil. Landlords continued to eschew any connection to piped water supplies despite the free installation of rising mains from 1881, because increased water usage would necessitate the reconstruction and more frequent emptying of cesspools if their properties remained separate from the sewer system. Powerful resistance to draining human waste into the Seine also came from the "fanatics of Seine water" who advocated the continuing use of the river for drinking water, along with Louis Pasteur and other influential microbiologists who feared the public health effects of contaminating the Seine with cholera and typhoid.[48] What was distinctive, however, about the immediate post-Haussmann era was the coexistence of a number of competing conceptions of the most appropriate means to regulate the flow of water in the city.

It was not until 1894 that the link between private dwellings and the sewer system was finally made obligatory, as rising water usage (which doubled between 1870 and 1890 in spite of the recalcitrance of private landlords) and the cholera epidemics of 1884 and 1892 eventually overwhelmed the traditional reliance on cesspools.[49] The fact that tenants themselves were in increasing numbers seeking out properties that were connected to the city's water and sewer system must also have increased the pressure to complete the modernization of urban infrastructure.[50] Furthermore, technical and scientific opinion was beginning to shift decisively toward the *tout à l'égout* solution in recognition that combined sewers for storm water and human waste were being increasingly adopted by other European cities.[51] By the end of the nineteenth century, the combined pressures of disease and growing water usage, along with the advent of inorganic fertilizers and growing public aversion to human waste, had overwhelmed lingering premodern conceptions of urban order and introduced a new set of relationships between water and urban society.[52]

## 1.3   Water, Modernity, and the Bashful Civic Body

The eventual cross connection of Haussmann's storm sewers to take human feces, the *tout à l'égout* solution to public health, reflects a complex shift in attitudes toward the use of water. During the medieval and early modern period there had been little use of water in Europe for personal hygiene and hence little need for sewers to drain water away from private dwellings. In the premodern period the use of water for washing remained predominantly a collective endeavor and was often therapeutic or recreational.[53] Evidence suggests that group bathing in Europe began to decline from the fifteenth century onward in the wake of the Counter-Reformation and changing moral codes regarding public nudity. In the eighteenth century, however, we find a rediscovery of bathing which undermined existing conceptions of the relationship between water and the body.[54] The sensuous flow of water and even the advocacy of cleanliness in readiness for sexual pleasure struck at the heart of conflicting concerns with moral purity, hygiene, and social order. Washing had long been associated with pagan sensuality in early Christian belief, and for most of the nineteenth century the bathroom was restricted to the homes of the rich, tourist hotels, and luxury brothels. Consequently, the associations of water with opulence, debauchery, and pre-Revolutionary court society persisted into the modern period.[55]

A decisive change occurred with what the historian Alain Corbin terms the "olfactory revolution" whereby the bourgeois sense of smell became newly sensitized to body odor, leading to an increasing desire for private space. It was not so much that conditions had suddenly changed as that there was a new intolerance under the sensory realignment of modernity.[56] The emergence of new standards of cleanliness brought individuals into contact with their own bodily smells and contributed to the emergence of a "new narcissism." By the end of the nineteenth century there were firmly barred doors to washrooms and bathrooms and an elimination of the "old promiscuity in defecation and the jumble of excremental odors."[57] As the places for washing and defecation became separated, this led to the increasingly complex design of private space and the interior of buildings.

During the later decades of the nineteenth century, more and more towns and cities across Europe became integrated into comprehensive water supply and sewerage systems in order to accommodate the increasing demand for personal use of water.[58] With the growing use of private washrooms, the smell of human excrement began to lose the last semblance of its rural associations with fertility: from now on it was to be indicative of disorder, decay, and physical repulsion. This is reflected in a survey of Parisian smells published in 1881 which recorded that cesspits, along with refuse and sewers, offered the three most unpleasant odors and that the proliferation of regulations for the construction and operation of cesspits had proved utterly futile in solving this problem.[59] The newfound bashfulness concerning bodily functions in bourgeois French society emphasized the association of sewers with excrement. With the growing involvement of the state, under the guise of public health reform, the management of excrement became an increasingly rationalized activity, resulting in a steady decline in the use of cesspits, the activities of night soil collectors, and communal places for defecation. Henceforth, the "regimes of the alimentary" were to be increasingly confined to domestic space under a new set of relationships between the body, technology, and urban architecture.[60]

One of the consequences of the reconstruction of the sewer system was that all wastewater was now discharged into the Seine at just two points along the river: Asnières and Saint-Denis. Unlike contemporary integrated sewage treatment systems, Haussmann's sewers were only intended for storm water and lacked any means for pollution control. With the increasing quantities of human and industrial wastes entering the sewer system, these two outlets left stretches of the river "a cauldron of bacteria, infection and disease."[61] With declining water quality, the irrigation systems at Gennevilliers were abandoned and the premodern "organic economy" was gradually lost to the demands of modern pollution control. With the loss of the organic continuities of premodern nature, a "modern" nature was being constructed through the planting of trees, the building of parks, and new transport links that enabled excursions to the city's hinterland. The Haussmanization of Paris was, above all, a process of redefining nature in metropolitan terms, of inscribing new patterns of social and spatial order within which nature was to be increasingly a focus of leisure and convenience rather than of material necessity. In Georges Seurat's *Bathers at Asnières* (1884), for example, we find a set of figures relaxing by the Seine (figure 1.5). This postimpressionist scene has for many observers served as a poignant critique of the emerging isolation and ennui of late nineteenth-century urban life. A succession of art historians and critics from Félix Fénéon onward have seen this type of work as the epitome of a kind of scientific realism that grew out of the prevailing rationalist and positivist ideologies of the time. Meyer Schapiro, for example, conceived of Seurat's vision as a counterpart to the technical and engineering outlooks of the Parisian lower middle classes. Schapiro even went so far as to suggest that Seurat's pointillist technique represented a combination of rationalist aesthetics with existential social critique.[62] Ranged against these materialist readings of this work are those critics who have conceived of the *Bathers* as lying closer to idealist and symbolist traditions, wherein art is not

1.5 Georges Seurat, *Bathers at Asnières* (1884). National Gallery, London.

predicated on the faithful mimesis of social reality but is a means to access a higher order of creative perfection.[63] The problem with relegating Seurat to some form of transcendent aesthetic universe, however, is that it may overlook the cultural significance of the work in both reflecting and reinforcing changing attitudes toward nature.

A more fruitful line of argument is to suggest that Seurat successfully captures a new kind of mediation between society and nature in post-Haussmann Paris: we are presented with a regularized, stylized, and commodified imagery based on leisure, spectacle, and the semblance of salubrity. In the place of the organic continuities of the past lies a new kind of nature for individualized leisure and consumption. A pictorial genre of the "urban pastoral" presented the outskirts of Paris as a harmonious interplay between nature and industry within which real labor was left invisible.[64] The changing place of water emerges as a central element in the shifting boundary between premodern and modern conceptions of nature. Yet the rationalizing impulse of modernity could never completely erase the surviving elements of a mythic urban space within which metaphors of bodily and social disorder could powerfully resurface to haunt the newly regulated urban society. Just as the water in Seurat's *Bathers* swirled with unseen bacteria, the underground city continued to provide a source of anxiety and fascination for metropolitan society.

## 1.4 Sewers and the Urban Uncanny

In one of the earliest surveys of the Paris sewers in 1824, the public health activist Alexandre Parent-Duchâtelet (1790–1835) prepared a detailed olfactory topography of underground Paris based on a series of specific smells such as *l'odeur fade* (insipid) and *l'odeur putride* (putrid).[65] For Parent-Duchâtelet sewers afforded the opportunity to combine aesthetic, moral, and scientific discourses, and formed an integral element in the meticulous documentation of the realms of urban life beyond the reaches of everyday bourgeois experience. In particular, the hygienist doctrine promulgated by

Parent-Duchâtelet and his successors emphasized an explicitly gendered conception of the relationship between sexuality and urban disorder:

> Prostitutes are as inevitable in an agglomeration of man as sewers, cesspits and garbage dumps; civil authority should conduct itself in the same manner in regard to the one as to the other: its duty is to survey them, to attenuate by every possible means the detriments inherent to them, and for that purpose to hide them, to relegate them to the most obscure corners, in a word to render their presence as inconspicuous as possible.[66]

The taming of nature through the new technologies of modernity carried with it an implicit echo in the social sphere.[67] In bourgeois French society women were relegated to a dichotomous olfactory universe of the "foul" and the "fragrant," which became manifest in the cultural and aesthetic discourses of urban design above and below ground. The relegation of women to an opposite world of nature and unreason had an increasingly powerful hold over the prevailing political and intellectual outlooks of nineteenth-century Paris, where the dichotomous cultural representation of women reached its apotheosis in relation to the flow of water through urban space. The public face of water in the lakes and fountains of imperial Paris was to be a celebration of the female form for the pleasure of the male citizen. Water-based sculptures and architectural forms allowed a symbolic continuity with classical themes centered around water, nudity, and human physical perfection. By the 1870s the Renaissance emphasis on the male nude as ideal human form was increasingly supplanted by the female nude and a new body aesthetic.[68] The ornamental public fountains of Haussmann's Paris, in the most expensive Belle-Époque neo-Fontainebleau style favored by Napoléon III, exemplified the combination of water with the control of women's sexuality.[69] Yet if these fountains, lakes, and other ornamental features represented the charm of virginal innocence, then the sewers continued to represent the dangerous obverse to female sexuality.

The association of women with impurity is not of course an invention of modernity; it is the reworking of premodern beliefs in the context of capitalist urbanization that is of interest here. In Second Empire Paris the repression of bodily functions in bourgeois society became increasingly manifested in a fear of women and the poor. Ideological readings of nature, which drew liberally on the latest scientific literature, contributed toward sharpening gender differences, with a newfound emphasis on the domestic ideal and the promotion of complementary gender roles. In reinforcing innate conceptions of "gender as nature," we find a convergence of the ideas and writings of Jules Michelet, Auguste Comte, Ernest Legouvé, and a panoply of other leading nineteenth-century scientists, writers, and intellectuals.[70] Underground urban infrastructure became a kind of repository for untamed nature, within which the innate tensions behind capitalist urbanization became magnified and distorted through the lens of middle-class anxiety.

The sewer has been consistently associated with what we might term the "urban uncanny" as a spatially defined sense of dread in modern urban societies. In order to understand the peculiar complexity of the "sewer" as a recurring spatial locus of the uncanny, we need to unravel how bodily metaphors have become transposed in urban space. Within Western intellectual traditions, Sigmund Freud's essay on "the uncanny" published in 1919 has served as a focal point for myriad debates concerning the connection between the psychological and spatial domains of modern societies.[71] The uncanny is best conceived as a "boundary aesthetic," with its spatiality rooted in anxieties of displacement and disorientation. Conventional accounts of the uncanny suggest that in passing from the world above ground into that below we are crossing between zones of the rational and irrational, culture and nature, male and female, visible and invisible.[72] Yet these dichotomous metaphors tend to conceal more than they reveal, obscuring the flows and interconnections that constitute material reality behind an illusion of stasis and symmetry.

Recent feminist scholarship has reinterpreted the Freudian reading of the uncanny in order to dispense with more simplistic gender-based conceptions of spatial disorientation and anxiety.[73] Rather than conceiving of the uncanny as a form of urban Gothicism in the Burkean aesthetic tradition, we are better served by reinterpreting it as an outcome of the complex intersection between the human body and the built environment. Within this schema, sewers represent a metaphorical space of defilement and confinement: their poignancy stemming from the connection between the private space of the home and the public *mélange* of urban infrastructure. The metaphorical grid surrounding the experience of the uncanny is ultimately a mystification of material reality in its implication that urban origins lie concealed beneath the surface of the city, rather than constituted through the more distant sets of social relations and spatial interconnections that sustain capitalist urbanization. The urban uncanny is a spatial fetishism of absence, a mythological response to the unseen and the unknown which weaves together popular misconceptions of how cities function with dominant ideological responses to urban disorder.

The new urban infrastructures of nineteenth-century Paris unsettled existing metaphors of urban space: with the breaking of the organic cycle, human excrement took on an intensely abject quality as part of a multiplicity of flows that integrate the body and urban society into an uneasy whole.[74] The "fear of touching" and the withdrawal from intimacy or curiosity toward strangers forms part of the atomization of social life under modernity.[75] In this context, the relationship between organic and social metaphors becomes problematized, since the city can no longer be meaningfully conceived as a holistic or autarchic entity but emerges as a dynamic intersection of the circulatory processes based around the exigencies of economic exchange. The "olfactory revolution," emboldened by the new discourses of the medical sciences, set in train an irreversible shift in water usage, the cultural significance of bodily smell, and the demand for private space. The urban transformation created a city in which social and economic differences were not only widened but were much more keenly felt.

The separation and reorganization of space set in motion an increasing dichotomy in the olfactory experience of the urban environment between the middle classes and the laboring poor (who were considered indifferent). The old vertical separation of the classes in the apartment houses of premodern Paris was to be gradually supplanted by a new emphasis on horizontal segregation. Edmond Texier's diagrammatic cross sections through apartment houses in his *Tableau de Paris* (1852–1853) show that the poor were clustered toward the higher stories of buildings.[76] With the construction boom of the Second Empire there was a progressive concentration of the middle classes in the central and western parts of the city. The quest for profit strengthened the social distribution of odors as the cleansing of the city involved a simultaneous relocation of the working classes and industry to the urban periphery.[77] Haussmann's leading critics, such as Louis Lazare, repeatedly drew attention to the mass displacement of people from central Paris and the emergence of the new slums and *faubourgs* at the city limits. The obverse to the rational city was not to be found beneath the streets but in what the art historian T. J. Clark has termed the "melancholic *banlieue*," a muddle of suburban sprawl, smallholdings, and displaced communities on the outskirts of Paris.[78]

## 1.5 Symbiotic Modernities

The provision of light, mobility, energy, and water form part of an urban palimpsest in the progressive rationalization of urban space.[79] From the early nineteenth century onward, advances in medicine, chemistry, and demography began to reveal the high death rates of towns. As a consequence, the progress of science and the administrative needs of the state developed in a symbiotic fashion.[80] Yet the development of new urban infrastructures also stemmed from the demand for greater privacy under the intensified self-awareness of modernity. The increasing aversion to communal washing facilities and the smell of excrement led to the growing use of water for washing and cleaning, which then had to be drained away.

This combination of factors behind the reconstruction of nineteenth-century cities led to a contradictory response on the part of urban planners. Though much of the literature on nineteenth-century Paris portrays Haussmann as a figure who faced rather than avoided the consequences of modernity, the actual sequence of events presents a far more complex picture. The creation of a modern metropolis introduced new sources of disorder that conflicted with existing conceptions of urban form and the premodern circulation of water in cities. We saw, for example, how Haussmann resisted the use of his new sewers for human feces. His advocacy of spatial order was rooted in a desire for a holistic and organic union of the city in all its parts, predicated on a conception of public health that owed more to the neo-Hippocratic doctrines of the past than to the latest advances in scientific thought.[81] The eventual integration of private dwellings into the city's sewer network in the post-Haussmann era was driven to a greater extent by changing attitudes toward the use of water than by any putative triumph of microbiological rationality over competing conceptions of public health.

The contradictory rationale behind the reconstruction of the Paris sewer system challenges simplistic tautologies that simply equate modernity with the process of Haussmannization.[82] Haussmann was unable to reconcile his conception of urban order with the disengagement of urban design from explicitly organic metaphors. The tensions within Haussmann's conception of water flow in urban space stemmed from an "uneven modernity" which extended across the physical, engineering, and medical sciences. Eighteenth-century conceptions of the elemental purity of water persisted despite the gathering pace of technological and scientific advances.[83] In the context of water and urban design, we need to differentiate essentially premodern holistic and geometric visions from the powerful exigencies of capitalist urbanization operating at successively wider spatial scales. Under capitalist space and time the corporeal unity of the premodern city was to be irrevocably altered, exposing an innate tension between function and perfection in the design of Second Empire Paris. The reconstruction of Paris under Haussmann was founded on a peculiar political medley of state intervention, liberal deference to powerful economic elites, and a mix of aristocratic and imperial visions for the French metropolis. In his memoirs Haussmann frequently compared his reconstruction of the Paris water and sewer system to that of imperial Rome. In Pliny's description of Rome, for example, the city's sewers are singled out as "the most noteworthy achievement of all," and parts of the original Roman aqueducts were actually incorporated into Paris's new water supply system from the Dhuys, the Vanne, and the Marne in the eastern Paris basin.[84] Yet the imperial pretensions behind the rebuilding of Paris were shattered by a succession of foreign policy failures in Crimea, Italy, and Mexico, culminating in the defeat of the Second Empire in the Franco-Prussian War of 1870. While these episodes have been condemned by posterity, the legacy of urban and infrastructural reconstruction has met with admiration by a succession of twentieth-century historians such as Richard de Kaufmann, Wladimir d'Ormesson, Raoul Busquet, and André Monzet.[85]

Despite the physical transformation of sewers over the modern period, they have never entirely lost their earlier associations with danger, disorder, and threatening infestations. Through their various gender-laden, seditious, and mutagenic permutations sewers have come to symbolize the particular fears of each successive phase of bourgeois society. They have consistently been portrayed as a focal point for political threats to social and political stability, both real and imagined: during the 1870 siege of Paris in the Franco-Prussian war there was apprehension that the Germans might secretly enter the city through the sewers, and the city authorities actually sealed the Collecteur Général d'Asnières in order to assuage public fears.[86] The sewer has consistently been portrayed as a symbol of the unclean city, a metonym for what William Godwin termed the "entire excrementation of the Metropolis."[87] The "cesspool city" of the nineteenth century was a place where metaphors of disease and moral degeneration mingled with the threat of women and the laboring classes to middle-class society. Even the most progressive and perceptive of nineteenth-century commentators on urban life, such as Friedrich Engels, Charles Dickens, and Charles Baudelaire, failed to look beyond their dichotomous urban worlds of "dirt and cleanliness."[88] To understand the enduring association of the subterranean city with the "urban uncanny," we need to transcend these dualistic metaphors and develop a richer appreciation of how human bodies and urban form interact. In technical terms the Paris sewer system is both accessible and transparent: it remains the only sewer system in the world whose entire network can be directly inspected.

By tracing the flow of water through the "urban alimentary system," we can discern a series of tensions and contradictions that underlie the development of the modern city. Water provides a powerful link between the body and the built environment within which competing conceptions of public health and spatial order have become entwined. The very fluidity of water as both a biophysical and a symbolic agency serves to disrupt and challenge simplistic understandings of how complex urban societies function, and the degree to which social and spatial order can ever be achieved under the contradictory dynamics of capitalist urbanization. The Haussmann era, as we have seen, was both contradictory in its inception and also highly uneven in its practical impact. It was not until the 1930s, after all, that the whole of Paris was finally integrated into the city's water and sewer system as a "public service" in distinction to the more differentiated and private approaches of the past.[89]

# 2

## Borrowed Light: Journeys through Weimar Berlin

... everything is much more complicated than one believes, and also much simpler.

*Andor Kraszna-Krausz*[1]

Die Schönheit einer Stadt liegt entweder in der Funktion ihrer zweckerfüllten Arbeit oder in der Funktion ihrer (scheinbar) zwecklosen Ruhe und Erholung.

*Martin Wagner*[2]

The German Weimar Republic, which lasted from 1919 until 1933, has long been viewed as the unstable precursor to the rise of Nazism, its cultural achievements flourishing in exile in the United States and elsewhere. More recently, however, this turbulent period has been critically reassessed in order to consider the precise ways in which the Weimar era, with Berlin at its center, played a leading role in the development of twentieth-century cultural and political thought, and of the multifaceted phenomenon of modernism in particular.[3] One of the less studied aspects of this distinctive cultural and intellectual epoch is the place of urban nature in architecture, planning, and urban design. Water is integral to the emerging dynamics of metropolitan nature: no longer restricted to functional or ornamental domains, a new set of interactions between water, the human body, and urban space can be discerned. Most significantly water became the focus of new cultures of nature, including the growing popularity of bathing at the urban fringe, thereby establishing new connections between public culture and these distinctive types of seminatural landscapes. In this chapter we move through a series of architectonic discourses in relation to metropolitan nature, culminating in the cinematic representation of an excursion to a popular bathing lake.

During the 1920s Berlin emerged as an "experimental city" at the leading edge of new developments in architecture, art, cinema, photography, theater, and other fields.[4] The Prussian imperial city of parade grounds, with its characteristic Wilhelmine facades, was transformed into the capital of European modernism. "If the new movement in the arts is going to produce a Utopia," wrote the British art historian William Gaunt after visiting in 1930, "that Utopia will be found in Germany and the center of it in Berlin."[5]

In the early part of the Weimar era, in the immediate aftermath of the First World War, Berlin was a city in crisis, plunged into economic and political turmoil. In the art of George Grosz, for example, the city is shown as chaotic and violent, bathed in an eerie red light. Grosz's imagery, both of central and peripheral spaces in the city, denotes a brutally transactional landscape extending to both human flesh and the built environment.[6] Similarly in the expressionist cinema of F. W. Murnau, Georg Wilhelm Pabst, and Robert Wiene, a sense of foreboding and despair is captured through a fascination with the macabre.[7] In architecture the expressionist impulse became reflected in fantastical and escapist designs for the metropolis of the future: Bruno Taut's *Alpine Architecture* (1919) depicts a vertiginous city nestling among mountain peaks, while sketches by Erich Mendelsohn resemble architectural exoskeletons, as if the buildings had emerged out of the ground like elaborately shaped concrete fungi.[8] The organic theme is also reflected in the ebullient sketches and drawings of Hans Scharoun, who would later become one of the pivotal exponents of architectural modernism for the city.[9]

By the mid-1920s, however, after the introduction of a stable currency in 1923, the German economy began to recover, and Berlin in particular saw the growth of electronics, optics, publishing, filmmaking, and other sectors. The most comprehensive Weimar-era survey of Berlin, written by the geographer Friedrich Leyden, presents a detailed picture of what was by the late 1920s the fourth largest city in the world, after London, New York, and Tokyo, with over four million inhabitants. Using an array of data sources, Leyden depicts the city as increasingly polarized between densely packed inner-urban districts such as Kreuzberg, Wedding, and Friedrichshain and leafy suburbs such as Steglitz and Zehlendorf.[10]

As its economy recovered, Berlin emerged as an international center of migration for artists, writers, intellectuals, and political dissidents. By 1924, for example, the Russian community alone had grown from 50,000 to 300,000, including figures such as Vladimir Nabokov, whose first novel, *Kamera Obskura* (later translated as *Laughter in the Dark*), is set in Weimar Berlin, using cinema as its central metaphor. In literature and the visual arts the experience of modernity was often captured by the use of montage to represent the dynamism and simultaneity of urban life. In Otto Möller's painting *City* (1921), for instance, Berlin is represented as a kaleidoscope of brightly colored symbols such as numbers, windows, and street signs, whereas in the collages of Hannah Höch we find fragments of found images, such as photographs in magazines, that are mixed together to acerbic and often comical effect.[11] As for the experience of time, Alfred Döblin's novel *Berlin Alexanderplatz* (1929) can be compared with James Joyce's *Ulysses* (1922) in its complex layering of simultaneous events.[12] The contemporary *Großstadt* was not simply seen as the acme of industrial modernity but also as an ideological counter to antiurban sentiment, providing a social milieu within which women, sexual minorities, political dissidents, and others could enjoy unparalleled freedom.[13]

For many writers Berlin was the archetypal modern metropolis at the leading edge of social and cultural change. Indeed, the sense that the city exemplified aspects of modernity certainly predates the Weimar era, as captured in the turn-of-the-century essays of Georg Simmel. Simmel found in Berlin the distillation of the "money economy," the incessant regularization of time, and the intensified psychological effects of modernity.[14] By the 1920s the earlier sociological insights of Simmel were supplemented by the historical materialism of Walter Benjamin, the economic history of Werner Hegemann, and the astute critical observations of Siegfried Kracauer.[15] Studies such as Hegemann's *Das steinerne Berlin* (The stony Berlin), published in 1930, revealed the impact of land monopolies on working-class housing conditions that had produced the characteristically dense urban structure of "rental barracks" (*Mietskaserne*).[16] The functional dynamics of the capitalist city were now subject to rigorous critical analysis that drew attention to how patterns of ownership and control affected urban development.[17]

During the 1920s there was a change of emphasis in the city's visual arts toward the depiction of everyday life in photography, documentary filmmaking, and portraiture. Dubbed the "new objectivity" (*Neue Sachlichkeit*) by the art historian Gustav Friedrich Hartlaub in 1923, this new movement held implications far beyond the visual arts.[18] The photography of László Moholy-Nagy, for example, revealed new possibilities for the portrayal of the human body, everyday objects, and architectural aspects of the city. Other photographers such as Martin Munkacsy produced work that included the detailed and unflinching portrayal of urban poverty. The "cellar apartment" photographs of Munkacsy, reminiscent of the New York images of Jacob Riis, drew attention to the dank overcrowded conditions in which many people lived. Similarly, innovative documentary films such as Slatan Dudow's *Wie der Arbeiter wohnt* (How the worker lives) (1930) and Albrecht Viktor Blum's *Im Schatten der Weltstadt* (In the shadow of the world city) (1930) promoted a progressive political agenda for housing reform, better working conditions, and the alleviation of poverty.[19]

The post-Loosian conceptions of design, pioneered by the Bauhaus school, stripped away unnecessary ornament and sought to make explicit the connections between objects and society.[20] The Wilhelmine facades that hid "dark and airless dwellings" were to be replaced by an open city filled with "sunshine, light and air."[21]

The landscapes of Weimar Berlin were marked by a series of contrasts: whereas the inner city was becoming increasingly overcrowded, toward the city's edge there were extensive undeveloped areas of forests, lakes, and other remnants of the original landscape. As the city grew, the need for better access to nature for the working population became more pressing, drawing together disparate ideas including didactic conceptions of the civilizing effects of parks, and in particular a series of modernist architectural interventions that explored the possibility of creating a new synthesis between nature and culture. The earlier "open space politics" (*Freiflächenpolitik*) of allotments, park design, and other fields became much more complex and wide ranging, extending to youth groups, trade unions, and tenant organizations, in order protect spaces of nature in and around the city.[22] The distinctively lake-studded woodland landscapes that encircled the growing city became the overriding focus of new political campaigns to protect these remaining spaces of "wild nature" from uncontrolled development: these locales were being encountered by more people at the precise moment of their intensified erasure through speculative development. The industrial city generated an intensely modern sensibility toward nature that encompassed an aesthetic longing for "unspoiled" landscapes, a yearning for tranquillity, and a new enthusiasm for outside sports.[23] Water in particular offered possibilities for extending and elaborating earlier conceptions of urban nature, since it not only intersected with the engineering of the modern city but also provided opportunities for new forms of corporeal freedom.[24] Urban lakes such as Wannsee were suddenly and dramatically thrown into the middle of these emerging debates as a focal point for the protection and reconstruction of metropolitan nature.

CHAPTER 2

2.1   RECONNECTING THE CITY

With the extension of Berlin's boundaries in 1920 the population was doubled at a stroke, and the city could now be governed as an integrated metropolitan region for the first time. The expanded city incorporated already built-up areas such as Köpenick, Spandau, and Steglitz, along with significant tracts of lake-dominated forest landscapes. Under the progressive administration of mayor Gustav Böß, from 1921 to 1929, greater attention was now placed on the strategic needs of the city in fields such as housing, transport, and leisure. His appointment of Martin Wagner as the city's building commissioner (*Stadtbaurat*) in 1925, with responsibility for planning, proved of crucial significance, since Wagner had already developed a series of sophisticated proposals for the city.[25] The newly created office for planning headed by Wagner was located within the city's housing department, so that housing and planning were integrally linked from the outset.[26]

The 1920s saw significant improvements in transport and urban infrastructure for Berlin, along with the implementation of an 1918 law against light-limiting features for new housing developments such as "back courtyards" (*Hinterhöfen*) or "side wings" (*Seitenflügeln*) that characterized the worst of the overcrowded *Mietskaserne*. Although the complete reconstruction of the inner city was impossible—the Böß-Wagner partnership could never wield the same power as Haussmann in Second Empire Paris—there were important opportunities for new projects in less-developed outer parts of the city. Under Wagner, the engineering-led vision that had underpinned the Hobrecht plan for Berlin of 1862 was replaced by an expanded conception of the city: no longer dominated by public health concerns, the purview of planning now extended to human happiness in its broadest sense.[27]

Under Wagner's influence Berlin emerged as a focal point for new approaches in architecture, planning, and urban design as he brought together an unparalleled range of contemporary talent, enabling a series of innovative proposals to be implemented. Examples include the Britz or "horseshoe" estate (1925–1933), designed by Wagner and Bruno Taut, and Siemensstadt (1926–1933), also known as "the ring," designed by Wagner and Hans Scharoun, with further contributions from Walter Gropius, Hugo Häring, Paul Rudolf Henning, Otto Bartning, Leberecht Migge, and Fred Forbát.[28] These new public housing projects were marked by high-quality building standards, spacious floor plans with extensive "double aspect" to ensure plentiful light, and the rejection of historicist ornamentation to hide or distort reality. The projects often placed emphasis on communal facilities such as washrooms or kindergartens along with shared gardens to produce new kinds of living spaces (*Lebensbereiche*).[29]

Wagner developed an integrated conception of relations between work, leisure, and urban space. His emphasis on the need to extend greater social control over banking, housing, and land markets marks the most distinctive and enduring element in his political vision for the city. Wagner emphasized how questions of land ownership and finance shaped the city: it was not enough to merely design better homes; it was also necessary to facilitate new patterns of investment through, for example, the creation of worker-controlled building societies.[30] In his essay "Das neue Berlin" (The new Berlin), published in 1929, Wagner bemoaned the power of speculators to undermine the public interest through their "capital connections" (*Kapitalverflechtungen*). Without greater control over capital and land the role of the urban planner would necessarily remain restricted, and problems such as access to better housing would not be resolved.[31]

With his neo-Weberian interpretation of the modern city as a complex machine, Wagner perceived an elevated role for technical expertise in guiding social and economic development toward "the rationalization of happiness."[32] He argued that the "integrated beauty and efficiency of the whole city" could only be realized by the rationalization of space and time.[33] The technological innovations that underpinned rising industrial productivity held the potential to liberate people from unnecessary work, in an echo of sentiments expressed by late nineteenth-century utopian thinkers such as William Morris and Edward Bellamy. If machines could run more efficiently, argued Wagner, then people would have a shorter working day and more time to enjoy life.[34] "The mechanization of work in the modern city," argued Wagner, "will enable the emergence of a more sophisticated and inclusive body culture."[35]

Wagner's conception of "open space politics" (*Freiflächenpolitik*) was founded on the enhancement of recreation opportunities for the urban majority who had little access to nature. His vision for Berlin drew on various sources, including his early collaboration with Leberecht Migge for park design in Rüstringen, the garden city designs of Ebenezer Howard, Martin Mächler and others, the Berlin city building exhibition (Berliner Städtebau-Ausstellung) of 1910, which paralleled similar events in other German cities such as Cologne and Hamburg, and the emergence of social movements for housing reform and access to nature.[36] In an essay published in 1916, for example, Wagner suggested that whereas the previous thirty years had been dominated by a politics of gas and water, the next thirty would see "a politics of light and air."[37] Precedents for Wagner's approach can also be found in the work of other landscape designers for Berlin such as Ludwig Lesser, founder of the German People's Park Association in 1913, who argued against narrowly didactic approaches to park design, and still further back in the nineteenth-century writings of Peter Josef Lenné, who lamented the worsening environmental conditions facing industrial workers.[38] The idea of improved health and happiness through direct contact with nature also stemmed from the emerging interest in heliotherapy and "sunshine cures" for tuberculosis and other diseases associated with urban poverty in the pre-antibiotic era, when architecture played a more direct role in public health discourse.[39]

Water lay at the center of Wagner's vision for Berlin, since most of the city's forests were interspersed with lakes that could become the focal point for the recreational needs of the whole city (figure 2.1). He argued that Berlin had become increasingly disconnected from water as a sphere of work, transport, and leisure; there had been a historic turning away (*Abkehr*) from water. The river Spree, for example, had largely lost its functional role

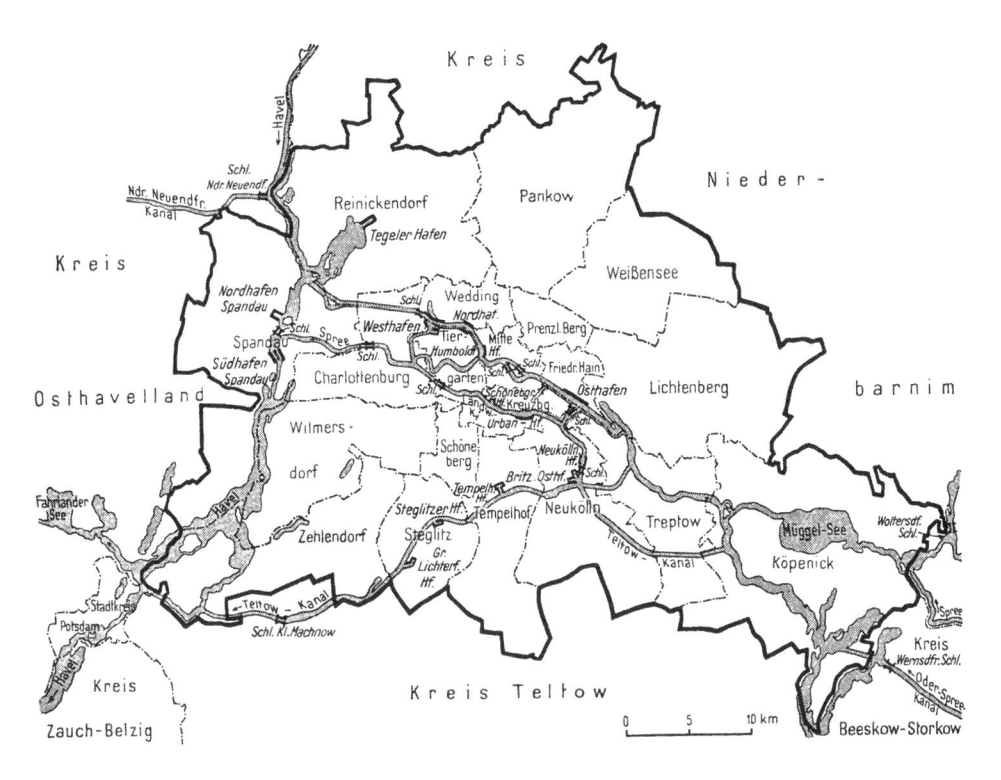

2.1 The principal rivers, lakes, and canals in Weimar Berlin. Source: Friedrich Leyden, *Groß-Berlin. Geographie der Weltstadt* (1933).

to railways, roads, and new types of industries, but had not yet acquired a new relationship with the city.[40] Unlike earlier uses of water in urban design, which tended toward decorative elements, Wagner was interested in ways of increasing direct contact with water as a source of leisure.[41]

The growing popularity of outdoor bathing for leisure marks a contrast with earlier metropolitan appropriations of water such as river baths (*Flußbäder*) or other supervised forms of washing. The completion of a modern water supply network for Berlin was accompanied from the 1890s onward by the construction of a series of working-class bathhouses across the city that reflected the prevailing emphasis, through both the Wilhelmine and Weimar periods, on the need for greater cleanliness as a facet of social order. The Wilhelmine bathhouses, exemplified by the architecture of Ludwig Hoffmann, presented an ornate extension into a public setting of the idealized interior space of the private bathroom. Under the impetus of modernist urban design in the 1920s, however, the earlier hygienist emphasis on clean bodies was embellished by a concern with aesthetic simplicity and the disbandment of the messy eclecticism of traditional or imperial architectural forms. The Weimar era instituted a far more complex conception of the body than the highly regularized corporeal freedoms associated with the architectonic manifestations of the early twentieth-century hygienist movement.[42] If the modern bathhouse served as a supplementary dimension to the limited domestic availability of water for the urban masses, then the role of the bathing lake can be interpreted as a welcome release from the claustrophobic gloom of the overcrowded *Mietskaserne*.

In addition to the growing emphasis on paddling pools, lidos, and other water features within municipal parks, there was a parallel emphasis on widening access to the hundreds of natural lakes in and around the city.[43] Before the enactment of a law in 1922 to help secure public access to "bank paths" (*Uferwegen*) alongside lakes, there had been no legal protection for most of the woodlands near Berlin: an earlier Prussian law of 1915 to improve forestry practices had proved of little use in an urban context.[44] As early as 1892, for example, the city of Berlin had made an unsuccessful attempt to protect the Grünewald forest, a former Prussian hunting ground to the southwest of the city, in order to create a new park. In 1910 the architect Hermann Jansen had presented an influential plan for a "greater Berlin" that would protect remaining spaces of nature as a woodland and meadow belt (*Wald- und Wiesengürtel*) for the city. The political momentum to provide greater protection for nature had also been emboldened by the widening of the political franchise and growing popular interest in nature. An emerging synthesis can be discerned between concerns with nature conservation and the strategically oriented social goals of urban planning in terms of health, land use, and the rationalization of space.[45]

Berlin's lakes had become indispensable for the relaxation (*Erholung*) of the city's workforce, yet they were being steadily encroached on by private developments so that many could only be seen over the "rooftops of villas." The building of luxury homes marked a continuation of trends under way since the nineteenth century, when lake-dominated landscapes on the urban fringe had become a focus of speculation for the indebted Prussian state, which had begun selling off plots of land to create "millionaire's colonies," accelerating the drift toward an increasingly inaccessible landscape.[46] Since many riverbanks remained "in private hands with restricted access," argued Wagner, "the tenement-dwelling masses can only enjoy urban nature through cruises and private cafes that take their 'good money'."[47] After taking over as Berlin's *Stadtbaurat* in 1925, Wagner moved swiftly to protect over 40,000 hectares of lakeside woodland from encroachment by private development in order to create a new kind of metropolitan nature on behalf of the city as a whole.[48]

The Wannsee bathing complex in the southwest of the city marks perhaps the most significant illustration of Wagner's ideal for urban nature. Completed in collaboration with Richard Ermisch in 1930, this expansive urban beach, with its pristine changing facilities and viewing platform, is a striking example of modernist architecture (figure 2.3).[49] For the architectural historian Ludovica Scarpa the Wannsee project exemplified Wagner's conception of the city as a machine for the production of happiness: here was a new public landscape at the leading edge of technology and design. The project proved an immediate success: in June 1930 alone there were some 360,000 visitors, compared with 160,000 during the same month in the previous year.[50] This surge in popularity for recreational forms of nature echoes similar developments in Robert Moses-era New York such as Jones Beach State Park, but in Berlin access was to be by public transport rather than car. The goal of mass use (*Massenbenutzung*) for new leisure facilities was facilitated by improvements in transport, principally through the extension of train lines to outer suburbs, enabled by close collaboration between Wagner and the city's transport commissioner Ernst Reuter.[51] By 1928, for example, the electrification of the entire rail network had been completed to create the largest integrated public transport network in the world. With the completion of new lines to the suburbs there were now cheap and fast connections from the densely populated inner parts of the city to Wannsee, Schlachtensee, and other popular bathing lakes toward the edge of the city.[52] These new transport connections facilitated working-class access to nature but also accentuated tensions with private landowners, as land values rose in anticipation of further development.

With the completion of the Wannsee bathing complex Wagner had extended the earlier architectonic advances of the hygienist movement associated with figures such as Robert Koch, Max von Pettenkofer, and Rudolf Virchow. The emphasis on engineering, public health, and the science-led rationalization of space had only engaged to a limited extent with new forms of corporeal freedom that had developed under the impetus of modernity and greater demands for cities to be spaces of leisure as well as work. New interactions between water and the body facilitated by advances in plumbing, new codes of personal hygiene, and the advent of municipal baths were now increasingly transposed to an open-air context, enabling direct contact with nature. In biopolitical terms Wagner loosened the ties between water and earlier disciplinary discourses regarding the body and emphasized instead the enhanced possibilities for pleasure in the modern city. Wagner, like many of his contemporaries, envisaged an expanded role

2.2 Model of proposed Wannsee bathing complex by Martin Wagner and Richard Ermisch. Photo by Arthur Köster (1929). Courtesy of the Baukunstarchiv, Akademie der Künste, Berlin.

2.3 The Wannsee bathing beach in 1930. Photo by Arthur Köster. Courtesy of the Baukunstarchiv, Akademie der Künste, Berlin.

for architecture, planning, and urban design in public policy, not simply in terms of reshaping urban space but as part of a nexus of expert interventions that could improve the health and happiness of society.

## 2.2  *Ausflug* Culture

From the nineteenth century onward, spaces of nature beyond the edge of the growing city became increasingly popular destinations for day trips. A distinctive excursion (*Ausflug*) culture began to emerge fostered by lakeside taverns for visitors, improvements in transport connections, and also enticing depictions of these locales in art, literature, and the popular press.[53] The Berlin-based artist Adolph Menzel provides some of the most detailed representations of the nineteenth-century city and its changing environs. In *Berlin-Potsdamer Eisenbahn* (The Berlin-Potsdam railway) (1847) Menzel presents a striking image of a train in the foreground speeding toward us, with the Berlin skyline in the distance. This dynamic scene juxtaposes the rapidly growing city with the relative tranquillity of the surrounding countryside. The composition does not suggest a neoromanticist critique of industrialization but rather a new synthesis between city and country.[54] The popular writer Theodor Fontane also exemplifies an emerging modern sensibility toward nature, describing his many walks around the edge of the city and in the nearby landscapes of Brandenburg.[55] In the novel *Irrungen, Wirrungen*, first published in 1888, Fontane describes a view from a guest house balcony in the west of the city that takes in two distinctive facets to urban nature: both the Zoological Garden (which opened in 1844) and the northern tip of the Grünewald forest.[56] Similarly, the paintings of Walter Leistikow, a founder of the Berlin secession movement in 1892, with links to European symbolism, also drew attention to landscapes on the urban fringe, but these highly stylized late impressionist representations of lakes and forests tend to downplay any human or technological presence.[57] Indeed, Leistikow's painting of the Schlachtensee lake near Wannsee, *Abendstimmung am Schlachtensee* (1895), shows no people at all.

The lakeside pine forests that were pivotal to the open-space politics of Weimar Berlin were not a fragment of "wild nature" but a distinctive kind of cultural landscape that had been extensively modified by centuries of human activity.[58] In the eighteenth century, for example, a "rational forest economy" emerged in Grünewald and other seminatural woodlands near Berlin that concentrated on exporting oak, especially for British shipbuilding. Over time the forests became dominated by less economically valuable trees such as pine, which lent them their characteristic look.[59] Although Grünewald has increasingly been regarded as a "natural" forest during the twentieth century (it gained protected status in 1979) it remains an extensively altered landscape that has acquired a distinctive role in open-space politics.

The politics of nature in Weimar Germany encompassed a wide spectrum of ideological vantage points ranging from a conservative disdain for metropolitan "inauthenticity" to modernist architectural visions of a synthesis between nature and culture. Progressive impulses included the articulation of a "right to nature"—connecting the whole city to light, air, and water—as well as new forms of "human nature" derived from scientific and technological interactions with the body in medicine, engineering, and other fields.[60] Running through these debates, however, was an ambivalence about what constituted German nature or landscape. To what extent, for example, could new technological networks such as roads or railway lines form part of a coherent cultural palimpsest? Modernist design lay at the heart of these tensions, since its avowed internationalism and clear break with tradition, exemplified by the projects of Wagner and his colleagues, illustrated a different set of potential relations between nature and society, and by implication a different type of social formation.

In the early twentieth century we can find some of the first references to a "natural right" to outdoor bathing that encouraged demands for wider public access to Wannsee and other lakes in and around Berlin. The creation of the Berliner Freibäder-Verein (Berlin Open Air Bathing Club) in 1907, for example, links with the new popularity of bathing as part of a wider set of reform efforts geared toward health and leisure. After 1900 the teaching of sports, including swimming lessons, became a common feature in school curricula, which further encouraged the enjoyment of water.[61] The legalization of bathing at Wannsee in the summer of 1907 provoked delight as a challenge to antiquated moral codes of behavior and the pervasive surveillance of everyday life.[62] Popular bathing beaches such as Wannsee were scenes of protracted conflict over bathing decorum, as the cultural legacy of Prussian prudery gradually gave way to a new cultural politics of the body.[63] The growing popularity of bathing in freshwater lakes was mirrored by the rise of seaside vacations in Europe that displaced the more austere, class-defined, and spa-oriented use of water in the past, in which contact with water was emphasized as a form of hydrotherapy rather than a simple source of pleasure.[64] Shared activities such as swimming had a carnivalesque dimension that could affirm forms of collective identity, while temporarily softening or inverting social differences and hierarchies.[65]

In Weimar Germany the human body itself became a focal point for the shifting ideological contours between the natural and the "unnatural." "German culture between 1910 and 1930," suggests Karl Toepfer, "cultivated an attitude toward the body unprecedented in its modernity, intensity and complexity."[66] From dance, theater, and the visual arts to architecture and urban design, an emphasis on the "modern body" can be discerned, which became reflected in new cultures of bathing. In more secluded places nude bathing began to flourish, including the designation of open-air pools (*Freibäder*) or "air baths" (*Luftbäder*), sometimes just forest clearings, where nudity was now officially tolerated.[67] There was an ascetic drive to affirm "body and soul" against exploitation at work or leisure (including criticism of the "alcohol industry"). Yet this politics of the body carried a tension between progressive ideals and more reactionary impulses associated with the ideological promulgation of ideal body types or Teutonic cults of nature worship.[68] The new forms of corporeal freedom contrasted with an emphasis on the "healthy body" in service to the needs of the nation (and especially the military).[69] Yet the corporeal dimensions to

German modernism, which reached their apogee in the "hygiene, sport, nature" slogan adopted under Ernst May, who between 1925 and 1930 held a position in Frankfurt equivalent Wagner's in Berlin, should not be interpreted as an inevitable precursor to the "geometrization of the masses" and the spectacle of fascist rallies.[70]

## 2.3  A Cinematic Interlude

One of the most vivid depictions of life in Weimar Berlin is the silent film *Menschen am Sonntag* (People on Sunday) (1930), which was shot largely on location during 1929. Directed by Robert Siodmak and Edgar J. Ulmer, the film has a simple narrative built around four young people, all played by nonprofessional actors, who meet on a Sunday to go for a picnic and a swim at Wannsee.[71] Rather than a prepared script, the screenplay by Billy Wilder drew on the use of reportage to create a series of interlinked scenarios based on the Sunday habits of Berliners, from which he pieced together a structure for the film based on the idea of a small excursion to the city limits. Wilder was clearly influenced by the reportage of Franz Hessel, Joseph Roth, and other contemporary writers, as well as Siegfried Kracauer's vignettes about everyday life in Berlin, which were published in the *Frankfurter Zeitung*.[72]

We first meet the protagonists in their real occupations—film extra, wine salesman, shop assistant, and taxi driver. The city is depicted as a collage of contrasting elements: street sweepers; a child's face being washed under a water hydrant; crowds surging out of railway stations; advertising hoardings by the roadside. Scenes of children playing in rubbish-strewn alleys are interspersed with shots of Wilhelmine facades, expensive cars, and shop window displays. The ubiquitous *Mietskaserne* are represented by a lingering shot looking into the gloomy depths of an inner courtyard with rotting windows and peeling paint. The one-room apartment which Erwin, the taxi driver, shares with his girlfriend has only a tap and a small washbasin, the incessant dripping of which disturbs his dinner. Although modern Berlin has now been plumbed, with a fully functional sewer system, the private bathroom remains a luxury for most of the working population, and the only privacy afforded by their room is a small dressing screen in the corner.[73]

The gray and dusty center of Berlin where the characters meet provides a poignant contrast with the lush landscapes of Wannsee on the outskirts of the city. We see their train speeding through pine forests toward a newly completed station, from which they walk through the woods and choose a picnic spot near the water's edge (figure 2.4). Brigitte, the shop assistant, places a portable gramophone player on the sand and the scene is set for flirtation, jealousy, and mirth. As they laugh at Erwin flipping sausages ever higher into the air, their faces are interspersed with other laughing faces from the city. A restored scene, which had been removed by the censors, shows a naked mannequin in a shop window, which cuts directly to stacks of gravestones for sale: sex and death are comically juxtaposed as our real-life protagonists enjoy their day out.[74] When Brigitte and Wolfgang steal away from the others to be alone in the woods, their postcoital languor is intercut by shots of rubbish scattered among the undergrowth. A sense of the fleeting nature of social encounters in the modern city is underlined. With the closing intertitles we are reminded of the significance of Sundays as the only regular respite from routine: "And then on Monday, back to work, back to everyday life. ... Four million people waiting until next Sunday."

The journey into nature is the setting for an eroticization of relations among the protagonists. The lakeside setting allows the characters to lose themselves beneath the trees: as they shed their clothes they cast aside their class differences and workplace identities.[75] The ensemble of scantily clad figures, with their simple picnic, also evokes Edouard Manet's *Le Déjeuner sur l'herbe* (1863), a painting that mocked classical allegories involving the representation of the human figure in idyllic naturalistic settings. The bathing figure is a recurring motif, but in the modern era there is a break from classical or religious allegory. The depiction of bathers in German art from

2.4  Still from *Menschen am Sonntag* (1930). Directed by Gregory J. Ulmer and Robert Siodmak. Courtesy of the British Film Institute, London.

around 1910 onward shows a steady move away from more straightforward figurative representations toward expressionist motifs in, for example, the work of Ludwig Kirchner and Max Pechstein.[76] The modern figure of the female bather also begins to dispel essentialist or mythological associations between gender and nature. As Linda Nochlin suggests in her reinterpretation of Renoir's *Bathers* (1887), we know little about "the position of the putative *female* spectator" in an era when the female body remained not only hidden from view but also largely cut off from nature, sunshine, and the pleasures of outdoor bathing.[77] *Menschen am Sonntag* evokes certain parallels with Seurat's depiction of bathers in the Asnières district of Paris, yet in this case the protagonists are brought together rather than stranded in a state of anomie (see chapter 1). Indeed, Hans Feld, writing in 1930, refers to the film as "pointillist" in reference to its experimental approach to cinematic realism.[78] Like Seurat's bathers, however, *Menschen am Sonntag* is closer to a mockery of classical motifs: these are not nymphs but ordinary people swimming; this is no classical allegory but a picnic with beer and sausages.

The film *Menschen am Sonntag* is interesting in terms of its complex collaboration between the directors, Robert Siodmak and Gregory Ulmer, the screenwriter Billy Wilder (working from the idea of Kurt Siodmak), the cinematographer Eugen Schüfftan, and the producer Moritz Seeler, who founded "Studio 1929." The production parallels aspects of the collaborative architecture and design of the Weimar period and has been referred to as an "open, collective text," thereby dispelling any straightforward auteurist reading.[79] The aesthetic influence of the "new objectivity" movement can be discerned in the deployment of nonprofessional actors, the emphasis on everyday life, and the use of experimental developments in photography.[80] Schüfftan's cinematography, for example, uses natural light as a spotlight through trees to illuminate faces (a device also deployed by the photographer Moholy-Nagy), as well as freeze-frame techniques reminiscent of Dziga Vertov.[81] The use of location settings also lends a certain verisimilitude to the film as a precursor to Italian neorealism in works such as Luchino Visconti's *La terra trema* (1948) or Roberto Rossellini's *Germania anno zero* (1948).[82] The inclusion of intersecting narratives in *Menschen am Sonntag* has parallels with other so-called "city symphony" or "cross-section" (*Querschnitt*) films such as Walter Ruttman's *Berlin, die Sinfonie der Großstadt* (1927) and *Melodie der Welt* (1929) (which also features Wannsee) and Berthold Viertel's *K 13 513—Die Abenteuer eines Zehnmarkscheines* (The adventures of a ten-mark note) (1926). Other significant influences include Russian avant-garde cinema and expressionist elements derived from Fritz Lang, Ernst Lubitsch, and F. W. Murnau (with whom both Ulmer and Schüfftan had previously worked).[83] The experimental aspects to the film are poignant because they coincide with silent cinema's attempt, at the moment of its demise, to differentiate itself from increasingly commercial studio productions and affirm the continuing significance of cinema within the cultural avant-garde.[84] For contemporary writers on Weimar visual culture, the subtlety of movements and gestures attainable within silent cinema affirmed its status on a level with the most sophisticated forms of literary or theatrical expression.[85]

Whereas films such as Ruttmann's *Berlin, die Sinfonie der Großstadt* present Berlin as a "city of work," *Menschen am Sonntag* presents a "city of leisure" in which social distinctions are more diffuse or ill-defined. A number of films toward the end of the silent era deploy Sunday as a contrast with the routine of the modern city, including Murnau's *Sunrise* (1927), Jean Gourguet's *Un rayon de soleil sur Paris* (1928), Marcel Carné's *Nogent—eldorado du Dimanche* (1929), and Jean Vigo's *A propos de Nice* (1930).[86] The aimlessness of a summer Sunday lends an element of melancholy to *Menschen am Sonntag*, since, in Andor Kraszna-Krausz's words, the characters "will perhaps meet again or perhaps not."[87] Interviewed in 1931, one of the actors, Brigitte Borchert, suggests that the film was really intended for their work colleagues in the north and east of the city, but that they would rather have seen a kitsch movie instead.[88] This cultural disjuncture illuminates the complexities of what constituted public culture in Weimar Berlin. The idea of film as art, or at least as a field of aesthetic experimentation, stood in an ambiguous relation with the development of mass culture and commercial cinema.

Critical reactions to the film were mixed: the Hungarian art critic Alfréd Kéményi, writing under the pseudonym Alfred Durus for the Communist Party's newspaper *Die Rote Fahne*, approved of its production outside the studio system and the influence of Russian film, but criticized its "petit bourgeois neutrality" and failure to instill a clear revolutionary message.[89] In contrast, Hans Feld, writing in the *Film-Kurier*, described the film as "a picture of life in the metropolis" and "city people in nature."[90] Similarly, Herbert Jhering, in the *Börsen-Courier*, delighted in its "magical lightness of movement," seeing the film as a German counterpart to King Vidor's depiction of New York in *The Crowd* (1928).[91] Other leading Weimar-era film critics such as Rudolf Arnheim, writing in *Weltbühne*, focused on the experimental feel of the film, and praised the young filmmakers whose mistakes "were a thousand times more important than what a troop of dexterous commercial filmmakers could make."[92] Siegfried Kracauer, however, strikes a more ambivalent tone. In his survey of German cinema written later in exile, Kracauer praises the influence of the "new objectivity" movement but worries that the group of young people appear to have no political convictions, leaving the film as "noncommittal as the other cross-section films."[93] The essentially white-collar ensemble, save the taxi driver, was part of a "rosy veneer" that hid a darker and more complex truth.[94] For Kracauer, writing in retrospect, the film evokes the fragility of the Weimar era and serves as a ghostly portent of a destroyed future.

A bittersweet sense of melancholy pervades *Menschen am Sonntag*, since it was filmed in 1929 during the last "carefree" summer of the Weimar Republic.[95] With its ironically interspersed elements such as military parades and Prussian statues, the film hints at some of the tensions behind the Weimar era.[96] By the early 1930s the new public bathing beaches at Wannsee witnessed violent political clashes—even the design of swimwear was used to denote political affiliations.[97] Martin Wagner, as a supporter of the Social Democratic Party (Sozialdemokratische Partei

Deutschlands, SPD), was among the first public officials to be dismissed by the Nazis when they seized power in 1933.[98] Even the most trivial or strategically insignificant social organizations such as choirs or sports clubs were targeted by the Nazis to eliminate any possible sources of opposition.[99] After the Nazi takeover of the Wannsee bathing beach administration in 1933, the director Hermann Clajus, also an SPD supporter, committed suicide. There were immediate attempts to ban Jewish people from swimming that were enshrined at a national level from the mid-1930s onward in a series of increasingly draconian decrees to prevent "racial mixing" in parks, restaurants, swimming pools, and other public spaces.[100] As for the makers of *Menschen am Sonntag*, all left Germany for exile in the United States, where Robert Siodmak and especially Billy Wilder would go on to enjoy glittering Hollywood careers.

## 2.4 TRACES AND FRAGMENTS

Weimar Berlin has left a distinctive legacy in terms of architecture, planning, and urban design that remains critically significant for the history of twentieth-century modernism.[101] The emphasis on connecting working-class Berlin with nature led by Wagner went far beyond the more decorative aspects of nineteenth-century beautification or the conventional parameters of municipal park design. His modernist vision was internationalist and rooted in a utopian reading of technology. Water played an integral role in the development of twentieth-century modernism in Berlin, not just in terms of the metabolic prerequisites of the city but also in the production of a "rational synthesis" between nature and culture. This emerging relationship between water and the modern city encompassed not just public health but also leisure, happiness, and the creation of a more democratic urban landscape in a poignant counterpoint to the ideological distortion of nature as a source of nationalist aspirations.

In a joint essay written in 1943 Martin Wagner and Walter Gropius called for cities to be shaped by an "orchestra of specialists," likening the master plan to a musical score.[102] Unlike the aesthetic and sanitary preoccupations of earlier planning traditions, Wagner's expert-led vision was rooted in an analytical critique of capitalist urbanization that centered on questions of housing, land ownership, and working-class quality of life. Yet there remained tensions in the thinking of Wagner and his contemporaries between neo-Weberian conceptions of the modern city as a machine—influenced by North American patterns of urbanization and the rise of the automobile—and more avant-garde approaches rooted in a utopian synthesis between nature and culture.[103]

Wagner's political vision rested on a wide-ranging critique of capitalist urbanization: not its violent overthrow but its radical reorganization. The advent of scientific urbanism, the impact of Taylorism on industrial productivity, and other elements of a technologically mediated modernity opened up prospects for a wider definition of the role of planning in public policy. The term "organized capitalism," introduced by Rudolf Hilferding, posed the possibility that capitalism was little more than an organizational system that could be modified at will.[104] Yet this reformist position accentuated political tensions on the left that led to a fragmentation of antinationalist forces, exemplified by mutual distrust between the SPD and the German Communist Party (Kommunistische Partei Deutschlands, KPD).[105]

For neo-Marxian architectural historians such as Aldo Rossi, Manfredo Tafuri, and Karel Teige, the shattered legacy of twentieth-century modernism showed the limits to utopian socialism under capitalist urbanization.[106] By the early 1930s Wagner and his colleagues found that the legal and political impediments to their work were increasingly difficult to overcome.[107] Wagner's growing frustration with the political aspects to urban planning—he resigned from the SPD in 1931 in protest at their evident complicity in land speculation—led to his adoption of an increasingly technocratic perspective in the remainder of his career.[108] Yet Wagner's failure lay ultimately not in the realm of architecture, planning, or urban design at all, but in the inability of social democracy to prevent the rise of fascism.

# 3

## Mosquitoes, Modernity, and Postcolonial Lagos

With the fall of mosquitoes came the rise of modernity.

*Jorg Sieweke*[1]

Mosquitoes are omnipresent in Lagos. Lagos mosquitoes, like people living in Lagos, are smart and intelligent. You will be wasting good time if you think you can get rid of them by adopting measures.

*Amby Njoku*[2]

If Berlin bathing cultures represent one facet of the relationship between water, modernity, and urban nature, a very different set of developments is revealed by a focus on the epidemiological dimensions to urban space. Although the health-threatening effects of water rapidly receded from the cities of Europe and North America, with the establishment of the "bacteriological city" and standardized patterns of infrastructure provision, the picture in the colonial cities of the global South is markedly different. The focus of this chapter is not waterborne disease associated with contaminated sources of drinking water but the socioecological dimensions to urban topography and the persistent threat of malaria.

We turn to a specific facet of the recent history of Lagos, Nigeria, now the largest city in sub-Saharan Africa. Despite its size and historical significance Lagos is a comparatively understudied city: the historiography of the colonial era remains heavily descriptive, with only limited engagement with more recent debates in anthropology, political science, and other fields.[3] Key dimensions to everyday life in the colonial city, including public health, have been neglected in relation to more familiar accounts of the gradual emergence of nationalist aspirations. Yet the politics of public health, and of malaria in particular, provides insights into the limits, priorities, and evident contradictions of colonial governmentality that continue to shape the contemporary city. The question of malaria is especially interesting because we can detect a fundamental shift in epidemiological discourse at the precise moment of the denouement of the colonial project and the opening of a new phase in the city's history. This chapter considers the shifting relationship between malaria and modernity in Lagos with particular emphasis on the 1940s, which marked a transitional phase to the postcolonial era. The term "modernity" is deployed in this context as a contradictory amalgam of different impulses ranging across fields such as architecture, law, and medicine, encompassing both the forcible rationalization of space and also demands for the improvement of human welfare.

Malaria remains the biggest cause of ill health in the global South, with around 90 percent of malaria-related deaths occurring in sub-Saharan Africa.[4] Though the disease has been recorded throughout human history, the word "malaria" is derived from the eighteenth-century Italian *mal'aria*, contracted from *mala aria* or "bad air."[5] Before the cause of malaria was fully understood it was widely linked with "miasmic" conceptions of disease caused, for example, "by breathing the damp air rising from marshy areas after sunset."[6] In 1880, however, the protozoan *Plasmodium* parasite causing malaria was observed for the first time at a French military hospital in Algeria; and in 1898 research in India and Italy showed that these parasites were present at an intermediate stage of development in the salivary glands of some mosquitoes from the genus *Anopheles*.[7] Thus the disease was revealed to be a complex interaction between three organisms: humans, mosquitoes, and the *Plasmodium* parasites. The impact of these scientific discoveries was immediate and had far-reaching effects on the control of malaria: levels of mortality associated with the construction of the Panama Canal, for example, were immediately reduced, and the emerging science of malariology became a key element in colonial public health discourse.[8]

The earliest known European description of Lagos Island by the Portuguese explorer Ruy de Sequeira in 1472 suggests an extremely inhospitable landscape of swamps with a profusion of biting insects. The term "mosquito coast," used originally in relation to Nicaragua and Honduras (where it derives from the name of the indigenous Miskito), is perhaps more apt for the coast of West Africa, where the extensive mangrove swamps provided ideal breeding grounds for mosquitoes. Though Sequeira named the partially submerged island Lago de Curamo—possibly after the Portuguese port of Lagos—there is also evidence of indigenous settlers from at least the sixteenth century who referred to the island as Eko.[9] By the early eighteenth century contact with Europeans became more frequent, but a more diversified local economy extending to palm oil and other commodities did not emerge until the second half of the nineteenth century after the end of the slave trade. Successive British colonial administrations after 1861 sought to transform Lagos into the "Liverpool of West Africa," but attempts to improve urban conditions were hampered by lack of financial support from the British Treasury, regional political instability, and wider economic perturbations affecting the price of commodities such as cotton and palm oil.[10]

3.1   FEVER CITY

The city of Lagos is dominated by water: its presence and the disruptive effects of flooding; its absence and the dangers and indignities experienced in the "infrastructure-free" parts of the city; and its role in the transmission of disease, which includes cholera, dysentery, and many other causes of ill health. The metropolitan region has gradually expanded outward from the original settlement of Lagos Island to incorporate Victoria Island, Apapa, Ajegunle, Ikeja, and more recently the Lekki Peninsula and parts of neighboring Ogun State, to produce an extended coastal conurbation with a current population between 12 and 15 million according to various estimates.

Water has a specific relationship with malaria since the larvae of the *Anopheles* mosquitoes, which act as vectors for malaria, develop in water. The *Anopheles* genus contains some 460 recognized species of which over 100 can transmit human malaria, but fewer than 40 commonly act as vectors in areas where malaria is endemic. (There are over 3,500 species of mosquitoes in all, and other genera such as *Aedes* and *Culex* also play a role in the transmission of diseases such as dengue fever and encephalitis.)[11] In the case of West Africa, it was not until the 1940s that the different species of mosquitoes involved in malaria transmission and their precise breeding habits were understood. Two forms of the critical vector *Anopheles gambiae* were recognized: *Anopheles gambiae gambiae* which breeds intermittently in rain-filled puddles and ditches, and *Anopheles gambiae melas* which breeds continuously in the brackish water of tidal swamps and presents the main threat to human health in Lagos and other coastal settlements.[12] In the 1960s *Anopheles gambiae* was further refined as a "species complex," subdivided into seven morphologically indistinguishable species which remain the most important group of vectors for malaria in sub-Saharan Africa. Unlike other *Anopheles* vectors such as the *A. funestus* complex, the *A. gambiae* complex displays a diversity of behavioral adaptations which were acknowledged as an impediment to the effective use of insecticides as early as the 1950s, before resistance to DDT became widely recognized. The evolutionary plasticity and ecological versatility of *Anopheles gambiae* remain a significant focus for epidemiological research and the complexities of malaria control: evidence suggests, for example, that mosquitoes have adapted toward biting their human hosts outside to avoid indoor use of insecticides and other protective measures.[13]

Although the earliest medical records for Lagos did not distinguish malaria from other forms of "fever," the evidence suggests that this fever was likely to be predominantly malaria because of the "low lying and swampy nature of the town and abundance of water and pools for the breeding of mosquitoes."[14] It is not until the 1940s, however, that we begin to have a more complete picture of the prevalence of malaria in Lagos. The first systematic studies of malaria incidence in children in the early 1950s, for example, found that around 98 percent of cases were due to *Plasmodium falciparum* and that a third of women about to give birth showed signs of malarial anemia.[15] Of the six principal *Plasmodium* species that cause malaria in humans, *Plasmodium falciparum* remains the most dangerous, accounting for some 80 percent of illnesses and 90 percent of deaths (over 200 species of *Plasmodium* are now recognized, but these mainly affect other animals).[16] Rates of infant mortality in Lagos showed a fall from over 390 per thousand in the 1890s—a staggeringly high figure—to around 140 by the early 1930s (which is comparable to Britain in the late nineteenth century). Analysis of autopsy data in Lagos from the period 1935–1950 suggests that around 9 percent of deaths of infants and 14 percent of deaths of children aged between one and four years were due to malaria.[17] The precision of the figures is tempered, however, by the fact that many children suffered from a combination of sources of ill health, so that malaria, tuberculosis, or other diseases become recorded only as the final cause of death. "The cause of death," writes the pediatrician Derrick B. Jelliffe, "is the result of an accumulation of disease rather than due to any single entity. The immature, anaemic baby, living in overcrowded unhygienic surroundings, becomes relatively malnourished in the second semester of life. His anaemia and nutrition deteriorate still further as a result of persistent malaria, leaving an attack of broncho-pneumonia or gastro-enteritis to add the last straw to his pathological burden."[18] The medical literature of the time uses the odd-sounding term "reproductive wastage"—introduced by the Royal Commission on Population (1950)—to encompass all forms of death associated with abortions, stillbirths, and the death of infants aged below one year.

Yet this ostensibly scientific terminology—it even refers to the mother as the "maternal organism"—obscures any direct engagement with the everyday lives of the people who were caught within the purview of the colonial medical services.[19] The degree to which different forms of suffering might be connected with the colonial project itself remains hidden.

The scientific discovery at the end of the nineteenth century that malaria was spread by mosquitoes which had bitten infected individuals had a profound impact on public health discourse. The focus of anxiety switched from the mosquito itself to the infected population, since it was the combination of insect vectors and human reservoirs for the *Plasmodium* parasite that caused the disease. Prominent malariologists such as Roland Ross, professor of tropical medicine at the University of Liverpool, immediately called for the segregation of Europeans in order to prevent the spread of malaria, along with other measures such as swamp drainage. Although Ross reflected on how the "generally very poor and semibarbarous" indigenous population might be protected against malaria, his main concern was with the protection of Europeans in order to defend British colonial interests.[20] The recognition of what the Lagos governor Sir William MacGregor termed an "imperial responsibility with regard to malaria" always held an innate ambiguity between narrow self-interest and wider public health ambitions.[21]

Ross and other colonial public health advocates repeatedly emphasized the specific challenges of West Africa, in comparison with India and other colonies where a greater degree of residential segregation and administrative order had already been achieved. In 1901, for example, the Reverend W. H. Findlay commented that "in regard to structure and location of houses, to dress, food and drink, personal habits … there is far less uniformity, carefulness, [and] adaptation to tropical conditions among Europeans in West Africa than in India."[22] The emerging emphasis on racial segregation conflicted with the heterogeneous characteristics of early twentieth-century Lagos, where Europeans lived among the indigenous population to a much greater extent than in the colonial cities of India, Singapore, and elsewhere.[23] The Isale-Eko district, for example, became known as the Brazilian quarter during the nineteenth century and was marked by Portuguese architectural influences and an influx of Yoruba emigrants from Brazil: freed slaves who sought to return to their country of origin.[24] Using Maryse Condé's definition of *créolité* as a form of cultural mixing that transcends narrowly Caribbean or Francophone associations, we can argue that Lagos shared important similarities with the more heterogeneous forms of urban culture encountered in the New World.[25]

From 1901 the British government adopted a policy of racial segregation, led by Colonial Secretary Joseph Chamberlain and based on the scientific findings of William Ross, Joseph Everett Dutton, John Stephens, and others.[26] Yet the governor of the Lagos Territory and Protectorate, Sir William MacGregor, vociferously rejected the new policy. He wrote to Chamberlain arguing that "residential segregation would be disastrous," and that "there is at present in this colony no racial question."[27] At a lecture given to Glasgow University in 1902 MacGregor again returned to the "segregation question":

> I make no attempt to put these separatist principles in practice, because to my mind they are impolitic, and unscientific. In a place like Lagos it would be most unwise to introduce any racial distinction of that kind. ... If we fear infection from them, then we can protect ourselves by curing them.[28]

MacGregor also clashed with Sir Frederick Lugard, high commissioner for Northern Nigeria from 1900 to 1906 and then governor of Nigeria from 1912 to 1919, who had enthusiastically sought to implement this new policy of racial segregation in northern Nigerian cities such as Jos and Kaduna.[29] In the British colony of Sierra Leone, the move toward residential segregation in Freetown had led to greater emphasis on racial differentiation within the civil service itself and attempts to "purify" the colonial administration. The new racism posed a fundamental challenge to the historically important position of high-ranking mixed-race people within Sierra Leonean society. In time, therefore, epidemiological insights into malaria became subsumed within a wider discourse of "scientific racism" that gathered momentum in the early decades of the twentieth century.[30] The "scientific racism" of Ross and his colleagues stemmed from a mix of cultural, political, and epidemiological arguments that served contradictory sets of interests, underlining the fragility and illegitimacy of the colonial project itself. The architecture of prophylaxis through racial segregation instituted in Freetown and other cities could not be applied to Lagos because the city was already too large, too complex, and too politically powerful to be reorganized against its will.

Though Ross insisted on the "desirability of segregating Europeans from natives in the tropics," he also recognized that the costs involved in building an "entire new cantonment for the Europeans" could better be spent on other measures that "might benefit the whole public rather than only a small section of it."[31] Also unclear in the epidemiological science of racial segregation was any indication of how far infected mosquitoes might travel to nourish themselves on European hosts or find suitable sites to lay eggs: the behavioral aspects to entomology, as explored by Ross and his contemporaries, had the somewhat paradoxical effect of shifting attention from the threat of human reservoirs of infection toward the significance of potential breeding sites.[32] In these and other writings Ross seems to acknowledge that the scientific findings, for which he gained the Nobel Prize in 1902, had complex and potentially iniquitous implications. It is to these "other measures" that we now turn.

### 3.2 Draining the Swamps

The British emphasis on residential segregation in the early twentieth century formed part of a wider program of antimalarial policies that included the destruction of breeding sites for mosquitoes, the application of larvicides, and the use of nets. The free distribution of quinine—to cure people directly—had been pioneered by the bacteriologist and physician Robert Koch in German-controlled Batavia (now Jakarta), Java, and in New Guinea, but this option of universal treatment had never been promoted by the British colonial authorities on the grounds of cost.[33] In Lagos there was some distribution of quinine to children, partly under the auspices of the Lagos Ladies' League formed in 1901, but this was never extended to the population as a whole.[34] Consequently, the focus in Lagos remained on the control of mosquitoes, either through the use of physical barriers or through more systematic attempts to destroy mosquito larvae and their breeding grounds.

The first attempts to drain the mangrove swamps around Lagos had begun in the late nineteenth century. The notorious Kokomaiko swamp, for example, had already been transformed into "a pleasant ground instead of being a manufactory of deadly disease."[35] By 1901 the Lagos governor, Sir William MacGregor, was calling for swamp reclamation to be a "political dogma in Lagos, from which there should be no relapse":

> There is no reason for clinging to the doleful prognostications so often uttered with regard to the hopelessness of ever rendering Lagos a fairly healthy city. When its swamps are filled up, when the island is surrounded by a sea wall, when its rainfall is utilised, and its sewage regulated, then Lagos will be sufficiently healthy to become the proud Queen of West Africa, the greatest emporium of trade in this part of the continent.[36]

MacGregor also recognized that the technical means to transform Lagos were straightforward and all that was lacking was the money to do it: "The principal factor in the sanitation of Lagos is nothing else than money. The scientific knowledge required to successfully deal with fever, for example, is of the simplest. The one great question is that of finance."[37]

It is not until the 1940s that we find the first comprehensive efforts to transform the "disease topography" of the Lagos coastline. The background to this shift is the emergence of the Royal Air Force base at Apapa opposite Lagos Island as the most important staging post for Britain in Africa during the Second World War. The military base was infested with mosquitoes, and military personnel were quickly becoming ill: a survey in 1942 found that every one-man tent was filled with 500 "blood-fed females" each morning and that 80 percent of all military ranks contracted malaria within three months of arrival.[38] To counter this threat, malaria control at Royal Air Force stations across Nigeria was organized under a Malaria Control Board in July 1942, funded initially by the military before receiving support from the Colonial Office.[39]

During the 1940s the city of Lagos and its inhabitants became an "experimental centre" for malariology, with entomological and topographical surveys carried out by the Malaria Field Laboratory of the Royal Army Medical Corps. The decision to embark on an extensive swamp drainage program to protect the Apapa airbase was driven in part by the realization that the use of larvicides such as the highly toxic Paris green (a copper aceto-arsenite) would be ineffective "over so vast an area," and also by the fact that oil (used to suffocate larvae) was in short supply because of the war. A first challenge for the comprehensive swamp drainage scheme was the need to produce some kind of accurate map of the region. Early surveys found that it was the brackish "ecological zones"—this modern-sounding term is used—where the mangrove *Avicennia nitida* and swamp-grass *Paspalum vaginatum* predominated that formed the main breeding sites for the *Anopheles* mosquitoes. For these surveys "a small party of Africans, armed with machetes, cleared tunnels through the gloom of the mangroves, and chain-and-compass survey laboriously delimited the swamp area and mapped out the ecological zones":[40]

> Progress through the tangled Rhizophora is slow, hot and difficult; not infrequently chest-deep streams must be forded and for hundreds of yards at a stretch the surveyor may sink to his knees in mud at every step.[41]

The period between 1942 and 1947 involved the comprehensive drainage of nearly all the major swamps in the Lagos metropolitan region (figure 3.1). More than 6 square miles (15.6 square kilometers) of swamps were drained—comprising a quarter of the urban area—with an immediate effect on the prevalence of malaria (figure 3.2).[42] The drainage scheme drew on methods already used by colonial authorities in Batavia, Freetown, Havana, the Malay peninsula, and other locations.[43] By the spring of 1943 a clear fall in the incidence of malaria at the Apapa military base had already been observed. "A most satisfying experience," writes the malariologist Alan Gilroy, "is to see the sun-cracked dry bottom of a lake that a week before was crossed by a canoe" and "land over which one formerly tramped laboriously, knee-deep in mud, [that] can be strolled across in light shoes."[44]

Where the swamps consisted of land submerged by tides, these areas were considered "property of the Crown"; but where there might be conflict with private owners, a form of eminent domain was deployed through the Destruction of Mosquitoes Ordinance of 1945 that enabled the governor of Nigeria to carry out work on any privately owned land. This law also gave the responsibility for the subsequent maintenance of drainage works to private owners, which provided a further pretext for indefinite intervention and control.

Although the pace of swamp clearance had dramatically increased in the 1940s, there remained problems with administrative fragmentation and rivalry between different colonial authorities. The Mosquito Control Board, for example, that was formed under the 1945 anti-mosquito ordinance carried out drainage work requested by the Director of Medical Services. Meanwhile a parallel body had been established in 1939, the Lagos Drainage and Swamp Reclamation Board, "formed to try and combine the rival activities of the Swamp Reclamation Committee, the Drainage Board and the Lagos Executive Development Board," yet "the Town Council was left out of the picture."[45] It seems likely that mutual suspicion between different agencies was spurred in part by emerging intersections between swamp drainage and land speculation in the midst of increasingly shrill

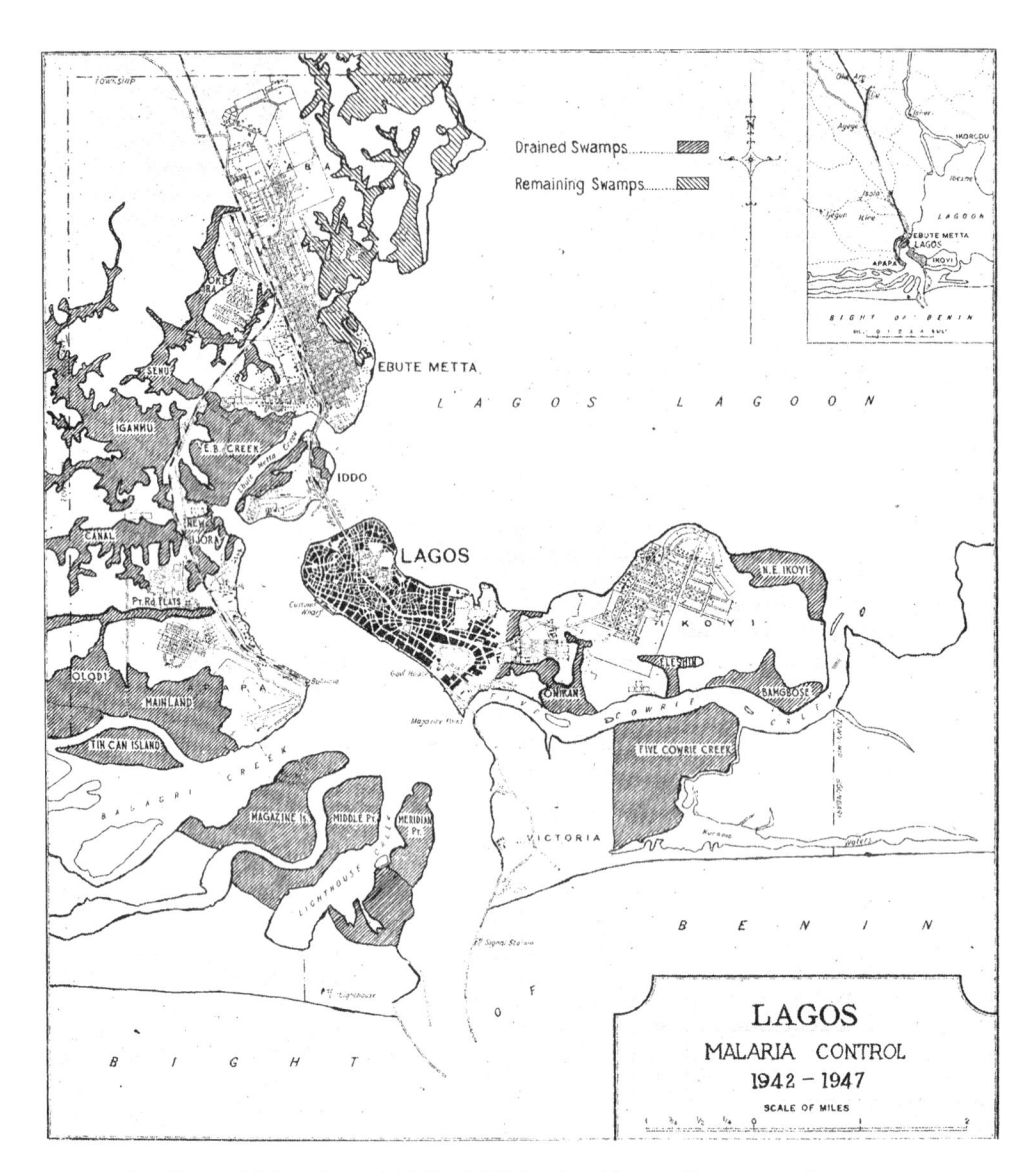

3.1 "Lagos, Malaria Control 1942–1947." Land and Survey Department, Lagos. Courtesy of the Wellcome Library, London.

3.2 Aerial view of swamp drainage scheme for Lagos with the characteristic "herringbone" pattern (circa early 1940s). Courtesy of the Wellcome Library, London.

condemnation of the unplanned character of Lagos, which was variously blamed by colonial authorities on women, "indiscipline," rural migration, and the persistence of a "village mentality." The main focus of tension lay between the Lagos Town Council, which strove toward being a "fully responsible municipality," and the Lagos Executive Development Board that held power over land and development rights.[46]

The swamp drainage scheme systematically extended control over land by the colonial state and became an integral part of the land development strategy of the Lagos Executive Development Board, which had been formed in 1928 in response to outbreaks of bubonic plague. This powerful, secretive, and unelected body proved to be a significant spur to the independence movement in the 1950s and came to symbolize the worst aspects of colonial administration.[47] In 1950 the first democratic elections took place for the Lagos Town Council, and the political tensions with the unelected Lagos Executive Development Board sharply intensified. The political character of Lagos was now rapidly evolving as "elements of the urban population," in particular the sizeable Yoruba middle class, became "active participants in urban politics." Other groups entering the political stage for the first time in the 1950s include "Muslims, indigenes, Ibos, women, members of the urban lower class, and traditional chiefs," so that "for the first time in its history, the Town Council boasted a heterogeneous and representative character."[48] Maladministration by colonial authorities was now being openly debated: in 1954, for example, the Nigerian National Democratic Party highlighted "the surreptitious attempt to evict the indigenous population of Lagos from their ancestral heritage to insalubrious surroundings" and called for the "immediate liquidation" of the Lagos Executive Development Board.[49]

The initial intersections between nationalist politics and malaria revolved principally around questions of land development, but the Ross doctrine of racial segregation did play a role in the Lagos scheme: the swamp drainage program initiated in July 1942 involved the removal of all villages within a one-mile zone of the Apapa military base. The indigenous population was now referred to as "the reservoir of infection," with particular anxiety over the "native child population" which had the highest incidence of malaria. Under this new landscape of control, no Africans were "permitted to remain at night within the one-mile zone," though "African-baited catching-stations" were to be deployed for research purposes beyond this exclusion zone.[50] This policy of "nocturnal segregation" had emerged in the British colonies on the pretext that the danger of racial mixing did not affect colonial officials by day but that at night a safe distance had to be maintained.[51]

The city of Lagos had benefited indirectly from the military-led efforts to control malaria, and over time it became clear that there was a confluence of interests between different fields of engineering, public health, and colonial administration. In 1948, for example, the director of public works for Nigeria, Sir Hubert Walker, praised "the value of the work to the war effort and the general health of Lagos." Walker suggested that it was fortunate that the field of malariology "trespassed on the engineer's domain" and deployed "the scientific approach, unhampered by preconceived ideas or conventional methods."[52] In 1948 the completed works were handed over to the Lagos Town Council's Public Health Department, which assumed responsibility for managing the newly drained land, and in the same year the first national ten-year program for malaria control in Nigeria was set up under the Colonial Development and Welfare Act of 1940, which marked a shift of emphasis toward social aspects of development, if only to shore up the fading legitimacy of the colonial project.[53] The drainage of Lagos marks a manifestation of malariology as a global science in the making, one that drew on the geopolitics of imperial self-interest but also connected with a nexus of locales for scientific research, such as the University of Liverpool, the London School of Hygiene and Tropical Medicine, and the nascent university system of Nigeria itself. The control of malaria also gives insight into the Nigerian state—at the meso level of the Lagos metropolitan region—as it sought to impose its own conception of spatial rationalization from above.[54] But hidden behind the confident inventories of swamps drained, rising hospital admission rates, and the latest scientific advances, we can find evidence of a degree of ambivalence about the scale of the task and the fading rationale for the colonial project. "There is no problem in West Africa that is not part of some other problem," reflected the Lagos-based malariologist Leonard Bruce-Chwatt; "all are intertwined; nothing can be treated in isolation."[55]

Descriptions of the drainage project set out clearly defined "ratios of control," suggesting that an optimum work efficiency was gained by "one European supervisor to 300 Africans."[56] Detailed accounts of the scheme suggest that only minimal managerial or technical responsibility was to be entrusted to non-Europeans (figure 3.4). The drainage of the Lagos swamps reveals a far more tightly centralized and state-directed mode of government control than the forms of "indirect rule" that had hitherto characterized the colonial project in Nigeria.[57] Large-scale drainage works have had an ambiguous relationship with the ideological dimensions to twentieth-century modernity: in fascist Italy, for example, the elimination of malaria through the eradication of swamps became a symbol of progress; and in German-occupied eastern Europe aspirations for land reclamation went hand in hand with race-based theories of Slavic primitivism.[58] In the case of Nigeria the drainage of malarial swamps emerged as a strategic priority for the British state that intersected with the wider ambiguities of colonialism as a project geared toward the disciplining of both society and nature. In any case, significant antecedents for European fascism lay in the "scramble for Africa" and the various experiments in governmentality that sought to impose radical conceptions of cultural and racial superiority.[59]

## 3.3   Historiographies of Absence

The detailed accounts of conditions affecting military personnel contrast with the relative absence of recorded information in newspapers, literature, or other sources on the experience of malaria within the indigenous Nigerian population. As Raymond Williams observes in relation to the

3.3 Surveying the swamps. The malariologist Leonard Bruce-Chwatt (right) and an unidentified colleague (circa early 1940s). Courtesy of the Wellcome Library, London.

3.4 Drainage works under construction for the Lagos swamps (circa early 1940s). Courtesy of the Wellcome Library, London.

"pastoral" and other idealized representations of life in rural Britain, it is difficult to uncover authentic records of the lives of the poor, who were often illiterate or had few opportunities to describe their own experiences.[60] The limits to historical representation have also been taken up by postcolonial scholarship, where the question of "subalternity," to use Gyan Prakash's term, goes further than Williams in acknowledging the limits to knowledge.[61] In a sense, the subaltern silence in relation to the threat of malaria contains two elements: first, the powerful and growing presence of a preexisting discursive framework for epidemiology, medicine, and other fields that was inseparable from the colonial project and framed all aspects of the written record in relation to it; and second, the limited degree to which indigenous populations, in all their heterogeneity, could remedy their living conditions, even in the heightened political moment of the run-up to Nigerian independence. Consequently, the stories of the urban poor in Lagos are largely absent from representations of everyday life in the city, and we must contend with "the systematic fragmentation of the record."[62] Even the burgeoning Nigerian literature of the 1950s seems to eschew direct engagement with the living conditions of the poor, perhaps because this literary circle was drawn to a large extent from social and cultural elites, and also because much of this literature owed more to drama than to realist fiction. In Cyprian Ekwensi's classic novel *People of the City*, for example, first published in 1954, Lagos is little more than a backcloth to the social and cultural tumult of the time, although in his later novel *Jagua Nana*, dating from 1961, the persistence of malaria, along with leaking roofs and open latrines, is used as evidence of political malfeasance.[63]

A slightly different slant is adopted in Chinua Achebe's novel *No Longer at Ease* (1960), set in Lagos during 1957, in the run-up to independence in 1960, where malaria makes a marginal appearance in terms of an individual act of kindness by a senior colonial official named Mr. Green toward an employee who is ill rather than "lazy."[64] Achebe narrates this incident through the eyes of an ambitious university-educated young Nigerian, Obi Okonkwo, who works for the civil service:

"I know," said Obi. "He will make a very interesting case for a psychologist. Charles—you know the messenger—told me that some time ago the A.A. wanted to sack him for sleeping in the office. But when the matter went up to Mr Green, he tore out the query from Charles's personal file. He said the poor man must be suffering from malaria, and the next day he bought him a tube of quinacrine."[65]

For Obi, the figure of Mr. Green is "a most intriguing character" who appears to devote himself to a country he does not believe in out of a sense of duty to "some great and supreme task":

Yes, a very interesting character. It was clear he loved Africa, but only Africa of a kind: the Africa of Charles, the messenger, the Africa of his garden-boy and steward-boy. … In 1900 Mr Green might have ranked among the great missionaries; in 1935 he would have made do with slapping headmasters in the presence of their pupils; but in 1957 he could only curse and swear.[66]

As for Lagos itself, Achebe suggests a twin city: one of electric lights, smartly dressed women, and high-life music, the other of gloomy slums. The city is divided between the newly emerging educated elites and the rest of the population, the main theme of the novel being not just the imminent departure of the British but incipient signs of a corruption that would mar the postcolonial state. Apart from the occasional mention of stinking or blocked drains, however, the conditions of the city itself are almost absent from the novel: slums, poverty, and ill health are little more than a shadowy and incidental presence to the main events. It may be that the occurrence of mosquitoes and malaria formed a continuous and ultimately unremarkable hazard of everyday life that appeared scarcely worthy of literary or journalistic attention: it is only with the first large-scale slum clearance programs of the 1950s that a distinctive Nigerian voice in relation to public health discourse becomes discernable in the public arena.

## 3.4  Fractured Modernities

The impetus toward modernity in an urban context has been bound up with social and spatial order: a dynamic that extends from the public arena to the design of individual homes. The realization that the mosquito was the vector for malaria instituted a flurry of design proposals in the early twentieth century for the "mosquito-proof room" and the "mosquito-proof house." The use of nets and screens became increasingly widespread, so that some kind of physical barrier against insects became a routine feature of wealthier homes. In 1902, for example, Sir William MacGregor, the Lagos governor, describes how he transformed his home from a "place of torture" by employing a full-time "mosquito-catcher" whose "sole duty it is to find and kill every mosquito that enters the house."[67]

Architectural responses to malaria and insect-borne disease revealed the limitations of a design-focused approach to public health. The threat of biting insects presented something of a conundrum for exponents of "tropical modernism": questions of regional adaptation, the use of natural ventilation, and other architectural innovations had to contend with the threat of malaria. In their book *Tropical Architecture in the Humid Zone*, published in 1956, Maxwell Fry and Jane Drew suggest that the "imaginary Mies van der Rohe house," with its optimum design, would make a "fully screened house possible" along with the elimination of standing water outside.[68] Yet these prestige projects, with their elaborate combination of European modernism with elements of vernacular African design, had little relevance for the wider population.[69]

Although swamp drainage had eliminated most of the mosquitoes associated with brackish water, other types of mosquitoes associated with standing water in the city itself remained relatively unaffected. Open sewers, for example, continued to provide breeding conditions for mosquitoes across the city despite attempts to modify their design.[70] In the 1950s banana trees were even banned from the city center because their leaves could retain small bodies of stagnant water favored by some mosquito species, principally the *Anopheles*, *Culex*, and *Aedes* genera, including vectors for lymphatic filariasis and yellow fever as well as malaria (see figure 3.5).[71] The Lagos City Council's mosquito control officials also had the power to fine residential compounds where any larvae could be found; consequently many buildings lacked roof gutters since these were often blocked by fallen leaves to produce small repositories of standing water.[72] Early studies of insects living in close proximity to human habitation revealed a diversity of vectors not just for malaria but for other dangerous or debilitating diseases.[73] The city's poor infrastructure—and especially its inadequate drainage system—continued to pose a grave threat to public health. A study of malaria prevalence among Lagos children in the late 1960s found that "the anopheline vectors of malaria are still commonly encountered in all parts of the area."[74]

Malaria control has also been hampered by the behavioral adaptability and "remarkable plasticity" of the mosquitoes themselves.[75] In colonial Egypt, for instance, Timothy Mitchell shows how the independent agency of nature—in this case represented by the spread of malaria-carrying mosquitoes in response to irrigation policy—should be woven into the analytical frame.[76] This is not to argue that mosquitoes are "actors" in the sense of an undifferentiated causal network that downplays human sentience, but to highlight the active role of biophysical factors in historical processes.

The legacy of malaria control in colonial Lagos ranges from "scientific racism" to humanitarian concern: it is no accident that leading figures involved in the colonial fight against malaria in the 1940s such as Leonard Bruce-Chwatt would later take on key roles within the World Health Organization or leading international research centers. The immediate postcolonial era in Nigeria reveals a high degree not only of administrative continuity with the past but also of the persistence of European influence in fields such as architecture, urban design, and medical science itself. Though the term "neocolonial" usefully captures the massive impact of oil extraction since the late 1950s, geopolitical meddling behind the Nigerian civil war in the 1960s, or IMF-imposed "structural adjustment programs"

3.5 Entomological survey work to check for the presence of insects in water hollows near to human habitation. The equipment being used is called a pooter. Courtesy of the Wellcome Library, London.

in the 1980s, there is another layer to the postcolonial transition that is both more subtle and more pervasive.

In 1955 the World Health Organization instituted a worldwide campaign to eliminate malaria with DDT, and the Second African Malaria Conference was held in Lagos. There was now optimism that malaria might be eradicated from Africa altogether. Nationalist demands added a further spur to existing medical discourse with a new emphasis on sharing public health benefits across the whole population.[77] By the late 1960s, however, a number of problems were emerging: there was growing evidence of mosquito resistance to DDT and other insecticides, even before the insecticides' environmental and health effects became recognized; more worrying still was the decreasing effectiveness of drugs such as chloroquine and pyrimethamine against malaria itself; and public health advocates had begun to recognize that inadequate primary health care systems and lack of resources would make the implementation of long-term solutions virtually impossible.[78] Even regions that had been temporarily cleared of the disease would be constantly under threat of reinfection from elsewhere.

The story of malaria in Lagos is unfinished. During the 1990s a significant increase in its incidence was recorded across the city.[79] The historical perception of malaria as a predominantly rural disease in sub-Saharan Africa is now changing. Urban populations are at significant risk for a combination of reasons: the use of irrigation in urban agriculture provides extensive pools of standing water; the prevalence of poor housing hampers the use of nets or screens, even if these protective measures are available; problems of misdiagnosis and limited access to health care increase the risk of death or serious illness; and dilapidated infrastructure systems provide ideal breeding conditions for mosquitoes (figure 3.7).[80] Indeed, the persistent relationship between malaria and inadequate urban infrastructure reveals how modernity itself, in its bifurcated colonial and neocolonial manifestations, has become part of the epidemiology of the disease.[81] As post-independence Nigeria morphed into a dysfunctional petro state, both physical infrastructure and civil society atrophied.[82] During the 1990s Lagos

3.6 Billboard warning against the danger of mosquitoes, Oshodi, Lagos (2003).
Photo: Matthew Gandy.

3.7 The open sewer system provides ideal breeding sites for mosquitoes. Amukoko, Lagos (2003). Photo: Matthew Gandy.

faced a combination of economic collapse, capital flight, and spiraling rates of poverty, which further contributed to the spread of malaria and other threats to public health.

### 3.5 Political Ecologies

Malaria is intimately connected with the interstices of the modern city: ramshackle structures, open drains, half-finished construction projects, and the "micro-ecologies" of the urban landscape.[83] Malaria forms part of the complex articulation between the urban and the rural, including the expanding frontiers of the city itself. Modernity has a paradoxical relationship with disease epidemiology: it has provided access to distinctive types of knowledge or medical expertise, but it also forms part of a larger picture of socioeconomic transformations operating at different scales, ranging from that of the individual household to the global economy. The disease and its insect vectors have undergone genetic changes and behavioral adaptations that further complicate the potential scope and impact of human interventions. The use of insecticide-treated bed nets has been widely promoted as an effective technological barrier to infection, but there remain significant problems in terms of their access and affordability.[84] In any case, the use of nets and screens for indoor protection forms part of an existing pattern of ad hoc measures that neglects to explore either the topographic characteristics of urban space or the ongoing behavioral adaptations of mosquitoes themselves. The public health threat of malaria is now far more intractable than that envisaged in the early postcolonial era, when confidence in medical and technological interventions was at its zenith.

The experience of malaria for the African majority in Lagos remains almost undocumented, beyond medical statistics. The lives of the urban poor were only recorded indirectly through other sources: charts, tables, and statistics compiled with increasing precision during the twentieth century give some insights into changing patterns of health in the colonial and postcolonial metropolis. It is only with large-scale ruptures such as slum clearance that some of these other voices are momentarily heard. More recently, instances of slum clearance do appear in literature—the destruction of Maroko is described in Maik Nwosu's *Invisible Chapters* (2001) and Chris Abani's *Graceland* (2004), for instance—but the earlier clearances of the 1950s and the experiences of the displaced bear no literary trace.[85]

Lagos, like many other cities in the global South, is faced by a contradiction: the poor are needed as labor in the urban economy, but the state is unable or unwilling to provide for them and the middle classes do not wish to live near them. The relationship between disease and segregation established in the colonial era persists in terms of middle-class intolerance: a "miasmic disdain" for the olfactory proximity of the poor is coupled with a neo-Haussmannite drive toward "urban redevelopment" and access to lucrative land. Lagos entered a new phase of technocratic administration after the transition to democracy in 1999, but it remains to be seen how these structural contradictions facing the rapidly growing city will be played out.

In the chapter's opening epigram from Amby Njoku we encounter a laconic use of the term "measures." A "measure" is defined as the "means of achieving a purpose," but in Njoku's sense we contend with the idea of the "measure" as something that is piecemeal, ineffectual, or doomed to failure. The idea of the "measure" has a degree of spontaneity or uncertainty about it. We could argue that Lagos has a long history of "measures," ranging from improvised strategies for everyday survival to autocratic attempts to impose control over the city such as the ill-fated "war against indiscipline" introduced by the Buhari military regime in 1984. What Lagos has consistently lacked, like many other cities in the global South, is a fully funded effort to transform urban infrastructure and health care provision. The elimination of malaria in Cuba, Italy, and elsewhere in the twentieth century was not the result of ad hoc "measures" but of a coordinated and long-term commitment to protect public health that extended to education, housing, nutrition, and many other areas of public policy. The experience of Lagos reminds us that poverty is not just a matter of economic insecurity or exposure to violence but also extends to other forms of corporeal vulnerability.

# 4

## Water, Poverty, and Urban Fragmentation in Mumbai

> Cities like Bombay—now Mumbai—have no clear place in the stories told so far that link capitalism, globalization, post-Fordism, and the growing dematerialization of capital.
>
> *Arjun Appadurai*[1]
>
> The nature of the municipal tap is feudal and bureaucratic. … You left the tap open before you went to sleep. When the water sputtered and splattered at three, four or five a.m. and sometimes not at all, was when your day began.
>
> *Kiran Nagarkar*[2]

In a dusty office of the municipal buildings in B ward in downtown Mumbai is a detailed map of the Paris water supply system, placed under a sheet of thick glass on the desk of the chief engineer. This "hydraulic decoration" acknowledges an attachment to a utopian vision of the perfect city: a striving toward a perfect synthesis of engineering science and urban modernity. The intricate arrangement of blue lines—varying in thickness and shading to depict the hierarchical structure of the city's water mains—is counterposed with the familiar bridges, boulevards, and radial subdivisions of the Parisian arrondissements. This striking cartographic representation of Paris is suggestive of a tension between the idea of the modern city as a visible manifestation of conscious design and the complex array of unseen networks extending beneath the city streets. Mumbai, like any other modern city, bears the imprint of successive generations of civil engineers and urban planners, yet its hydrological structure has never closely corresponded with a rationalized conception of urban space: time and again, ambitious plans and schemes have been only partially realized, leaving the material reality of the city far short of any technical ideal. In the case of water and sanitation Mumbai has never been able to provide the most basic services to all of its population; high infant mortality rates persist through poorer parts of the city, and the rudimentary sewer system is regularly overwhelmed by heavy rains. In the record monsoon of 2005, for example, over 400 people lost their lives in the city as makeshift dwellings were washed away, buried in landslips, or simply inundated by rising flood waters that could not disperse through blocked or absent drains. "The shanties of the poor," writes Somini Sengupta, "as well as the refuse of the rich blocked gutters and creeks."[3] The increasing frequency and ferocity of these extreme weather events—probably connected with anthropogenic sources of climate change and certainly exacerbated by the concretization and destruction of the city's natural floodplain—add another facet to the city's seemingly intractable challenge to modernize its water infrastructure.

Mumbai is a city of striking contrasts: the glittering towers and opulent hotels of Nariman Point, along with elite neighborhoods such as Juhu, Worli, and Malabar Hill, are encircled by informal settlements and armies of pavement dwellers who sleep in neat rows alongside city streets at night. By some estimates over half of the city's estimated 18 million people live in the *zopadpatti* or slums, which appear in city maps as amorphous gray areas clustered along roads and railways lines and extending into some of the most polluted and insalubrious spaces next to creeks and the remnants of once lush mangrove forests (figure 4.1).[4] During the nineteenth century Bombay became the largest and most economically lucrative city in the British Empire, and it remains the most important "global city" in the Indian subcontinent, with downtown property prices in the mid-1990s briefly exceeding those of Manhattan.[5] Changes in the Indian economy since the 1980s associated with market deregulation and trade liberalization have contributed toward a growing national and international role for Mumbai, which now accounts for as much as 75 percent of India's stock exchange transactions and at least a third of Indian tax revenues. Over the last fifteen years the relative importance of established administrative and industrial centers such as Chennai (Madras) and Kolkata (Calcutta) has declined in comparison with cities such as Bengaluru (Bangalore), Hyderabad, and Mumbai at the leading edge of expanding economic sectors such as financial services, software, and microelectronics. A widening segregation in both income and lifestyles is emerging between new global elites and the urban poor, whose continuing marginality is underpinned by intensified efforts to reclaim Indian cities from "encroachers and polluters."[6] Nominally public spaces such as parks and pavements or state-owned tracts of land alongside railway tracks or water distribution pipes have become the focus of bitter conflict between the middle classes and the urban poor: the "beautification" of Indian cities such as Delhi and Mumbai has involved renewed attempts to displace poorer communities and informal settlements from valuable land in order to bring about a forcible rearrangement of urban space.[7]

The inequalities and injustices that mark everyday life in contemporary Mumbai are exemplified by problems of access to water. The city's municipal water supply system is derived from six sources outside the city: Tansa, Modak Sagar, Upper Vaitarna, Bhatsa, Vihar, and Tulsi (figure 4.2). This network of dams and reservoirs links Mumbai to the vast interior of the state of Maharashtra—India's third largest—a region more extensive than Italy, marked by intense rural poverty, high levels of farmer suicide, and Maoist tribal agitation toward the east. Yet this vast transfer of nearly

# Chapter 4

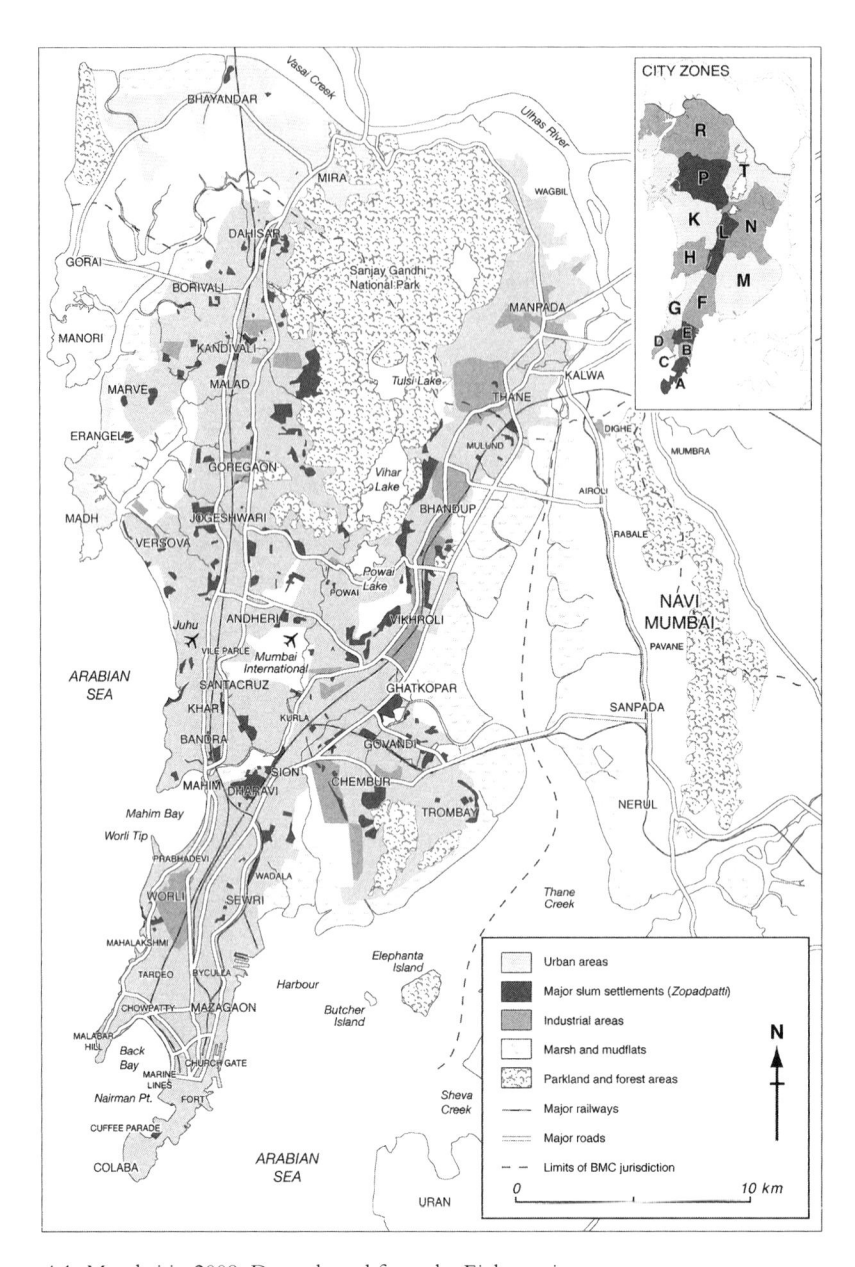

4.1 Mumbai in 2008. Data adapted from the Eichner city map.

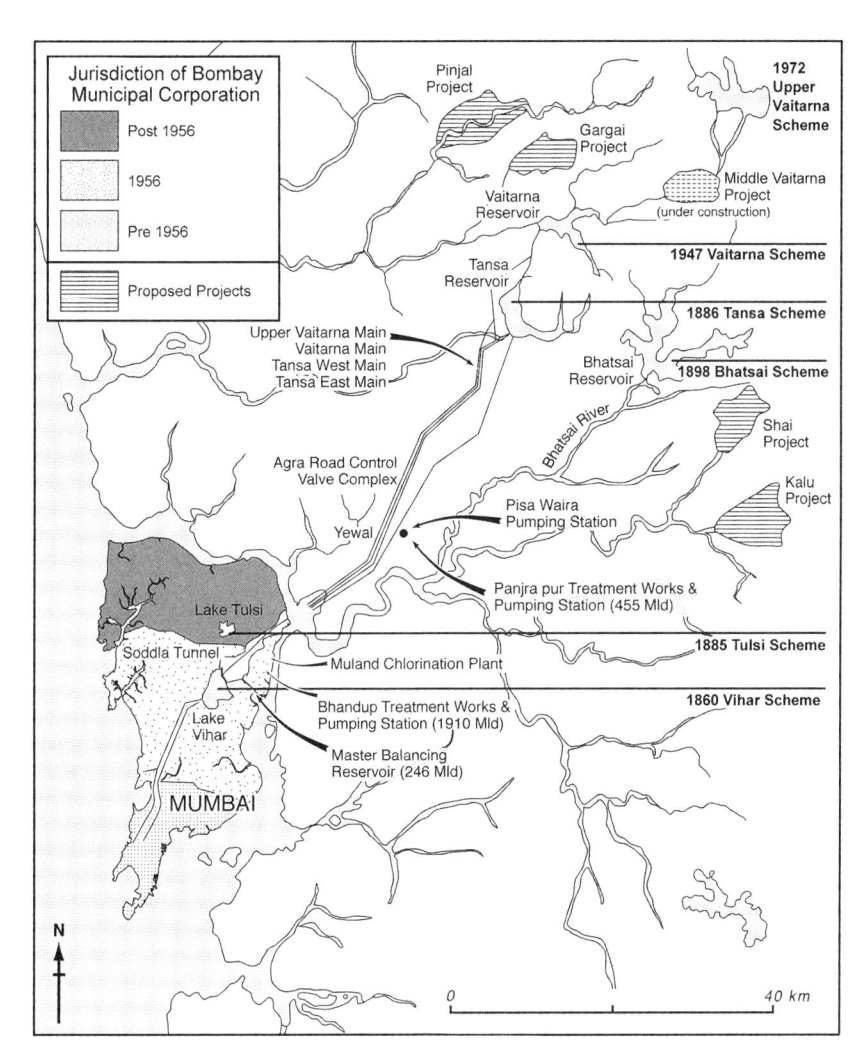

4.2 The development of the Mumbai water supply system. Adapted from original concept by Nayan Parekh and Rachana Sheth.

three billion liters of water a day from the jungles, lakes, and mountains of the state of Maharashtra does not meet the city's needs: many businesses and local communities rely on thousands of wells and bore holes scattered across the metropolitan region, along with hundreds of private tankers and innumerable illegal connections. Modern means of storing and distributing water in the city are also juxtaposed with complex patterns of water use originating in the precolonial era. A number of so-called "tanks"—elaborate bodies of water surrounded by stone steps—still exist in the city, but these are now mainly used for recreation or ceremonial bathing. The Bandanga tank, for example, located in the hilly and relatively prosperous Malabar Hill district of the city that overlooks the Arabian Sea, is an oasis of relative calm surrounded by a frenzy of construction activity; the cracked and faded paintings of elephants and other religious motifs that surround the water are indicative of a set of premodern and precapitalist forms of cultural engagement with nature that persist within the contemporary city (figure 4.3). Water is a dominant symbol in Hindu mythology: intricate distinctions and ritualized washing play a role in purification and social differentiation.[8] In many apartment blocks taps are separated by caste or religion, and in times of shortage lower castes are routinely "shooed away from their taps."[9] Although this "caste-based apartheid" is now weakening in comparison with rural areas, it has not yet disappeared and provides an added layer of social inequality to the religious, regional, and class-based tensions associated with the development of the modern city.

## 4.1   Colonial and Postcolonial Antecedents

Though contemporary Mumbai is widely considered to be an emerging megacity alongside other behemoths such as Dhaka or Lagos, it has in fact been of global significance in terms of its size and trading links since the early nineteenth century, so that its current predicament represents an accelerated continuity of its past. The epithet "maximum city," recently coined by Suketu Mehta in his vibrant survey of a city "reaching its own

4.3 Bandanga Tank, Malabar Hill, Mumbai (2002). Photo by Matthew Gandy.

extremity," could equally apply to the city's historical experience.[10] By the 1820s the colonial city of Bombay had a population of 300,000 making it the sixth largest city in the world. Yet the island city's rapid growth—spurred significantly by the opium trade with China—had been accompanied by a steady deterioration in urban conditions. A particular concern, as in many other nineteenth-century cities, was the unreliability of water supplies, with frequent shortages experienced during summer months. In recognition of the parlous situation, the Bombay Municipality appointed the city's first water commission in June 1845 to investigate alternative sources, and the commission recommended that a dam should be built to the north of the city at Vihar (figure 4.2). It was not until the severe droughts of 1854 and 1855, however, which necessitated the requisitioning of boats to bring in emergency supplies, that any political agreement could be found to raise property rates to cover the costs of the proposed water works.[11] The necessity of sanitary reform was also given added political impetus by the Indian rebellion of 1857 that underlined the ideological task of building a greater degree of loyalty to the colonial state. The construction of a functional infrastructure system forms a clear element in the building of political legitimacy for the modern state, yet the improvement of urban conditions in Indian cities had to contend with two sets of political dynamics: first, reluctance on the part of the British government to invest in their overseas colonies, for which any "internal improvements" had to be profitable;[12] and second, emerging tensions between colonial administrators and the Indian majority over any increases in taxes or property rates, particularly if the benefits of infrastructure improvements appeared to accrue mainly to European enclaves.[13] The struggling legitimacy of the colonial state in India was to a significant degree dependent on a gamut of infrastructure projects, ranging from water supply to extending webs of roads, railways, and communication systems, that might facilitate more devolved or bureaucratic variants of political control derived from bastardized elements of Victorian liberalism. Yet the emphasis on infrastructure in the classic Weberian model of a nascent modern state could only delay rather than efface the growing resentment toward colonial rule. The political turbulence of the 1850s, for example, added a sense of urgency to the more abstract and complex task of establishing a functional legal system and the extension of governmental interventions to hitherto neglected spheres such as the domestic arena, as part of a more systematized attempt to challenge sources of social or cultural resistance to colonial authority.[14]

When finally completed in 1860, the Vihar project for Bombay was the first municipal water supply scheme in British India, but access remained highly uneven, with only better-off households able to afford the smaller distribution pipes and taps within their homes.[15] By linking water supply with property taxes, however, the Vihar project soon became a key source of municipal revenue, as the number of individual house connections leapt from just over 200 in 1860 to over 6,000 in 1865 despite increases in taxes. The management of the new water system became part of a more complex set of municipal responsibilities and marked the emergence of a more technically oriented approach to handling the interrelated challenges of social and economic development.[16] In 1863, for example, the leading British civil engineer Robert Rawlinson called for a modern sewer system to be constructed in Bombay "according to true scientific principles," and in 1864 the first reliable census for Bombay was undertaken.[17] A stream of reports and expert recommendations now became part of the nexus of policy deliberation as engineers and public health advocates sought to promote the latest administrative and technical advances from Europe, yet this discourse of urban modernity increasingly conflicted with both the fiscal conservatism of colonial authorities and a sharpening cultural and demographic differentiation among segments of the urban population. This dilemma is illustrated by the publication of the influential report of Hector Tulloch, who argued that the critical issue facing the city was how to extend the "borrowing capacity and powers of the Municipality."[18] Like many other engineers Tulloch simply assumed that if the technical and financial issues could be resolved then an integrated hydrological system of the type adopted in London or Paris could also be implemented in Bombay.

By the 1880s, however, the earlier enthusiasm for the "new science" of sanitation became modified by emerging discourses of cultural and racial difference to account for widening disparities in living conditions.[19] In effect, the failures of technical modernization were increasingly ascribed to innate differences in sanitary practices or tolerance thresholds rather than to the limitations of liberal conceptions of urban government exposed by the colonial metropolis.

Although the city's first modern water system provided 32 million liters a day, this quickly proved insufficient: in 1872 the Vihar system was more than doubled in size, and the drought of 1879 forced the rapid completion of the Tulsi scheme (figure 4.2). A series of legislative changes in the 1870s and 1880s initiated a major reorganization of urban government with the creation of the Bombay Municipal Corporation: a structure that emulated the administrative scope of a Victorian town hall yet never enjoyed the same degree of effectiveness or legitimacy in tackling problems such as inadequate housing or sanitation. In the light of the Tulloch Report of 1872 and a range of other expert advice, the Municipality decided in 1885 to proceed with the far more ambitious Tansa scheme. This "magnificent project," effused the historian and colonial administrator Sir William Hunter, "will provide the city with an inexhaustible water-supply" that will "afford another splendid proof of the public spirit of the citizens of Bombay and the skill of English engineers."[20] The completion of the first stage of the Tansa scheme in 1892 brought a further 77 million liters a day to the city, followed by the completion of further stages in 1915, 1925, and 1948, adding a further daily capacity of 331 million liters. At first, however, the Tansa scheme contributed toward a public health catastrophe: the improved water supply in the absence of adequate drainage in low-lying parts of the city left whole districts in a filthy waterlogged state. As a result of the deteriorating environmental conditions, a series of bubonic plague outbreaks began in 1896 which lasted until the First World War, with devastating social and economic consequences.[21] The panicked exodus of half of the city's population during the worst episodes of disease led to emergency measures including the creation of the Bombay Improvement Trust, which set about acquiring land and demolishing slums (as we saw in the case of Lagos in the 1950s). Yet these radical measures emerged not out of any philanthropic concern with intolerable living conditions but "in the paranoid fear of the city's elites of pestilence and disease spreading to their bungalows from poorly ventilated and overcrowded slums."[22] The city's predicament at this time emerged out of a mix of laissez-faire economic doctrine, social indifference among the city's elites, and an administrative inability to coordinate different processes of modernization for the benefit of the city as a whole: a dynamic that would play a significant role in spurring the development of nationalist political sentiments during the early decades of the twentieth century.

At the time of India's independence in 1948 the city's total water supply stood at 494 million liters a day for a population of around 2 million, but there remained significant shortages and inequalities in access. Surveys in the 1950s found that the previous three decades had seen accelerating rates of growth with unprecedented levels of overcrowding through much of the city. And beyond the overcrowded tenements there were now extensive slums housing around 15 percent of the total population: an example in the vicinity of Sewri comprised a cluster of shacks housing 850 people with only two taps.[23] The anticipated demographic and commercial development of Bombay necessitated a further development of the city's water supply system, but this technical challenge also provoked a parallel discourse concerned with the containment of urban expansion, the implication being that technocratic approaches to urban policymaking could be extended from civil engineering to the social arena.[24] The first major water project for the city after independence, constructed between 1949 and 1957, involved the transfer of water from the more distant Vaitarna River and the construction of a vast reservoir named Modak Sagar after the city engineer Shri N. V. Modak (figure 4.4). The so-called "Vaitarna-cum-Tansa" project used the latest advances in engineering science and was the first water project to be completed using only Indian technical expertise,

4.4 Modak Sagar Dam, completed in 1957, Maharashtra state (2006). Photo by Matthew Gandy.

thereby exemplifying Jawaharlal Nehru's conception of a science-led Indian modernity within which hydraulic engineering would play a leading role. The development of engineering, especially in the immediate post-independence era, exhibited a sense of technocratic omnipotence before the overextension of the state's capacities and aspirations had been widely recognized.[25] The state, in other words, despite its colonial lineage and compromised legitimacy, had become the container for an immense program of national transformation in the context of a weak and fragmentary civil society. In the fields of civil engineering and urban planning we find that the Indian state adopted a form of "imitative modernity," to use Sudipta Kaviraj's term, as the earlier technocratic vision of the colonial city was supplanted by a new kind of modernist synthesis between science, technology, and urban society. Over time, not only the technical scope of this flawed vision but also the seemingly uncontrollable social and political dynamics of cities themselves threw into doubt the earlier certainties surrounding the search for a distinctive kind of Indian modernity.

By the mid-1960s Bombay's short-lived post-independence technocratic honeymoon had come to an end. Just as acute water shortages had forced the construction of the Tulsi project in 1879 ahead of the Tansa scheme, the failure of the monsoon in 1966 necessitated completion of the Ulhas River scheme in 1967 to divert additional water into the Tansa distribution network. In recognition of the growing scale and complexity of the city's water system, a new municipal department was created in 1971— the Water Supply and Sewerage Department—that brought together a range of planning and operational functions for the first time but did not include storm water drainage or regional water resources management. The use of an integrated treatment system at source was to be the model adopted for the much larger Bombay I–III projects under construction between 1980 and 1997, which included the completion of South Asia's largest water treatment complex—the vast Bhandup plant in the north of the city (figure 4.5). The completion of the Bombay I–III projects marked a significant expansion in the city's reliance on bilateral loans from the World

Bank, whose involvement in the financing of the city's water infrastructure dates from 1973. These projects faced an array of technical and logistical challenges ranging from disputes over land acquisition to continuing shortages of key materials such as cement, sand, and steel.[26] By the late 1990s, the three projects combined had added more than 1.3 billion liters a day of capacity to the city's water supply system, but demographic pressures left the city with a continuing inability to meet demand. Less than 70 percent of the city's inhabitants currently have access to piped water within their homes (and then for only a few hours a day), while the rest are dependent on shared standpipes, wells, tankers, and illegal connections.[27] And the rich—partly out of fashion and partly out of fear—have begun to consume increasing quantities of bottled water as part of "a new glamorous world of beauty and wealth" that seeks disengagement from the "disconnected segments" of the city.[28]

The UN predicts that Mumbai will be the third largest city in the world by 2025 (after Tokyo and Delhi)—with a population of over 26 million—and in anticipation of the city's inability to meet future needs, a range of new projects are either under way or at the planning stage.[29] The so-called Bhatsa (Mumbai IIIa) project has recently been completed, yet the much larger Mumbai IV project has not yet started because of uncertainties over its funding in combination with technical complexities (the scheme is dependent on power-driven rather than gravity-fed distribution) and conflict with the Maharashtra state government over its environmental impact.[30] There are also longer-term plans to link the city to the more distant parts of the Vaitarna and Ulhas river basins, along with the Gargai and Pinjal rivers, to create one of the largest and most technically complex water supply systems in the world.[31] The city is set to extend its "ecological frontier" ever further into the mountains of Maharashtra state in the context of a sharply polarized post–Narmada dams era of hydrological politics. (The damming of the Narmada River, beginning in the early 1960s, had by the late 1980s escalated into one of the most contentious environmental disputes facing modern India.)[32] Although the technocratic paradigm of

4.5 The Bhandup water treatment works, Mumbai (2006). Photo by Matthew Gandy.

large-scale infrastructure development continues to dominate longer-term planning, the fiscal, managerial, and regulatory context is now in a state of extended flux. The politics of Mumbai's "second nature"—the elaborate reworking of relations between nature and culture in the metropolitan arena—remains in a continuous state of contestation as new political formations seek to control the city. The rise of "bourgeois environmentalism," for example, and the renewed assault on informal settlements led by middle-class interest groups, often working in partnership with the state or international agencies, suggests an urban vision in which the inadequacies of urban infrastructure are likely to be disguised rather than directly addressed. If water infrastructure is to be conceptualized as a particular facet of the social production of nature, then we need to explore how specific manifestations of this "second nature" relate to the majority experience of the city as a space in which access to basic necessities is brutally circumscribed. The Indian experience of capitalist urbanization—through both colonial and postcolonial eras—has been marked by consistent ambivalence toward the modern city driven not only by material realities but also by the capacity of urban dysfunction to expose the limitations and weaknesses of nationalist aspirations to create a functional modernity. The large-scale water engineering projects that serve cities such as Mumbai are emblematic of the power of the Indian state to enforce its will on the regional hydrological cycle, while the dynamics of urban growth have long eluded the grasp of urban technocrats.

## 4.2 Hydrological Dystopias

Contemporary Mumbai faces massive inequalities in access to water: while most downtown districts receive water for at least short periods every day, there are outlying parts of the city that remain largely unconnected to the city's water distribution network. These disparities in access to water are etched into the urban landscape: some of the slums with the worst service provision are traversed by giant water pipes that have been transformed into precarious networks of elevated walkways. At Mahim Creek, for example, local residents make their way along the tops of huge pipes, their makeshift dwellings only a few meters away from the edge of the heavily polluted and mosquito-infested Mithi River (figure 4.6). The spatial interstices of the city's water infrastructure form ribbons of extreme deprivation that connect some of the poorest communities in the city. The situation is most acute at the urban fringe in districts such as Bhayandar, Mira Road, and Thane where rapid growth has not been accompanied by adequate improvements in basic infrastructure.[33] Unregulated construction activity and the provision of illegal water abstraction licenses have created intolerable conditions for poor communities living in what the film director Dev Benegal terms "landscapes of disaster."[34] Since the mid-1990s there has been growing political unrest in many of these marginal communities over the demand for better services. In March 1999, for example, in the district of Bhayandar on the city's northern edge, thousands of residents rioted for three days because of extended interruptions in their municipal water supply.[35] The city's "water mafia" took advantage of the situation by raising their prices tenfold for tanker supplies, and railway lines were blocked by hundreds of protesters to prevent commuter trains from entering the city. The roots of this popular unrest can be found in long-standing arrangements between local politicians and the tanker lobby to permanently delay the construction of new water infrastructure.[36] In October 2000 water shortages also led to unrest in the suburbs of Borivali, Andheri, and Chembur and the spread of carnivalesque street blockades called *rasta roko* to parts of downtown Mumbai.[37]

The city's urban fringe exemplifies the disjuncture between the postcolonial state as a formal set of institutional practices and its vernacular modification at a local level through complex interactions between different sources of power and authority.[38] Even where municipal authorities have tried to improve services, they have faced problems of collusion between private water suppliers and corrupt local officials who seek to engage in rent-seeking activities by limiting access to piped water supplies

4.6 Mahim Creek, Mumbai (2006). Photo by Matthew Gandy.

for marginal communities. Yet the use of the term "corruption" in this context actually masks an array of different practices and needs to be more carefully differentiated. At a household level, for example, it tends to manifest in terms of "speed money" to get repairs or new connections done more quickly, the falsification of bills or meter readings, and the provision (or overlooking) of illegal service connections.[39] At a meso level the consequences become more complex and deleterious, such as the formation of price-fixing cartels, the political allocation of contracts (where legislators may even become de facto contractors), and the stifling of access to capital (which affects all forms of investment). Over the last fifteen years water provision has also become increasingly linked with criminalized networks as part of an intensification of political corruption in Mumbai associated with the Shiv Sena control of public institutions.[40] The interrelated problems of corruption and political party funding in a context of weak state formation worsened after the 1970s with the spread of extensive "pork barrel" practices by local political elites: a trend underpinned by the decline in workplace forms of political organization and the electoral mobilization of different social constituencies along the lines of religious or caste-based forms of identification. The promise of better water and sanitation is vulnerable both to large-scale political manipulation, as evidenced by the populist agenda of the Shiv Sena in the 1990s, and also to extending webs of control within slum communities, since "anyone can take charge of water and collect money."[41]

The problems of water access at the urban fringe are also exacerbated by an unregulated scramble for available sources. In fast-growing suburbs such as Andheri, Goregaon, Kandivali, Kurla, Chembur, and Ghatkopar there is extensive illegal groundwater extraction, including new bore holes for bottled water sellers, hotels, and the construction industry. Speculative real estate companies even advertise lavish condominiums that resemble luxury apartments anywhere in the world in almost every detail except for one thing: they have no water. In order to circumvent limits of supply, many hotels and wealthy apartment blocks deploy illegal booster pumps to enable them to suck more water out of the system, so that existing inequalities in access become intensified at a local level. The city's urban fringe is also the focus of conflict over regional water diversion, in which marginalized farming and tribal communities in districts such as Thane have found that their water is being increasingly diverted to meet the needs of new urban developments. The city's water engineering strategy has consistently focused on the needs of the city above any emphasis on regional water use, yet the rural water crisis and escalating rural-urban tensions over access to water resources are a significant spur to new waves of migration to the city. Rural water resources have also been exploited by companies such as Coca Cola which seeks to sell its Kinley brand of "packaged drinking water" in the city: their water abstraction operations at Wada, for example, undertaken in the face of lax or ineffectual regulation, are diverting groundwater reserves needed by nearby farms and villages.[42] Water politics in India is inextricably linked with the social and economic disintegration of rural communities and widening polarities in wealth between urban and rural areas: a dynamic that reaches its acme in the poisonous sectarian violence of neighboring Gujarat state, where the misappropriation of funds for water projects has been deliberately obscured by the organization of anti-Muslim pogroms by the ruling BJP (Bharatiya Janata Party) administration.[43]

The scramble to access groundwater resources because of the inadequacies of the city's water supply network have led to irreversible problems of saline incursion in bore holes and aquifers in coastal areas of the city. The municipal water system also suffers from periodic contamination—not from its mountainous and rural water sources, but from the corrosion and dilapidation of the water distribution system itself that fosters the spread of bacteria within pipes and enables dirty water to enter the network through cracks and fissures. The city's official statistics suggest that a quarter of the city's water is currently lost through a combination of leakage and pilferage (principally through illegal connections), yet the real figures for "unaccounted-for water" may be between 40 and 60 percent.[44] The city's poorly maintained network is dominated by old materials such as galvanized iron, and parts of the network in downtown Mumbai are over a hundred years old. The strongly seasonal patterns of precipitation (peaking during the monsoon) in combination with the effects of rising salinity, vibrations from road traffic, stray currents from rail networks, and road excavations by other utility companies have accelerated the rusting and disintegration of older pipes, so that contaminated groundwater can seep into the water distribution system during interruptions in supply.[45] The city also cannot undertake large-scale repairs of trunk mains because there are no alternative water distribution tunnels available to temporarily divert water (as in New York for example) and avoid extensive disruption, so that even minor problems can easily escalate into catastrophic system failures.[46]

These problems of dilapidated infrastructure also extend into the private spaces of the city. Crumbling tenements or *chawls* are characterized by rusting pipes, leaking taps, and filthy tanks for the storage of water necessitated by intermittent supplies—in a typical *chawl* at least fifteen families will share one tap for little more than two hours a day. Most rented housing has had extremely limited investment in repairs or improvements: a combination of rent controls and shadowy networks of ownership ensure that what Arjun Appadurai terms "spectral housing" predominates through much of the city.[47] Some 16,000 buildings in Mumbai are currently "cessed," with rents frozen at 1947 levels, and must rely on a municipal agency called the Maharashtra Housing and Area Development Authority (MHADA) to carry out minimal repairs that do little more than prop up these structures to prevent them from collapsing into the street (although more than 2,000 of them have fallen down in the last decade).[48] The extraordinary complexity of the city—both in terms of its morphology and ownership—presents a bewildering array of technical, fiscal, and legal obstacles to improvements in water supply extending from the large-scale reconstruction of the underground city to the plumbing of individual tenements.

The city's water system has been dominated by a supply-oriented engineering ethos since its inception in the middle decades of the nineteenth century. Apart from drought years such as 1992, water conservation efforts have been only a marginal concern. More recently, however, there has been an emphasis on measures such as rainwater harvesting as a rediscovery of traditional approaches to water management that contrast with the gigantism of established technomanagerialist paradigms in water engineering.[49] This emphasis on smaller-scale technological responses to water scarcity has quickly emerged as a point of intersection between social, ecological, and architectural discourses concerned with both water management and the need for local government reform. For the first time in its history the city's engineers have been engaged in educational outreach programs to encourage changes in household water use and the introduction of water-saving technologies. At the same time, the spread of more profligate uses of water by the city's middle classes, exemplified by the construction of exclusive leisure complexes, reflects changing patterns of consumption that serve to undermine efforts at water conservation. Since the early 2000s the main roads leading to the north of the city have been fringed by vast billboards not just for the latest investment opportunities or fashion accessories but also for exclusive water complexes such as the aptly named Water Kingdom, billed as "Asia's largest theme water park," whose hoardings routinely depict pale-skinned women enjoying the newly constructed swimming pools and other water features. In this instance luxury access to water is being used to reinforce existing caste-based social distinctions and gendered stereotypes geared toward aspirant urban elites.

Critical within this emerging political dynamic is the role of grassroots campaigns to extend citizenship rights to marginalized communities through the actions of nongovernmental organizations such as the Society for the Promotion of Area Resource Centres (SPARC), the National Slum Dwellers Federation (NSDF), and the women's organization Mahila Milan.[50] Schemes to construct new toilet blocks in some of the city's poorest communities have involved a partnership between the municipal BMC, the World Bank, and SPARC based on what Appadurai terms "countergovernmentality."[51] The deployment of repertoires of local knowledge has allowed some of the poorest communities in the city to become visible for the first time, not just as objects of strategic intervention by the state, but as legitimate social and political entities in their own right. These new developments involve not only a vertical dispersal of power from the state to the grassroots level but also the horizontal development of networks between cities: the National Slum Dwellers Federation, for example, whose membership includes at least a quarter of million households in Mumbai, is now active in over 50 Indian cities.[52] Examples of new administrative structures emerging in Mumbai include the Local Area Citizens' Committees or LACCs set up in each of the city's wards with the support of the state and influential NGOs such as Action for Good Governance and Networking in India (AGNI). Yet the local urban environmental concerns encompassed by these new patterns of public participation remain only tangentially linked to strategic agendas evolving at a regional level. The floods of 2005, for example, not only revealed the full extent of the city's infrastructure crisis but also raised complex fiscal and political issues that cannot be handled on an ad hoc local basis. Equally, the expansion in local initiatives to improve access to sanitation must be set alongside the gathering impetus to eradicate informal settlements throughout the city. As a consequence, the expansion of community-based forms of government administration remains vulnerable to the innate weakness of the state to effect urban change and the fragile intersection between different sources of power and authority within the city. An emerging focus on "knowledge capital" involves a far greater fluidity in localized initiatives through the emphasis on experimentation and sharing of information, yet the role of the state remains highly ambiguous, as evidenced by the recent attempts of political and economic elites to capture strategic policy agendas and stymie the perceived electoral threat of the poor. The growing emphasis on what Darshini Mahadevia terms the "enterprise state" has involved a far greater focus on individualized or small-scale forms of public participation while broader strategic agendas elude public scrutiny.[53] We can discern a conceptual tension emerging between a more upbeat and extensively improvised urban vision rooted in the generation of new forms of cultural capital and a more wary interpretation of the fragility and volatility of recent urban political mobilization.[54]

The contemporary politics of infrastructure provision in Indian cities is inseparable from the shifting legal, demographic, and territorial definition of the "slum city" and its ambiguous relationship with discourses of modernity, progress, and the reconstruction of civil society.[55] New forms of political activism have been necessitated by the glaring disjuncture between formal rights—as set out in constitutional and legalistic frameworks—and the material realities of social injustice experienced by the urban poor.[56] This tension between a universalist conception of citizenship and the reality of extreme and multiple forms of social stratification has led some scholars such as Partha Chatterjee to distinguish between "civil society" and "political society" in an Indian context. For Chatterjee a distinction can be drawn between citizens and "populations" whose demographic presence must be accommodated or at least tolerated to a certain degree in order to allow urban space to function effectively. Critical to this dynamic is the malleability of the state and its legal norms, so that a series of pragmatic and varied instances of state action, inaction, and interaction with other sources of power can be observed.[57] The relationship between the state, democracy, and the public realm, for example, is scarcely addressed in the recent elision between neoliberal reform and the imposition of "good governance" that has characterized much recent writing on the policy dilemmas facing the cities of the global South.[58] While the emphasis on "citizen engagement" in providing and managing basic services has had some impact in specific fields such as the provision of toilet blocks and other small-scale structures, the emphasis on a more radical diminution of the role of the state in some of the "governance" literature holds far more complex implications for long-term investment in the public realm, whether this be the maintenance of existing technological networks or the construction of major new infrastructure projects such as storm water sewers or the extension of trunk mains to outlying parts of the city.[59] The extension of NGO activity into former areas of municipal responsibility should not be confused with greater public accountability, since these organizations are themselves embedded in social power structures and cannot be removed by electoral means if they fail to fulfill grassroots expectations: state structures, however flawed, have a continuity that is not shared by NGOs, which can depend on the input and commitment of a relatively small number of highly motivated or powerful individuals.

During the last decade there has been a growing emphasis on attempts to secure greater cost recovery for water services through the introduction of metering, revised tariff structures, and higher charges (which have risen sharply since the 1990s). Part of the pretext for these changes is the recognition that current charges amount to less than a quarter of the actual costs of water provision.[60] The city faces a major dilemma over the financing of its water infrastructure: there are now attempts to disentangle the city from its reliance on bilateral loans from the World Bank and other international lenders, but the linking of local capital markets to infrastructure investment would require a range of complex institutional and regulatory reforms.[61] The scale of institutional change needed for municipal bonds, for example, contrasts with the higher rates of return (and punitive water charges) that would be demanded if the city were to rely on private equity.[62] Insufficient levels of investment in the existing system have been used as an ideological pretext to push for divesting the entire system to the private sector: a degraded public realm is thereby portrayed as evidence of state failure, juxtaposed with sophisticated lobbying activities from international water companies.[63] Since the mid-1990s there have been external pressures from the World Bank, working alongside European water companies, to pursue privatization, but the city is resisting calls to sell off its water system.[64] A pilot privatization of water services has been undertaken for one of the city's wealthiest wards in Andheri, and the French company Castalia carried out a feasibility study with the support of the World Bank.[65] Yet leading municipal engineers raise concerns that the imposition of a simplistic privatization model ignores the complexities of ground-level realities and the immense spatial differences across the metropolitan region.[66] These worries have also been heightened by a spate of privatization failures across the global South in cities such as Cochabamba, Buenos Aires, Dakar, Jakarta, and Manila, that have in some cases led to violent protests and bitter recriminations.[67] The shifting political context is also illustrated by the recent abandonment of many planned schemes, including a major World Bank–funded project for Delhi. Many private-sector water suppliers now recognize that there are few if any profits to be made in extending water services to urban slums and are actively seeking to reduce their exposure to financial risk: leading global water companies such as SAUR, Suez, and Veolia (formerly Vivendi) are now trying to disentangle themselves from complex and loss-making commitments.

### 4.3  ZONES AND ENCLAVES

The absence of a fully functional water and sewer network in contemporary Mumbai can be attributed to a number of factors. A long-standing issue is the extent to which the Indian state has been "captured" by the middle classes, so that its political agenda has been consistently diverted from the universal provision of basic services. The middle classes have long been the principal beneficiaries of the municipal water supply system, and they exert little pressure on the state to extend services to the poor.[68] Where there are deficiencies in the municipal system, wealthier households also engage in a variety of opt-out strategies by investing in pumps, storage tanks, and purification devices, so that any cohesive political response to governmental failure is countered by a myriad of individual responses. Despite the constitutional amendment of 1992 to widen political representation by women and scheduled castes, the political strength of the urban middle classes in India appears to be growing as part of a new discourse of "environmental improvement." Recent campaigns to clean up Indian cities have been framed around the appropriation of public spaces for the use of "respectable citizens," with a growing intolerance for informal markets, makeshift structures, and the spatial and cultural practices of the urban poor.[69] In tandem with this embourgeoisement of existing public spaces, there has been a parallel political maneuver to capture "empty" or marginal spaces such as former industrial areas for speculative urban development. At an ideological level this new middle-class self-confidence with regard to the city has been emboldened through the selective appropriation of a cultural and architectural legacy within which significant parts of the urban experience have been effectively erased.[70] The reconstruction of urban space is occurring within a weak civil society in which the practice of citizenship has been deeply coded by moralistic social distinctions rather than any clearly articulated conception of human rights. Burgeoning middle-class intolerance toward the poor as impostors within the city is being facilitated, it seems, by public-interest litigation that had once afforded a degree of protection to the most vulnerable sections of the urban poor such as pavement dwellers. The legal recognition of the right to shelter, for example, has been weakened by an increasingly narrow interpretation of the state's obligations to the poor and greater emphasis on cleanliness and "encroachment."[71]

A further factor behind persistent inequalities in access to water and sanitation is that, apart from dramatic exceptions such as the Surat plague of 1994 or the resurgence of the deadly *falciparum* strain of malaria during the 1990s, the public health crisis facing slum dwellers does not directly endanger middle-class residents.[72] Many cities face a paradoxical combination of increasing wealth and deteriorating public health through the disruptive effects of social and economic upheaval.[73] Severe disparities in public health can persist because of the array of technological, scientific, and architectural innovations that enable wealthy households to insulate themselves from the environmental conditions of the poor. The recent history of Mumbai has militated against the kind of progressive political movements that galvanized processes of sanitary reform in, for example, European cities during the second half of the nineteenth century and the early decades of the twentieth: though similar technical discourses of spatial rationalization are now active, the political and institutional structures to implement these sanitary ideals have been largely absent. The persistence of inequalities in the distribution of basic services in Indian cities is integrally related to the long-standing weakness of working-class organizations within the urban political arena.[74] The circulation of capital involved in the construction of water infrastructure networks reveals the intersection between the urbanization of nature and the exercise of social power.[75] The middle-class monopoly of public service provision in Indian cities stems from their effective control of the postcolonial state apparatus and their perpetuation of colonial dualities in patterns of urban administration. Yet the weakness of the state, particularly beyond middle-class enclaves, necessitates an expanded definition of power to account for the daily practices through which resources such as land, water, and shelter rights are actually allocated. We can usefully elaborate the concept of "social power" as encompassing not just the historical and legislative lineage of state formation but also other sources of power lying beyond formal institutional structures.[76] The city has been simultaneously shaped by officially acknowledged forms of state intervention and an expanding zone of local negotiations to produce a "shadow state," where the boundaries between different loci of political authority and legitimacy become extensively blurred.[77] Like other Indian cities, contemporary Mumbai is characterized by three primary and overlapping sources of authority encompassing the state, the community, and networks of powerful individuals capable of enacting a mix of "terror and generosity."[78]

The combination of economic restructuring and the intensification of religious difference presents a markedly different context for political mobilization than that of the more homogeneous urban proletariats of late nineteenth- or early twentieth-century Europe. The decline of Mumbai as an industrial city since the 1970s has been a critical factor in its political realignment of the last thirty years: the proportion of manufacturing employment in the city fell from 36% in 1980 to under 29% in 1990, while employment in the "informal sector" grew from 55% to 66% over the same period.[79] The Bombay textile strike of 1981–1982, for example, and the subsequent closure of 58 mills contributed to a weakening of working-class politics within the city and the disengagement of industrialization from nationalist aspirations. Deindustrialization and economic liberalization have fostered a vast growth in the informal economy and the spread of poverty and unemployment to previously relatively prosperous communities.[80] A combination of economic restructuring and political realignment has precipitated a breakdown in established patterns of urban public life, with a diminution in the political and economic power of the Parsi, Muslim, and Jewish communities as the Hindu nationalist Shiv Sena movement shifted its ire from migrant labor toward the much larger Muslim population within the city. In the wake of the devastating anti-Muslim riots of 1992, the Shiv Sena steadily increased its political control of the city, transforming what Christopher de Bellaigue describes as "a liberal city called Bombay" into "an ugly, disturbing shrine city called Mumbai."[81]

The "rebranding" of the city as Mumbai in 1995—all traces of the former name were to be erased from official language—marks the symbolic capture of the city by this new political formation. Though the Shiv Sena, in alliance with the Hindu nationalist BJP, only gained political control of the Maharashtra state government between 1995 and 1999, they have continued to exert extensive influence through business deals and local networks.[82] The Shiv Sena and their allies have promoted a distinctive form of "saffron capitalism" through their combination of nativist Marathi politics and a tactical alliance with corporate global giants in fields such as energy and telecommunications. The Shiv Sena came to power in 1995 on the back of grandiose promises to provide "free housing" for four million slum dwellers, yet by 1999 less than 80 apartment blocks had been completed amid confusion, recrimination, and political disarray. High-profile defections, bitter internecine feuding, and the erosion of support from women, the working poor, and other vulnerable groups have added to a sense of disorientation in Hindu nationalist politics.[83]

The susceptibility of the Indian state to capture by authoritarian political movements is illustrative of the intersection of wealth redistribution and electoral mobilization in the context of a degraded or largely absent civil society.[84] This political dynamic also underlines the degree to which teleological readings of the evolution of the modern city have failed to appreciate the fundamental fluidity and instability of the postcolonial urban arena. The secular alliances of the post-independence era have weakened to produce a more fragmentary urban political constellation, yet this instability in the cohesion of civil society stems in large part from the peculiar circumstances of state formation under Indian modernity. As Partha Chatterjee and other scholars have shown, historic injustices have persisted under even the more reformist Nehruvian conceptions of the postcolonial state, as have conceptions of the public realm as a "colonized outside" to be spurned in preference to the apparent authenticity of the "community" or the domestic realm. Though modern political practices emerged through a combination of colonialism, liberalism, and capitalism, there has been a consistent countercurrent of antipathy toward governmental forms of sovereignty reflected in the defense of social norms that are inscribed within fixed or cosmological belief systems whose provenance lies outside the modern state.[85] The urbanization of nature, and of water in particular, is inscribed with these rival sources of social and political legitimacy: a fragile secular modernity must contend with powerful ideological conceptions of nature and culture as essentially fixed and nonnegotiable entities, so that the power of nature within Indian modernity is deeply contradictory. The marketization of water, for example, evokes opposition not just on grounds of social justice as articulated within a modern public sphere but also as it threatens traditional understandings of use, entitlement, and social organization.

Mumbai's municipal government has been undergoing a protracted process of hollowing out through the dispersal of responsibilities toward new institutional structures. Existing forms of urban administration have come under sustained pressure since the 1980s through the growing strength of Maharashtra state and the creation in 1975 of a new agency, the Mumbai Metropolitan Region Development Authority (MMRDA), to coordinate the planning of large infrastructure projects—an institutional precondition from the World Bank for expanding bilateral sources of finance. A hiatus has opened up between a corporate agenda led by a strengthened MMRDA and various forms of grassroots globalization with links to international campaigns around sanitation, shelter rights, and other issues. The MMRDA, which is independent of municipal jurisdiction, has been given a range of strategic planning powers for roads, bridges, and also slum clearance. This new planning agency represents the imposition of a different kind of governmental paradigm that conforms to the technocratic demands of the World Bank. The complexity of institutional structures involved in providing water and sanitation infrastructure, from the scale of individual homes or informal settlements to that of large-scale investment programs, raises different sets of issues than other urban technological networks such as power, transport, or telecommunications.[86] For the urban poor, state failure encompasses not just processes of deliberate exclusion from basic services but also a continual uncertainty over the use of violence to bring "invisible spaces" into the capitalist land market. In Anand Patwardhan's award-winning documentary *Bombay Our City* (1985), for example, the terrified inhabitants of a slum try to prevent the feared "municipality" from entering their settlement and destroying their homes. Tellingly, the clearance itself is undertaken by men employed from another slum at the behest of the municipal authorities. The ambiguous relationship between slum dwellers and the state is a key dimension to the urban infrastructure crisis, because it illuminates the limited extent to which citizenship rights have been extended to participants within the urban labor market.

Over the last forty years we can identify three distinct phases in political discourse concerning the provision of basic infrastructure for the poor. First came an intensified or exaggerated form of "authoritarian governmentality" in the wake of Indira Gandhi's declared state of emergency from 1975 to 1977, which resulted in the brutal removal of many settlements in combination with programs of forcible sterilization. The political repression of this time involved an extensive assault on the power of trade unions and working-class political movements: a development that would have repercussions for the emergence of authoritarian populism in the 1990s. Second, there was a liberal interregnum beginning in the mid-1980s with greater emphasis on the ad hoc upgrading of basic services, tenure rights, and the extension of the political franchise, but with a highly uneven impact on urban morphology and social inclusion. This uncoordinated, incremental and "paralegal" extension of basic services operated within a structural tension between the illegitimacy of slum dwellers and their organizational threat to property rights and the need to accommodate cheap labor within the metropolitan economy. The Mumbai slums have emerged as a micro sphere of negotiation between state agencies and "discrete elements of the heterogeneous social," which culminated in an agreement in 2000 to provide greater security of tenure and additional municipal standpipes for settlements that were in existence before 1995.[87] And third, there has been a more recent reversion to a renewed mode of authoritarian governmentality driven by the recapitalization of the Mumbai land market. This latest phase forms part of a wider strategy to integrate the city more effectively within the global economy and marks a movement away from nationally oriented governmental paradigms, whether linked to the elimination or the upgrading of informal settlements. In recent years there have been intensified attempts to eradicate slum settlements in lucrative locations through the denial of basic services and various forms of harassment.[88] Since 2004 a more violent politics of "spatial elimination" has reemerged in order to reclaim much larger areas of commercially valuable land, such as the vast Dharavi settlement south of Mahim Creek. In the wake of the McKinsey report *Vision Mumbai* (2003)—which sets out a strategy to boost Mumbai's credentials as a global city—there has been a wave of slum demolitions leaving at least 350,000 people homeless.[89] This report, commissioned by the powerful corporate coalition Bombay First, sets out a neo-Haussmannite agenda of urban boosterism to forcibly transform Mumbai into a world-class city on a par with other emerging citadels of economic dynamism and prosperity in east and south Asia.[90]

4.4   Neo-Haussmannite Dreams

Mumbai's elites have long desired to transform their city into something else: the current preoccupation with Shanghai can be likened to the earlier aspirations of Rajiv Gandhi to reclaim the metropolis under the guise of "urban beautification" or Sharad Pawar's plan to create a "new Singapore." It is revealing that one of the most substantial improvements to the city's physical infrastructure in recent years has been the construction of a new elevated highway providing speedier connections between the international airport and the rest of the city.[91] The attempted Shanghaiization of the city through the subservience of urban political discourse to a globalized set of economic interests can improve the connectivity of "premium spaces," but the majority of the population may find themselves marginalized just as when nineteenth-century engineers ignored the "inferior" communities living beyond European enclaves. Contemporary Mumbai lies caught between the Nehruvian technocratic vision of the immediate post-independence era and a newly emerging urban discourse rooted in the spread of wireless technologies and the increasing political and economic power of the urban middle classes. Indeed, the very idea of the modern city has for much of the postcolonial era played an ambiguous role within Indian conceptions of national identity with its deep ideological attachment to rural life.[92] The desire to emulate aspects of Chinese urbanization by the city's most powerful political and economic strata marks the adoption of a new "urban imaginary" or global archetype, within which certain Indian cities—notably Bengaluru (Bangalore), Hyderabad, and Mumbai—appear to be playing a leading role.[93]

Although generations of architects, planners, and engineers sought to transform colonial Bombay into a modern metropolis—a modernist impulse exemplified by the abundance of art deco buildings clustered along Marine Drive and other prestigious locations—the reality was always better represented in the overcrowded *chawls*, tenements, and makeshift dwellings that increasingly characterized the majority experience of the city.

Foreign-trained engineers marveled at the sanitary revolution that had been achieved in the cities of Europe, but the fiscal and political realities of the colonial and postcolonial city militated against any comprehensive modernization of the metropolis. The episodic nature of the hydrological cycle in India—epitomized by the monsoon—also thwarted the imposition of an engineering model derived from the European experience, in which a universalist impulse occluded any close engagement with the cultural, political, and also biophysical complexities of other urban contexts.[94] The landscape architects Anuradha Mathur and Dilip da Cunha's recent conceptualization of "Mumbai-in-an-estuary" for the exhibition *Soak* (2009) marks a self-conscious break from both the global boosterism that ignores the biophysical setting of the city and the technomodern striving for environmental control.[95] The idealized model of the "bacteriological city" with its universal water and sewerage systems rests on the assumption that urban space is both relatively homogeneous and spatially coherent, which is at odds with the extreme forms of social polarization and spatial fragmentation experienced in many of the cities of the global South.

In the immediate post-independence era we find a new kind of technomanagerialist modernity in which the scale and complexity of water engineering became emblematic of a national self-confidence associated with a distinctive form of authoritarian governmentality. In the 1950s, for example, engineers and planners suggested using roadblocks to prevent further rural-urban migration, and the bifurcated strategies of colonial administration were reworked to produce a new geography of citizens and noncitizens. This governmental duality has persisted in subsequent decades so that division, inequality, and injustice have become defining features of the postcolonial city: the cosmopolitan elites who ruled post-independence Bombay were merely supplanted by an unstable postsecular *hindutva*-dominated political constellation. The current neo-Haussmannite impetus toward the physical reconstruction of Mumbai and other Indian cities marks only the latest phase in the struggle to capture governmental institutions and alter the dynamics of capitalist urbanization.

Alternative hydrological visions have struggled to free themselves from either an idealized rurality or various forms of "ecological thrift" rooted in a reworking of precolonial forms of water management. The Indian environmental movement has tended to draw its ideological sustenance from ecological or cultural metaphors rather than engaging with issues of social justice within the urban arena. A rural bias within much Indian environmental thought has served to obscure latent tensions between the urban poor and new forms of "bourgeois environmentalism."[96] An emerging middle-class militancy over aesthetic and quality-of-life issues threatens to disrupt politically inclusive responses to the urban infrastructure crisis and perpetuate many of the inequalities associated with the colonial city.

## 5

### Tracing the Los Angeles River

Reason told the Explorer that millions of Angelinos must cross the river daily, most of them two, four, six times. Yet nobody he *knew* had seen it, save one acquaintance who lived in far-off Glendale, and even *he* wasn't sure.

*Dick Roraback*[1]

There is a sense of impermanence about the whole place; and yet its threats are also in many and strange ways the making of its splendours.

*Reyner Banham*[2]

## Chapter 5

> The river, this concrete river,
> Becomes a steaming, bubbling
> Snake of water, pouring over
> Nightmares of wakefulness;
> Pouring out a rush of birds;
> A flow of clear liquid
> On a cloudless day.
> Not like the black oil stains we lie in,
> Not like the factory air engulfing us;
> Not this plastic death in a can.
>
> Luis J. Rodríguez[3]

The contrast between the Los Angeles River as absence, spectacle, and lived experience could not be more stark. Few Angelinos can trace the fifty-one-mile course of the river as it makes its way from Canoga Park through Burbank, Glendale, and then south through downtown Los Angeles toward Long Beach. For much of its course the river is now little more than a concrete channel hidden within the heart of the city: buildings face away, billboards obscure its location, and its banks are often inaccessible behind miles of fenced-off levees. The placid blue haze over Los Angeles belies sudden downpours in the river's mountainous catchment area that can produce dramatic and sometimes deadly flash floods. These torrents of storm water transform the dry channel into a fierce surge of power that threatens to sweep anything in its path into the Pacific Ocean.

So unknown had the Los Angeles River become that in the autumn of 1985 the journalist Dick Roraback, referring to himself as "the Explorer," undertook a series of walks along the course of the river for the *Los Angeles Times*, starting with its mouth in Long Beach. "A mere mile or so north of its mouth," remarks Roraback, "it narrows to a stream, then a creek, finally a joke."[4] Writing a series of dérive-style articles about his experiences as he wandered along its course, Roraback created a picture of what the modern river had become. His vignettes include loners sitting in abandoned chairs, tire tracks from roadster races across the concrete, and the "traditional LA marshland of cattails and shopping carts."[5] His wry commentary is also peppered with historical references to figures such as Juan Crespi, whose diaries from August 1769 provide the first descriptions of the original river:

> It is at approximately the North Broadway crossing that Crespi first laid eyes on the "lush, green, wide-spreading valley … a most handsome garden, countless rose-gardens, the soil all dark and friable."
>
> Were Crespi to cross the bridge today into Lincoln Heights, the only friables he would find—welcome though they may be—would be in the kitchen of La Pasadita, a pastel-blue restaurant redolent of hamburgers, hot dogs, huevos rancheros and chorizo.[6]

For Roraback, and many other writers, the river has served as a powerful symbol for the incongruity of Los Angeles, the disconnection between river and city serving as metaphor for other fractures and divisions. In the historian Robert Fogelson's account of the early growth of the city until the 1930s, it is the theme of fragmentation that drives his analysis of the contradictory social, political, and spatial dynamics that underpinned the transformation of a small agricultural settlement into a vast conurbation.[7] By the middle decades of the twentieth century the city had evolved into an urban patchwork without a definite center, appearing to defy the familiar parameters of urban legibility.[8]

For the architectural historian Reyner Banham, writing in the early 1970s, it was the experience of driving rather than any particular landmark that characterized Los Angeles. The city appeared to have achieved some kind of dynamic equilibrium between technology and modernity, since the newly completed freeway system was not yet in a state of gridlock. For

Banham, the freeway system or "autopia" denoted one of the "four ecologies" of Los Angeles he identified, along with the beach, the plains, and the surrounding hills. Banham used the term ecology to refer to the "human ecology" of modern lifestyles, rather than in the scientific sense of interactions between organisms or the emerging ecological critique of modernity. For Banham, Los Angeles was the ultimate "pop city" in which anything was possible: the Italian futurists' work, the Archigram projections, and a host of other avant-garde experimentations seemed to have found a natural home.[9] The city's exhilarating combination of space and technology appeared to offer a pathway out of the "impasse of modernist formalism."[10]

Banham's assessment of Los Angeles is strongly influenced by the German geographer Anton Wagner, who completed one of the first systematic studies of the city in the early 1930s. Wagner was particularly interested in the scale and speed of the city's development and gathered together a diverse array of sources, including maps and aerial photographs.[11] For Wagner, drawing on the European experience of urbanization, Los Angeles did not fit into established conceptions of urban development, not least because of its administrative fragmentation.[12] His interest in the varied landscapes of Los Angeles mirrored a series of developments in American geography focused on the "cultural landscape" as an alternative to crude forms of environmental determinism. In this sense Wagner's study marks a synthesis between the work of the American geographer Carl Sauer, who led the so-called Berkeley School, and similar regionalist approaches that had developed within German geography, including the work of his doctoral supervisor Oskar Schmieder, who had held a visiting position at Berkeley in the late 1920s while studying Spanish influences on the American landscape.[13] Yet Wagner took this interest in cultural landscapes further than most of the Berkeley geographers or his compatriots by focusing on urban rather than predominantly agricultural landscapes, including the distinctive forms of human artifice such as the "stage set cities" (*Kulissenstädte*) of Hollywood and Culver City.[14] His interest in the "drilling tower forests"

(*Bohrturmwälder*) of Los Angeles, for example, reworks metaphors of nature in an urban context and also resonates with an emerging fascination with the vast scale of American technological landscapes.[15] For Wagner, the central question guiding his study was the speed and scale of the city's development in relation to its inhospitable semiarid setting, and the paradoxical hydrological dilemma of having both too much and too little water:

> Under the third and largest boom that Los Angeles has experienced, the "water questions" [*Wasserfragen*] derived from the semiarid character of the climate have for the first time been regarded in a new light, arising from a combination of unwanted storm water and a growing shortage of groundwater. This poses new and immense tasks.[16]

The presence of "wild" or unpredictable nature, in its multifarious relationships with the sprawling Los Angeles conurbation, has been a repeated focus of attention. Flood, fire, drought, seismic threats, and unexpected ecological disturbances form part of an array of biophysical uncertainties facing the region. Mike Davis, for example, presents a landscape of anxiety driven by a combination of environmental threats, political paranoia, and socioeconomic tensions. Davis presents a cultural elision between the city's threatening environment and the anxious urban imaginaries that have accompanied its modern development.[17] Other writers such as the geographer Jennifer Wolch present a more upbeat assessment of the city's current prospects, delineating signs of an "ecological urban citizenship" emerging from a myriad of grassroots initiatives and new forms of public engagement with nature.[18] Water is a constant feature in many of these accounts: its absence, its unpredictability, its sudden power, and its tangled relationships with the social and political history of the city.[19] At a structural level Los Angeles is a city that has repeatedly sought to confound or ignore its own topography. Yet it is also a city whose dynamics have been shaped by the biophysical characteristics of its setting from the earliest period of its expansion.

## Chapter 5

Los Angeles presents a paradox. On the one hand it has frequently been evoked as exceptional or unique in comparison with other cities, while on the other hand it has at various points been held up as a paradigmatic city, a portent of things to come.[20] The apparent conundrum of the Los Angeles River can speak to both propositions: the problem of how to enable a city to grow within an erratic floodplain is not unique, and the channelization or enclosure of urban rivers has occurred elsewhere, yet the scale and symbolic significance the Los Angeles River and the range of cultural, technical, and political discourses engendered by its transformation stand out.

### 5.1 Extinguishing Nature

In order to preserve the city's life nerve [*Lebensnerv*] of further expansion, the people have cooperated in the face of a common danger. Together they have been able to undertake the greatest of nature-conquering [*naturbezwingenden*] tasks.

*Anton Wagner*[21]

The contemporary Los Angeles River bears almost no relation to the landscapes first encountered by the Portolà expedition of 1769, yet the need to adapt to the complex hydrological characteristics of the region has marked every phase of its human occupancy. Before extensive human modification, the Los Angeles River essentially comprised two different fluvial systems: a shallow multichannel river for much of the year, interspersed with an occasional torrent fed by winter storms that could alter the entire course of the river.[22] The original Pueblo de los Angeles was established by Spanish settlers near to what is now the Los Angeles River in 1781 and involved the gradual erasure of Gabrielino Indian villages which had based their living on fishing, hunting, and the trading of soapstone (steatite).[23] The Spanish used delta waters for irrigation canals or *zanjas* that stretched from the Santa Monica shore to San Bernardino in the east, and the settlement began to acquire a reputation as a "garden paradise" on account of its production of almonds, oranges, grapes, and other produce. Idyllic descriptions of Los Angeles, along with burgeoning wine production, added to its allure, and by the early nineteenth century Los Angeles had emerged as the most important agricultural settlement on the Pacific Coast. The city's first major phase of expansion was spurred by several further developments, including the aftermath of the Mexican-American War (1846–1848), the discovery of gold in 1848, and the admission of California as a new American state in 1850.[24]

By the late nineteenth century, however, there were demands to replace the polluted *zanjas* with safer piped water supplies. The city's first water commissioner or *zanjero* had been appointed in 1854 to oversee the use of the river ditches for irrigation. The last of the *zanjas* were closed down in 1904, but the allocation of water rights proved very problematic.[25] The Spanish legal emphasis on public ownership of water resources was ideally suited to equitable use of water in arid conditions, but this conflicted with growing commercial interests. In 1868, after extensive lobbying, the city leased its water rights to a private company which then transferred its rights to a subsidiary, in an ill-fated arrangement that proved incapable of providing adequate supplies to the fast-growing city. In a foreshadowing of current concerns over water privatization, the company was accused of charging exorbitant rates and distributing excessive profits to its stockholders rather than making adequate investments in its supply system. With the expiry of the company's lease in 1898 there were protracted arguments over the value of its assets; Los Angeles eventually regained control over its water supply in 1902 in the wake of growing public hostility to the so-called "water ring" and their practices.[26] The Los Angeles River was still used at this time for irrigation and the supply of aquifers for wells, but after the completion of the 233-mile Los Angeles Aqueduct in 1913, linking Los Angeles with more distant water sources in the Owens Valley, the city was no longer dependent on the river for its water.[27]

The original landscape of the river was derived from the complex mosaic of its floodplain, with woods, meadows, and shallow braided streams. Edward Ord's plan of 1849 clearly shows the river—still without any bridges—as a dividing line between cultivated lands and early road networks to the west and an expanse of largely uncultivated lands to the east (figure 5.1). William Mulholland, the engineer for the Los Angeles Aqueduct, remarked in 1877 that the original river was still a "beautiful, limpid little stream with willows on its banks," but this picturesque landscape belied complex technical and political challenges.[28] And even Mulholland's remarks are something of an anomaly, since the river and its riparian landscapes would already have been heavily degraded by the time he began work in the city. After the completion of the city's aqueduct in 1913 the river was increasingly used as an open sewer or dumping ground for refuse and became "the antithesis of the Arcadian ideal so central to the promotion of the region" (figure 5.2).[29]

At the height of the Depression in 1930, an ambitious park plan for the city was prepared by Frederick Law Olmsted Jr. and Harlan Bartholomew. The plan, funded but then rejected by the Los Angeles Chamber of Commerce, opposed the further channelization of the river and suggested that land at risk of flooding should be used to create a new park system for the city: "Such land would have to be acquired only once," argued Olmsted and Bartholomew, "yet would serve a double purpose—flood-control use and park use—not conflicting but positively benefiting each other."[30] They recognized that tax revenues were being consistently devoted to infrastructure projects that underpinned the continued expansion of the urban fringe rather than improving conditions within the existing city. If implemented, the plan would have brought greatest benefits to the working-class communities in the south of the city that had least access to parks and open spaces. Yet the realization of the Olmsted-Bartholomew plan would have involved removing swathes of land from the city's burgeoning real estate market, and the scheme was ultimately quashed by the city's political and business elites.[31] The report's publication only five

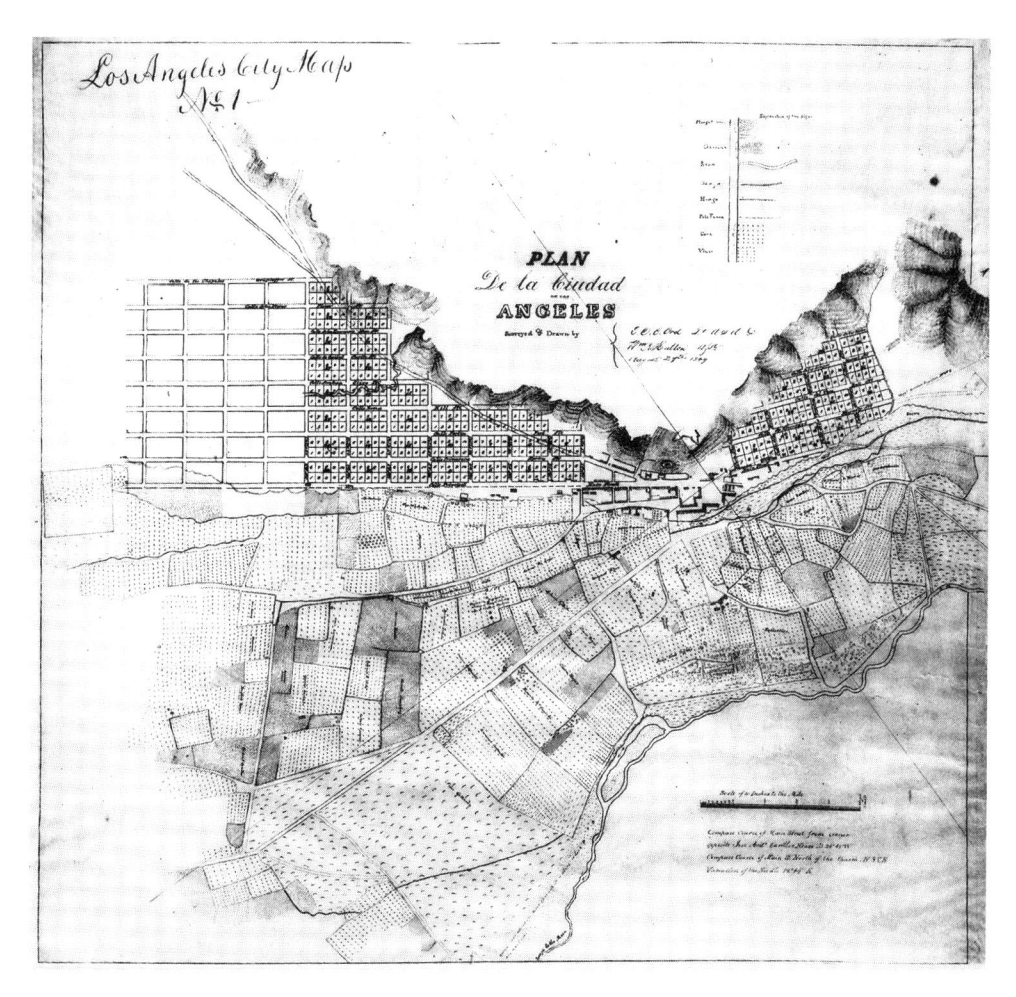

5.1 Edward Ord, plan of the city of Los Angeles (1849). Courtesy of the Los Angeles City Archives.

5.2 Side view of Macy Street Bridge (1925). Originally from the Security Pacific National Bank Collection. Source: Los Angeles Public Library Photo Collection.

months after the stock market crash that heralded the Great Depression further marginalized a project whose expenditure was focused principally on real estate acquisition rather than labor-intensive public works.[32]

The decisive issue proved to be the volatility of the river itself: the original river featured multiple channels spread over a typical "wash" landscape, but with increasing development the river's floodplain became further constricted, limiting its capacity to cope with variations in rainfall. Major floods took place in 1886, 1914, 1934, and most notably in 1938, when over 100 people lost their lives across the county, some swept away by collapsing bridges. In the wake of the 1914 flood there were calls for a radical solution to the flooding problem, and in the following year the Los Angeles Flood Control District was set up which favored a more ambitious technical solution to the problem. River channelization, so that flood control could be achieved with minimal loss of land for development, was pressed for by the major floodplain landowners, such as the Southern Pacific Railroad, which resisted the implementation of any "hazard ordinance" or other planning interventions that might limit their use of land within the river's extensive natural floodplain.[33] In Wagner's study of Los Angeles he recognized how the danger of "washouts" (*Auswaschungen*) had produced a distinctive kind of planning blight characterized by an increasingly disordered, polluted, and ethnically divided urban landscape:

> A fissure in the normal development of the inner city was caused by the appearance of railroads on both sides of the dry bed of the Los Angeles River, where industry began to settle, especially along the western bank between the railroad [*Bahnkörper*] and the riverbed. Slaughterhouses, breweries, steel foundries, gas works, and heavy industry, therefore, settled early on in connection with the railroad yards on the flood terrace between Alameda Street in the west and the upper east side in the east. The extended industrial city strengthened the division between the east and west sides, which had already developed

in relation to the dry riverbed. The residential areas that lay to the leeward side of industry consequently declined into a working-class district, in which immigrants predominated, Italians and Mexicans in East Los Angeles, Jews, Mexicans, Japanese, and Italians on the upper east side.[34]

The industrialization of the river and its floodplain delineated nearby communities more sharply in terms of class and ethnicity, so that the Anglo cultural disengagement from the river predated its physical reconstruction.[35] The creation of an industrial-riparian zone is thus inseparable from the wider political dynamics of race, politics, and the production of urban space in Los Angeles.[36] The uncertainties generated by flood risk accentuated existing forms of residential segregation and reflected a fragmentary political arena in which a more inclusive conception of the public interest had been consistently deflected or ignored.

Despite the centrality of flood risk to emerging patterns of urban development, there was growing evidence that a local solution to the flood control problem could not be found: there were bitter disagreements between engineers; key works were not carried out; and a series of critical bond issues to fund new infrastructure projects were rejected by voters. As late as 1933 some 86 percent of stream channels still had no levees or protective structures at all. In the face of this technical and political impasse the responsibility for flood control was eventually passed to the federal government.[37] From 1935 onward the city set about modifying the river with banks, levees, and drainage channels using newly approved funds under the Works Progress Administration (WPA) set up by the Roosevelt administration. The devastating flood of March 1938, however, showed the need for a more ambitious approach to flood control; in 1941 Congress eventually approved construction by the Army Corps of Engineers of a fifty-one-mile-long concrete channel, and other defense works, in place of the original river. The powers of the Army Corps of Engineers had also recently been extended under the Flood Control Act of 1936, in the wake of nationwide flooding, and Los Angeles County would receive more flood control expenditure than any other region in the United States.[38] This vast program of public works was initiated under the aegis of the New Deal with the promise of job creation and economic recovery. For a mix of fiscal, political, and technical reasons there was a move toward a technocratic solution to flood risk control in Los Angeles, in which the impetus now lay at the national rather than local or regional level.[39]

The US Army Corps of Engineers was faced with an extremely limited range of options, because there had been no zoning restrictions on development within the city's floodplain. Furthermore, the semiarid characteristics of the Los Angeles River and its catchment area meant that an alternative floodplain-oriented approach would have involved a vast and prohibitively expensive alteration in land use, perhaps beyond even that envisaged by Olmsted and Bartholomew. As the geographer Blake Gumprecht argues:

> Flood control officials in general and the U.S. Army Corps of Engineers in particular are blamed for turning the Los Angeles River into an eyesore, when in truth they were little more than undertakers, closing the coffin on a river that was by and large already dead. The destruction of the river had begun half a century before the first concrete was poured, when the river, so unlike the waterways back home remembered by residents in this city of perpetual newcomers, began to be viewed not as a giver of life or a thing of beauty, but as a dumping ground—for horse carcasses, petroleum waste, and the city's garbage.[40]

The main phase of the channelization took over twenty years to complete, deployed over 10,000 workers, and used some three million barrels of concrete. The original Los Angeles River was converted into an immense flood control system with no other purpose than the effective removal of storm water from the city (figure 5.3). Even the word "river" appears to lose its veracity: in amendments to the city's 1965 master plan, for example, what was once the site of the Los Angeles River is referred to as merely a series of "open flood control channels," in a case of historical erasure through technological transformation.[41] In December 1969 the newly concretized river faced its first real test: several days of heavy rain propelled immense quantities of storm water through the city, and the new flood defenses prevented what might have been another catastrophic flood.[42]

The creation of Banham's "autopia" also held more complex environmental implications than he and many other urbanists had anticipated: the increasing conversion of the city's surface into roads and parking lots served to amplify the potential impact of sudden downpours by directing runoff straight into the city's simplified network of concrete channels in a matter of minutes.[43] By the late 1970s the specter of deadly floods had reemerged: flash floods in 1978, 1980, and 1983 had left over thirty people dead and exposed the vulnerability of the city's drainage system. Forest fires in the river's mountainous catchment area, worsened by the spread of urban sprawl into semiarid hillsides, also began to intersect with heightened flood risk.[44] The marked increases in flood water levels since the 1970s led to proposals by the US Army Corps of Engineers, who retain responsibility for flood control in Los Angeles County, to raise the height of the existing channels. Their 1987 report indicated a potential return to the devastating floods of the 1930s and demanded a new program of public works to enhance the capacity of the city's flood defenses.[45] On the other hand, a high-profile scientific symposium convened at Caltech in the wake of the 1980 flood began to reveal significant deficiencies and uncertainties in the scope of technical expertise across many areas, ranging from meteorological forecasting to the hydraulic behavior of watersheds.[46] In contrast to previous reports, the Los Angeles River Master Plan of 1996, instituted by the Los Angeles County Department of Public Works, now indicated greater critical scrutiny of the technical logic behind further concretization and sought to accommodate a wider range of interests in the future of the Los Angeles River beyond a narrow focus on flood control.[47] By the late 1990s, however, the perception

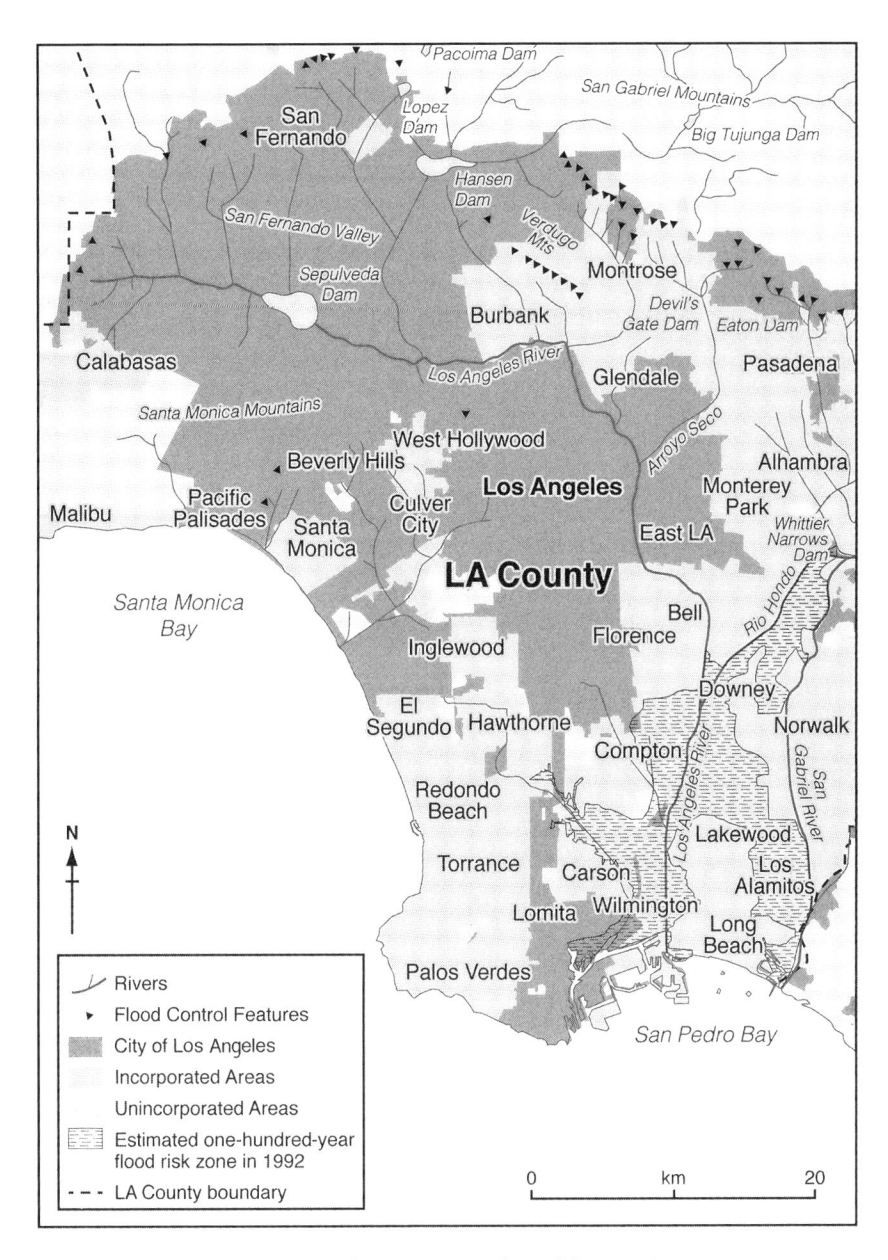

5.3 The contemporary Los Angeles River. Data derived from various sources.

of heightened risk within the river's hundred-year floodplain posed a further political dilemma, since the areas at greatest risk were in predominantly low-income neighborhoods. With the abrupt intervention of the Federal Emergency Management Agency (FEMA) in 1998, and demands to increase compulsory flood insurance premiums within a now expanded area deemed at risk, the tensions over future flood control policy began to escalate.[48]

The prospects for a continuation of the technocratic paradigm established in the 1930s have also been challenged by a more inclusive ecological vision for the whole metropolitan region, epitomized by the Southern California Institute of Architecture's call in 1990 for the creation of a democratically controlled Los Angeles River Valley Authority.[49] The difficulty is that, while more scenic tributaries of the Los Angeles River such as the Arroyo Seco, which passes through Pasadena, could be brought back into public consciousness, the rest of the river system remained dominated by concrete landscapes of neglect.[50] It would also be misleading to suggest that the local state, in the form of the Los Angeles County Flood Control District, had been impervious to the growth of environmental concern or to the lack of democratic accountability for infrastructure planning.[51] Even if only as a matter of political expediency we can detect a change of tone in the language of flood control by the early 1970s, so that the question of legitimacy began to be a necessary one to ask.

## 5.2 Concealed Landscapes

> The shadows crossing on the concrete below
> In a wry design of pediments like sentinels
> Invite my meditation on the fate
> Of these Spanish borderlands and their wild streams—delivered
> By the fractures of history into
> The hands and mechanized channels of
> Politics and the Army's corps of engineers.
>
> *L. A. Murillo*[52]

The threat of flooding in Los Angeles has produced some of the most striking and unusual landscapes of any American city. If we follow the San Diego Freeway a few miles south of Los Angeles International Airport, it begins to veer east toward Carson. The terrain becomes more disordered and is peppered with light industrial units and patches of waste ground. For a while the freeway is accompanied to the left by a waterway named the Dominguez Channel—a 15.7-mile-long (23.3 km) drainage channel largely obscured by billboards and embankments. Just north on the Long Beach Freeway there is an exit for Del Amo Boulevard—a dusty two-lane highway that traverses the Los Angeles River and heads east for the ethnically mixed working-class suburb of North Long Beach. The concrete bridge over the river connects with a grid of small suburban streets: the modest bungalows of a low-income neighborhood lying less than a mile from the Virginia Country Club and Bixby Knolls. From the middle of the bridge there is a dramatic view of the concrete channel of the Los Angeles River and its surrounding landscapes (figure 5.4). Although most of the river is fenced off to prevent public access, at the Del Amo Bridge there is a gravel track leading down to the debris-scattered banks of the channel. On the west side of the channel is an assemblage of shacks including paddocks for horses clustered under giant billboards. There are vegetable plots, chickens running around, and other signs of informal food production.

In the early 2000s these concrete channels, or "spreading grounds" as they are referred to in flood control terminology, seemed cut off from the rest of the city. In 2008, however, an unusual project was unveiled called the Dominguez Gap Wetlands Project, under the auspices of the LA County Flood Control District with additional funding from the State Water Resources Control Board, the California Coastal Conservancy, and the Rivers and Mountains Conservancy (figure 5.5). This experimental landscape, on the east side of the same stretch of the river, has several aims: groundwater recharge to help stem the salination of aquifers; water purification through the use of swale-type landscapes; the addition of a corridor-type ecosystem to enhance urban biodiversity; and the creation of a more

5.4 Los Angeles River, Carson (2002). Photo: Matthew Gandy.

5.5  Dominguez Gap Wetlands Project, Long Beach (2013). Photo: Matthew Gandy.

accessible type of riparian public space. The complexity of this designed fragment of urban nature is indicated by its own parallel system of intricately engineered hydrological management, including sprinklers and other kinds of "ecological irrigation" in what is an ostensibly semiarid landscape. The Dominguez Gap is a key demonstration project for the Los Angeles River Master Plan along with other schemes such as the Tujunga Wash Greenway and the River Restoration Project in the San Fernando Valley. In passing from Del Amo Boulevard to the Dominguez Gap there is an auditory transition from traffic noise to bird song; a network of paths and bikeways has opened up what was until recently a restricted type of strategic hydrological landscape, vestiges of which remain in the form of gated underpasses and other structures to discourage unwanted public access. In this contemporary example of an engineered urban landscape we actually encounter three different types of metropolitan nature: the dominant topography of concrete channels inherited from twentieth-century techno-modernity; the new socioecological synthesis of swale-type public landscapes; and the interstitial spaces of spontaneous nature emerging within and between these designed spaces but also increasingly predominant in the "nondesigned" utilitarian spaces beyond. This last form of metropolitan nature, including weeds and other vegetation, is interspersed with billboards, pylons, and other types of functional elements. The billboards are designed to be seen from the window of a moving car and do not invite motorists to interrupt their journey: they provide a hieroglyphics of consumption to be etched into the weary consciousness of the city.[53] Set further back from the billboards there are lines of pylons, so that the skyline resembles a giant spider's web suspended along the sides of the river channel (figures 5.6 and 5.7).

When Wagner completed his study of Los Angeles in the early 1930s, he was especially interested in the proliferating landscapes of roadside billboards (*Reklameschilder*) that depicted not just imaginary lifestyles but also idealized Californian landscapes. For Wagner, these facade landscapes emphasized the presence of "the incomplete" (*das Unfertige*) as a distinctive

5.6 Billboard and shacks next to the Los Angeles River, Carson (2002).
Photo: Matthew Gandy.

5.7 Landscapes adjacent to the Los Angeles River, Carson (2002).
Photo: Matthew Gandy.

type of spatial form, where planning uncertainty had generated a garish visual topography of billboards. Wagner was interested in reading the urban landscape (*Stadtlandschaft*) as the distinctive visual outcome of demographic patterns, building typologies, and different types of industrial activities. He observed that the clustering of billboards, then as now, was determined by a combination of limited development opportunities and the intensity of traffic flow, to produce a distinctive network of infrastructure corridors.[54]

The concrete channel of the Los Angeles River has often been likened to an empty freeway. In 1948 the Western Asphalt Association even suggested that the river be turned into a freeway—an idea that was mooted again in the 1970s as a solution to traffic congestion. Although the first LA freeway, the Arroyo Seco Parkway (now the Pasadena Freeway), opened in 1939, the main phase of freeway construction began in the 1950s in tandem with rapid urban growth.[55] In terms of design and materials the river channel can be considered as a hydrological counterpart or even precursor to the city's expanding freeway system. Whereas one network is open, visible, and part of the daily experience of the city, the other remains obscured, overlooked, and for many observers of uncertain function, even if it presents a parallel pattern across much of the city: the Ventura, Golden State, and Long Beach freeways all shadow the course of the river. The Long Beach Freeway alone follows it for some 16 miles from Maywood south to the ocean, creating a dense corridor of infrastructure including electricity pylons and other services running through the middle of Los Angeles County. Various other major roads also follow the route of channels and tributaries such as the San Diego Freeway, which runs parallel to the Dominguez Channel before the latter veers south toward Los Angeles Harbor. With the postwar building of new highways, which often flanked the river or were deliberately directed through poorer neighborhoods, an increasingly divided urban topography emerged marked by poverty, racism, and environmental blight. As highways became instruments of "slum clearance" during the 1950s, the marginalization of working-class communities in East Los Angeles and elsewhere accelerated through redlining, the abandonment of public housing programs, and intensified forms of neglect.[56]

The landscapes of the Los Angeles River are a quintessentially utilitarian space comprised of successive layers of human activity, where the technocratic paradigm of flood control has been gradually supplemented by highways, pylons, and other infrastructure networks. The unfolding dynamic between mobile and fixed elements of the urban landscape remains one of its most characteristic features. Whereas the pop-up landscapes of gas stations, swimming pools, and billboards depicted by the LA-based artist Ed Ruscha, for example, have an ephemeral or even unreal quality about them, the contemporary river has involved decades of complex physical reconstruction. Ruscha's Los Angeles is a cityscape rather than a landscape, in which the artificiality of everything is ironically accentuated. In *Twentysix Gasoline Stations* (1963), *Every Building on Sunset Strip* (1966), or *Some Los Angeles Apartments* (1967) Ruscha emphasizes the banal dimensions to urban space by presenting a morphological typology of the city in which sameness and difference become blurred. These projects based on a series of images of buildings, signs, or other structures are accessible, recognizable, and ostensibly straightforward, yet over time they have acquired additional significance as a form of conceptual art that unsettles existing relationships between design, function, and quotidian urban landscapes. The self-conscious preoccupation with the mundane poses questions about the nature of the art object and its relationship with urban visual culture.[57] To understand the cultural significance of the concrete landscapes that traverse the city, we need to interpret these utilitarian spaces in relation to the "kinetic geographies" and ephemeral spaces that have also characterized Los Angeles.

Few artists have tackled the landscapes of the Los Angeles River directly; an interesting exception is James Doolin, whose "saturated photo realism" is perhaps best known to academic readers through his paintings used for the covers of influential texts on the city.[58] Doolin, whose work has been described as "illusionistic," is known for his precise attention to minor details of material landscapes such as abandoned Styrofoam cups, rust stains on concrete, or visual nuances created by different forms of artificial lighting.[59] Like late impressionist artists such as Constantin Meunier, Claude

Monet, and James Whistler, Doolin has also recorded the vivid aesthetic effects produced by meteorological interactions with air pollution, along with the unusual vantage points created by the city's complex freeway system.[60] As Doug Harvey notes in *LA Weekly*:

> In his vertiginous depictions of negative social spaces—bus stops, empty billboards, the dry trough of the L.A. River, the concrete islands between freeway onramps—he managed an unlikely marriage between the lurid sublimity of California landscape tradition and the postindustrial apocalyptic melancholy of J. G. Ballard.[61]

In Doolin's *Study #4 for Connections* (1992) the river channel is shown as bleak and deserted (figure 5.8). The only trace of movement is a freight train on the left running in parallel to the channel and two vapor trails cutting across the sky. Apart from some scattered vegetation above the levee to the right, the lone sign of natural life is a flock of birds taking off in the distance. Doolin's urban landscapes are reminiscent of earlier American artists such as Edward Hopper, whose work evolved in a symbiotic relationship with cinematic noir; but in Doolin's art there is a greater emphasis on the functional rather than oneiric dimensions of urban space.

The poet Luis J. Rodríguez reminds us that the bleak landscapes of the LA River are both a place of collective memory and also a powerful metaphor for the abstractions of power that have wrought such injustice upon the city's poor. His poem "The concrete river," an excerpt from which opens this chapter, also makes reference to hidden worlds of refuge, despair, and imagination. The new ecological imaginaries for the river offer little place for the "spray-painted outpourings" that provide "a chaos of color for the eyes."[62] Among the most powerful cultural appropriations of the concrete river is Judith Baca's mural entitled *The Great Wall of Los Angeles*, located in the Tujunga Wash drainage channel, which chronicles key political moments in the city's history. Taking 13 years to complete,

5.8 James Doolin, *Study # 4 for Connections* (1992). Courtesy of the Koplin Del Rio Gallery, Culver City, and the Estate of James Doolin.

between 1976 and 1989, and involving more than 400 local residents, this vast public art project provides an alternate urban history inscribed into the physical infrastructure of the city.[63] To erase all trace of these quotidian landscapes is to ignore the other city that lies concealed behind the dominant cultural and political narratives of urban change, not just alternative spaces of collective meaning but also zones of cultural experimentation. "Though there is strong advocacy for the river's renewal and restoration," argues the landscape architect David Fletcher, "there is as yet little constituency for understanding the river as it is and as it will be in the future, for the infrastructural sublime, for the freakological, for the river as artifact."[64] Though explicit recognition of the river's parallel human and natural ecologies is made in earlier critical interventions, such as the Southern California Institute of Architecture's influential report in 1990, the leitmotif of already existing diversity is harder to discern in the more recent wave of policy-oriented documentation that emphasizes landscape design over urban politics.[65]

The concrete river has produced its own distinctive cultural and biophysical ecologies that combine aspects of the architectural avant-garde with spontaneous assemblages of fauna and flora drawn from across the world.[66] Grasses, rushes, and other plants have colonized parts of the concrete levees, and in three small sections of the channel where the bottom was not completely covered in concrete, there is a profusion of vegetation that has provoked alarm from flood control engineers (figure 5.9).[67] Various cultural projects have emerged which are interested in studying the river channels in their current state: the photographic study by Greg Ercolano, for example, entitled FOVICKS (Friends of Vast Industrial Concrete Kafkaesque Structures), seeks to record the river "the way it is."[68] In a similar vein, the photographer Camilo José Vergara, who has documented many aspects of urban change in American cities, asks whether "concrete is wrong."[69] Vergara emphasizes how these unusual landscapes intersect with alternative urban cultures to produce a "transitory nature":

5.9  Glendale Narrows, Los Angeles River (2013). Photo: Matthew Gandy.

Every time I am in the neighborhood, I head for the river's Sixth Street entrance. Here, I take in a vast vista: water flowing between steep concrete walls, or, in dry weather, the cracked cement bed; heavy trains crawling alongside; monumental bridges; and evenly spaced high-voltage towers.[70]

Concrete, so strongly associated with architectural modernism, has routinely provoked a mix of fear, hostility, or indifference. There is something threatening or unstable about concrete; its tendency to weather easily, or to become cracked and stained, has contributed to a degree of antipathy toward its presence in the urban environment.[71] Yet concrete is also a kind of modernity in and of itself, since it would be difficult to imagine a contemporary city in the absence of concrete. The double-edged connotation of the phrase "concrete jungle" evokes a sense of sociospatial marginality but also a lineage to antiurban sentiments rooted in naturalistic metaphors of wildness and disorder.[72] In the concrete jungle, as represented in popular culture, nature is largely extinguished beneath a dense labyrinth of human structures, including freeways, industrial installations, and high-rise buildings. Nature if present at all is incidental, functional, or in the case of surface water a potential threat, yet at the same time concrete is itself a material manifestation of nature transformed into human artifice. Concrete has also been the material associated with accelerated or exaggerated forms of twentieth-century modernity, its availability and versatility instilling a sense of control for engineers and politicians alike. Concrete is inextricably related to what the architectural historian Adrian Forty terms "urgent modernization," variously enlisted in technomodern projects of nation building or the Promethean fashioning of nature.[73]

The scale of LA's concrete river, like that of the freeways that often parallel its route, provides an exhilarated form of landscape experience that has often been captured in cinema, literature, and the visual arts. The strangely deserted riparian landscapes have served as a cinematic icon, especially from the late 1960s onward when location shooting became more commonplace in American filmmaking.[74] The city's hidden river has frequently served as a repository of urban malaise interconnecting with neo-noir evocations of crime and political intrigue. In John Boorman's *Point Blank* (1967), for example, the fresh-out-of-jail John Walker (played by Lee Marvin) goes in search of the former gang members who have double-crossed him. He is sent to a deserted stretch of the river channel to collect his money, only to realize just in time that he has been set up: the dollar-sized bundles of blank paper flutter away into the shallow water (figure 5.10). Walker has left prison to find that there is no "real money" anymore, just credit cards, wired transfers, and the inscrutable world of white-collar crime. *Point Blank* is notable for its emphasis on concrete, steel, and glass—not unlike Michelangelo Antonioni's depiction of Los Angeles in *Zabriskie Point* (1970)—where modernism has become little more than surface appearances and the interchangeable elements of the International Style. As Thom Andersen suggests, in his compendium of cinematic representations of the city in *LA Plays Itself* (2003), modernist architecture, and the pristine hillside home in particular, had by the 1970s become the ultimate criminal hideout, where polished exteriors hid the real machinations of wealth and power exerted on the city below.

The perception of the river as a place of danger has also been played out in geographies of fear within Los Angeles: residents in wealthy neighborhoods such as Griffith Park, for example, have requested that barriers to the channel be made higher, in an opposite move to the environmentalist demands for greater public access.[75] The creation of a "tent city" on the river's banks, in an effort to hide the homeless during the 1984 Los Angeles Olympics, reinforced the perception of the river as a marginalized landscape to be eradicated from the public imagination, but it also served as an ironic recognition of the river's long-standing role as a refuge for displaced people, including gypsies, South American transvestites, and other outsiders.[76]

5.10 Still from *Point Blank* (1967). Directed by John Boorman. Courtesy of MGM/The Kobal Collection.

5.3  Reconstructing Nature

The engineered landscapes of the Los Angeles River can be interpreted as a functionalist counterpart to more visible aspects of the city's topography: had the river not been channelized, subsequent patterns of development would have been significantly different. The historian Kevin Starr suggests that the river has been "transmogrified, like Los Angeles itself, into a postmodernist construction of symbolic engineering."[77] Yet the river does not correspond to any neat distinction between the modern and the postmodern: although the complete reconstruction of the Los Angeles River was initiated under the New Deal, it was not completed until the 1960s. Similarly, the river's more recent cultural and ecological appropriation marks an elaboration rather than a straightforward repudiation of the connections between engineering, flood control, and the public realm. The term "postmodern" has in any case been largely subsumed within the discourses of the modern, or the "late modern," despite the critical role of Los Angeles as an architectural and intellectual epicenter for many dimensions to postmodern urbanism in the 1980s and early 1990s.[78]

The politics of the Los Angeles River begins to change decisively in the 1980s. A first development, with strong links to the environmental justice movement, is exemplified by a grassroots organization called the Mothers of East Los Angeles (MELA), which was created in 1984 to campaign against polluting and health-threatening industrial facilities near to poor and predominantly Latino or African American residential neighborhoods. Beginning with opposition to the construction of a new prison, MELA then directed its attention to a proposed incineration plant, an oil pipeline, and hazardous waste treatment facilities. Their campaigns were not simply against environmental threats but also to protect the local labor market from the effects of unwanted land use.[79] There are historical parallels with the earlier, predominantly Chicano-led opposition to the building of freeways through working-class neighborhoods in the 1950s: the new freeways and the industrialized course of the river formed part of an increasingly divided and health-threatening urban landscape. Indeed, these early postwar struggles over infrastructure in the city reveal the "countervalent forces of modernity" which have been so extensively occluded in the popular imagination of Los Angeles.[80]

The working-class chicana/Latina character of MELA provides a stark contrast with a second dimension to river politics focusing on the landscape of the river channel, led by the Friends of the Los Angeles River (FoLAR). Founded in 1986, FoLAR has focused its energies on restoring the river to a more "natural state" along its entire course, and its campaigns can be viewed in the wider context of "ecological restoration," albeit with greater recognition of the artificiality of designed nature.[81] FoLAR, along with other groups such as the California Native Plants Society, has pushed for a "nativist" conception of riparian ecology that is at odds with more inclusive conceptions of the intersections between ecology and urban culture. Invasive plants have acquired ideological or even geopolitical resonance beyond their ecological impact on the native flora.[82] It was not so long ago, after all, that a faction within the local chapter of the Sierra Club had considered excluding African Americans as members, let alone recognizing the global characteristics of urban nature.[83] If the future of the river, along with its degraded landscapes, has become symbolic of the future of the city itself, there remain ideological ambiguities between different ecological imaginaries and wider conceptions of urban society.

The inadequacies of existing flood defenses to cope with future conditions, and spiraling insurance premiums for housing near the river, have exposed tensions between the exigencies of flood control and alternative visions for the city. In the late 1990s, for example, modifications to the concrete levees introduced by the Army Corps of Engineers in order to improve the city's flood defenses were subject to extensive legal challenges by a range of environmentalist organizations led by FoLAR.[84] Other recent events such as the flooding of the Mississippi River in 1993 have fostered skepticism toward reliance on physical flood defenses alone, opening up possibilities for different approaches.[85]

By the late 1980s a further strand of river politics had emerged which reflected wider trends toward the "upgrading" of postindustrial riverfronts.[86] In 1990 Mayor Tom Bradley set up the Los Angeles River Task Force in recognition of growing public interest in these neglected landscapes. Following the approval of a parks referendum by California voters in March 1999, a significant tranche of money was set aside for river-related projects; yet none of these initiatives has involved the removal or significant modification of existing flood defenses.[87] "The greening of the Los Angeles River," reflects the writer D. J. Waldie, "is a sobering demonstration of the limits of environmental restoration in an urban landscape."[88] Various plans have been prepared calling for the linking of green spaces along the river in order to provide an uninterrupted pathway for cyclists and pedestrians, but these "parallel landscapes" do not offer an alternative conception of flood control. Although these new landscapes are far removed from the "autopia" of the past, they nonetheless shadow existing technological networks. The future of the river has become linked with heritage-oriented interest in the city's original riparian landscapes, as well as part of wider strategies to revitalize inner urban neighborhoods. There remain tensions, however, between an emphasis on rebuilding the riverfront, including the historical reconstruction of *zanja* walkways, and more ecologically oriented proposals rooted in landscape restoration. The rezoning of riverside sites from industrial to residential use is already setting in train a form of "ecological gentrification" that will further marginalize low-income communities in the vicinity of the river as rents and land values begin to spiral upward.[89]

In 1999 one of the first major tests of a new approach to the river emerged in relation to a proposed industrial park on a forty-acre site known as the Cornfield (or the Cornfields in some accounts) near to LA's Chinatown. The former rail yard had been initially dismissed by the city's Cultural Heritage Department as having "no historical or cultural value worth preserving."[90] This dispute brought together an array of over thirty different groups, including FoLAR, the Chinese Consolidated Benevolent

Association of Los Angeles, Concerned Citizens of South Central Los Angeles, Friends of the Los Angeles River, Latino Urban Forum, Lincoln Heights Neighborhood Preservation Association, Los Angeles County Bicycle Coalition, and Mothers of East Los Angeles-Santa Isabel, reflecting a broadening of the earlier campaigns of the 1980s.[91] The creation of the Chinatown Yard Alliance from these disparate groups saw one of the most significant multiethnic grassroots campaigns in the city's history including a successful legal challenge against the city's largest real estate developer, which had intended to use the site for warehousing without any environmental impact assessment.[92] In the wake of high-profile debates during the city's mayoral race, the site was finally purchased by the state, in an expensive move that marked a reprise of the earlier difficulties in expanding park provision in the 1930s. Yet the site's identification in the recently completed Los Angeles River Revitalization Master Plan as an "opportunity area," rather than a public space under democratic oversight, is indicative of a continuing ambiguity in the relationship between ostensibly "empty" riparian sites and the wider dynamics of urban change.[93]

Long-standing problems of metropolitan fragmentation continue to militate against coordinated strategies for the river, as they did in the past. In 1941, for example, Samuel B. Morris, former chief engineer for Pasadena and dean of the Engineering School at Stanford University, noted that "the very complex water problems of the Los Angeles Metropolitan Area are made more difficult by the number of unco-ordinated agencies which have grown up to meet individual needs. Each city or community has its water department, water company, or irrigation company, its storm drains and sanitary sewers."[94] By the mid-1920s the city of Los Angeles had all but given up on annexing the other municipal districts of Los Angeles County (which now number 128 in total).[95] With the passing of Proposition 13 in 1978, which limited local property tax rates, the fiscal and political momentum toward further fragmentation became even stronger.[96] Regional public agencies such as the engineering-dominated LA County Flood Control District saw their budgets slashed and retreated into a minimal role of infrastructure repair before being subsumed within LA County's Department of Public Works in 1985.[97] The failure of the 1999 bill in the California legislature to effectively recreate a regional authority for managing the river—the Los Angeles and San Gabriel Rivers Conservancy—became inevitable because of fierce opposition from smaller municipalities.[98] As a result, the river continues to pass through thirteen different jurisdictions within the county and is managed by over thirty different agencies. Any plans for the river must therefore contend with a "crazy quilt" of jurisdictions and responsibilities, in which the many different authorities "seem uncertain of who is responsible for what."[99]

The eventual release of the Los Angeles River Revitalization Master Plan in 2007, after some five years of deliberation, sought to combine modern flood control with new demands for ecological restoration and public access, envisaged within a twenty-five- to fifty-year time frame:

> As a long-term goal, the River's ecological and hydrological functioning can be restored through re-creation of a continuous riparian habitat corridor within the channel, and through removal of the concrete walls where feasible. If one completely restored the River to a naturalized condition throughout its entire length, it would be very difficult to achieve flood control requirements and maintain current urban development.[100]

Though ambitious in its scope, and invoking key figures in the history of North American planning such as Daniel Burnham, the new plan also illuminates the limits to large-scale policymaking. Burnham is mentioned in the context of planning for the "next several generations," yet this extended temporal horizon obscures a series of socioecological tensions and continuing structural impediments to strategic planning for Los Angeles.[101] The plan envisages the creation of a vast network of interconnected riparian landscapes that would require enormous quantities of water to be realized. Looking at the question of water more broadly, Los Angeles has gradually expanded its water rights in a semiarid context to encompass control over a vast area, an urban catchment that may ultimately extend to northern California and the Sacramento River during the twenty-first century. Yet successive generations of engineers, including William Mulholland, never seriously questioned the high per capita consumption of water in Los Angeles evidenced by the proliferation of lawns, golf courses, private swimming pools, and other forms of nonagricultural irrigation. As for flood management, every effort to control development in the city's floodplain or set up a regional authority to manage the river system as a whole has been stymied.

### 5.4   Riparian Anomie

The Los Angeles River presents something of an anomaly: as a nonnavigable river which usually has little or no water, it appears to confound expectations of what an urban river should look like: the contrast with other urban rivers in North America such as the Charles or the Hudson could not be more stark. Yet without its transformation into a vast drainage channel the development of Los Angeles would have been seriously if not irrevocably disrupted. The strange concrete landscapes of the contemporary river reveal more about the underlying dynamics of the city than the many popular projections of Los Angeles in which traces of urban disorder are either effaced or dramatized.

Recent developments are suggestive of a new kind of synthesis between nature and culture in the public imagination. Yet the rediscovery of the river as an ecological facet to the city may in part be the paradoxical outcome of the success of the river channel in controlling the danger of flooding, so that this fragment of urban nature is no longer seen as threatening but domesticated and amenable to new forms of cultural appropriation. The emphasis on restoration tends to downplay the artificiality of the current river system, since its source for most of the year is no longer the mountains to the north of the city but treated wastewater from within the metropolitan region.[102]

The fragility of any new consensus in relation to the river depends in large part on how necessary improvements and modifications in flood defenses can be combined with wider social and ecological objectives. The recent focus on intensifying flood risk and the impact of El Niño events, for example, has raised suspicions that the river faces a social construction of risk as a discursive strategy on the part of the US Army Corps of Engineers and their allies to maintain their technical hegemony. Like the suspicious goings-on in Roman Polanski's film *Chinatown* (1974), in which reservoirs were being secretly drained at night to produce a water shortage, there remains a conspiratorial dimension to the ecological critique of hydrological engineering.

For their part the US Army Corps of Engineers and local flood control engineers have had difficulty keeping pace with the growth of the city. At certain points in history engineers have solemnly called for the city's expansion to be brought under control—an anxiety expressed by technical elites in other fast-growing cities in India and elsewhere. Yet "engineering" was never a monolithic block of power or opinion anyway: the views of engineers have sometimes coincided with the opinions of city boosters and their allies but not always. And engineers themselves have sometimes been in open conflict with each other about what to do, especially in the pre-New Deal era, when deadly floods were more frequent.[103]

The derision directed at the landscapes of the Los Angeles River tends to downplay the degree to which these marginal spaces have played a distinctive role within urban culture. The environmentalist disdain for the "concrete river" ignores cultural appropriations of the drainage channels that utilize these landscapes in unexpected ways. A functional space that had been the focus of spontaneous forms of social and ecological appropriation is now being reframed, if not partly reengineered, to meet a very different set of expectations. The imposition of an "ecological simulacrum" in the place of the concrete river, as part of a wider agenda of landscape restoration, raises its own set of social, cultural, and ecological assumptions that are invariably cloaked beneath rhetorical claims to present the public interest. The specter of "ecological gentrification" lurking behind the greening of Los Angeles suggests that the arguments for environmental justice, first articulated in the 1980s, remain vitally relevant.

# 6

## Fears, Fantasies, and Floods: The Inundation of London

> As the sun rose over the lagoon, driving clouds of steam into the great golden pall, Kerans felt the terrible stench of the water-line, the sweet compacted smells of dead vegetation and rotting animal carcases.
>
> J. G. Ballard[1]

In J. G. Ballard's novel *The Drowned World* we encounter a future London submerged under water where survivors must contend with searing heat, giant mosquitoes, and the constant threat of attack by rival gangs scattered across the flooded city. First published in 1962, the book explores what might happen to London in a postcatastrophe scenario. The destruction of Ballard's London is attributed to a dramatic change in climate so that the city is thrown into a reprise of the Jurassic era, complete with giant gymnosperms, aggressive iguanas, and screeching bats swooping out of the sky. For Ballard, it is the fragility of modernity—and the institutional threads that enable everyday life—that drives his ironic neo-Hobbesian vision of the future. Without these complex structures of organization and control, there is a return to small-scale human societies dominated by fear, violence, and the brutal reassertion of patriarchal authority.

At one level Ballard's gloomy synopsis can be read as an indicator of the exhaustion of utopian politics, but like the best science fiction writers he uses a hypothetical scenario to explore aspects of contemporary society.[2] Interviewed in the German broadsheet *Die Zeit*, Ballard conceded that the flooding of New Orleans in 2005 appeared to mirror many aspects of his vision of social breakdown in advanced industrial societies. What Ballard found most disturbing, however, was not that city's technological failure—which had in any case been widely predicted—but the virulent racism that posed an even greater threat to the survivors.[3] The entanglements of race, class, and vulnerability to catastrophic events traverse the realm of fiction, chiefly through the trope of middle-class fear and anxiety, as well as the material topographies of contemporary cities. If London really were to be flooded as Ballard describes, it would be mostly poorer communities in the east of the city that would first be inundated by rising water levels, rather than middle-class neighborhoods historically located on higher ground to escape the pollution of the nineteenth-century city.

London has frequently been the setting for imaginary visions of the future. In Richard Jefferies's novel *After London* (1885) he describes a city reclaimed by the swamps it was built on, in an early fantasy of urban disappearance. "It became green everywhere in the first spring, after London ended," writes Jefferies, "so that all the country looked alike."[4] Similarly, in John Wyndham's novel *The Day of the Triffids* (1951) we encounter images of an abandoned London where nature has taken over:

> Almost every building was beginning to wear a green wig beneath which its roofs would damply rot. Through many a window one had glimpses of fallen ceilings, curves of peeling paper, and walls glistening with damp. The gardens of the Parks and Squares were wildernesses creeping out across the bordering streets. Growing things seemed, indeed, to press out everywhere, rooting in the crevices between the paving stones, springing from cracks in concrete, finding lodgements even in the seats of abandoned cars.[5]

In addition to literary forays into the unknown, London has also been the focus of cinematic spectacles of future collapse. In Danny Boyle's *28 Days Later* (2002) a deadly virus called "rage" produces a city overrun by violent zombies, while in Alfonso Cuarón's *Children of Men* (2006), derived from the P. D. James novel, eighteen years of global infertility has led to a general psychological malaise and the free distribution of suicide pills. Cuarón's London in the year 2029 is shown as shabby and authoritarian, with most people existing in a state of fear. And in Tony Mitchell's contemporary disaster movie *Flood* (2007), based on the novel by Richard Doyle, London faces disaster from a huge tidal surge, wrongly predicted to be heading for the Netherlands. A mix of professional rivalries and meteorological ineptitude leaves the capital under water and over 200,000 people unaccounted for.

The flooding of Brisbane, Jakarta, Mumbai, Prague, and especially New Orleans in the 2000s, and of Bangkok, Beijing, Manila, and New York in the early 2010s, has dramatically revealed the vulnerability of modern cities to extreme climatic events. If London's flood defenses were to fail, a catastrophic inundation would endanger not just the metropolitan region but also the wider UK economy. In some imaginary scenarios one major breach of London's flood defenses might be followed by repeated inundations and large-scale abandonment of the city. If the entire city were to be deserted—as evoked by John Wyndham and some popular science literature—the "return to nature" would be surprisingly fast, with much of the city covered by woodland and dense vegetation within a couple of decades. In the science journalist Laura Spinney's account of an abandoned future London, further failures in crumbling flood defense infrastructure would enable repeated inundations, as buildings began to tumble under their tilting foundations and rising groundwater levels flooded the subterranean city. The Thames would begin to revert to its much wider and more unpredictable course within a greatly expanded floodplain. Tree roots would tear apart the surface of the city, further accelerating its disintegration. Fires caused by lightning strikes would burn uncontrollably until little flammable material remained. And within a few centuries there would be little trace of London left.[6]

Twenty-first-century London faces a dilemma in relation to increasing flood risk. On the one hand, a technocratic solution, rooted in a continuation of the engineering-led responses to flood risk used in the past, would inevitably point to a new generation of defensive structures far larger than those currently in use. On the other hand, various strategies of adaptation could be developed, within which an expanded role for London's floodplain would be pivotal. Proposals for adaptation range from the ecological restoration of the Thames estuary to various avant-garde scenarios in which large parts of the city remain permanently submerged.

### 6.1   Projections: Scientific and Otherwise

> Warming will cause the Thames to rise and London will flood. Don't think it won't happen by the way: it simply will. All the scientific predictions are that the Thames will flood very seriously.
>
> *Jonathan Glancey*[7]

The Thames runs a somewhat erratic course from its source in rural Gloucestershire in the West of England before passing through London to reach the North Sea. Over the last two thousand years the river and its estuary have been transformed by isostatic adjustments caused by the postglacial sinking of southeast England, eustatic changes from rising sea levels after the last Ice Age and more recent climatic variations, and a range of human modifications of the riverscape that have gathered momentum from the medieval period onward.[8] These include the encroachment, embanking, and bridging of the river in central London to produce a much more restricted flow, along with extensive land reclamation from salt marshes within the river's floodplain toward the east. As a result of these changes, the tidal range of the Thames has risen from less than a meter during the Roman period to over seven meters today to create what is effectively a tidal canal, with the most rapid increase experienced from the seventeenth century onward. The river's defended tidal floodplain now comprises some 35,000 hectares of low-lying land, extending from Teddington in west London and including large parts of central and east London, as well as significant coastal areas of Kent and Essex bordering the Thames estuary (figure 6.1).[9]

Nineteenth-century London saw a flurry of legislative and engineering responses to the increased incidence of floods and tidal surges.[10] During the twentieth century London experienced two serious floods: 1928 and 1953. The flood of 7 January 1928, caused by the combination of rapid snowmelt in the Thames catchment area and a storm surge on the east coast, was the last occasion that serious flooding was experienced in central London, with 14 deaths and thousands left homeless. The North Sea flood of 1 February 1953, caused by a combination of a high spring tide and storm surge, left 307 dead, mainly east of London in the counties of Essex, Lincolnshire, Norfolk, and Suffolk. In the Netherlands there was even greater destruction, with the loss of 1,836 lives and some 1,365 square kilometers of land inundated by seawater.[11] In response the UK government set up the somewhat inconclusive Waverley Committee in 1954, which eventually led to the so-called Bondi Report of 1967, written by the mathematician Professor Hermann Bondi. Bondi had been appointed by UK Prime Minister Harold Wilson, on the advice of his chief scientific adviser Professor Solly Zuckerman, to carry out a wide-ranging survey of London's future flood defense needs.[12] Reflecting on the possibility of an "exceptional disaster," Bondi considered that "such a problem always raises issues of an almost philosophical kind." He warned that "a major surge flood in London would be a disaster of the singular and immense kind. ... It would be indeed a knock-out blow to the nerve centre of the country." "We are at risk all the time," concluded Bondi, "and further delay only invites the disaster."[13] The Hamburg flood of February 1962, in which over 300

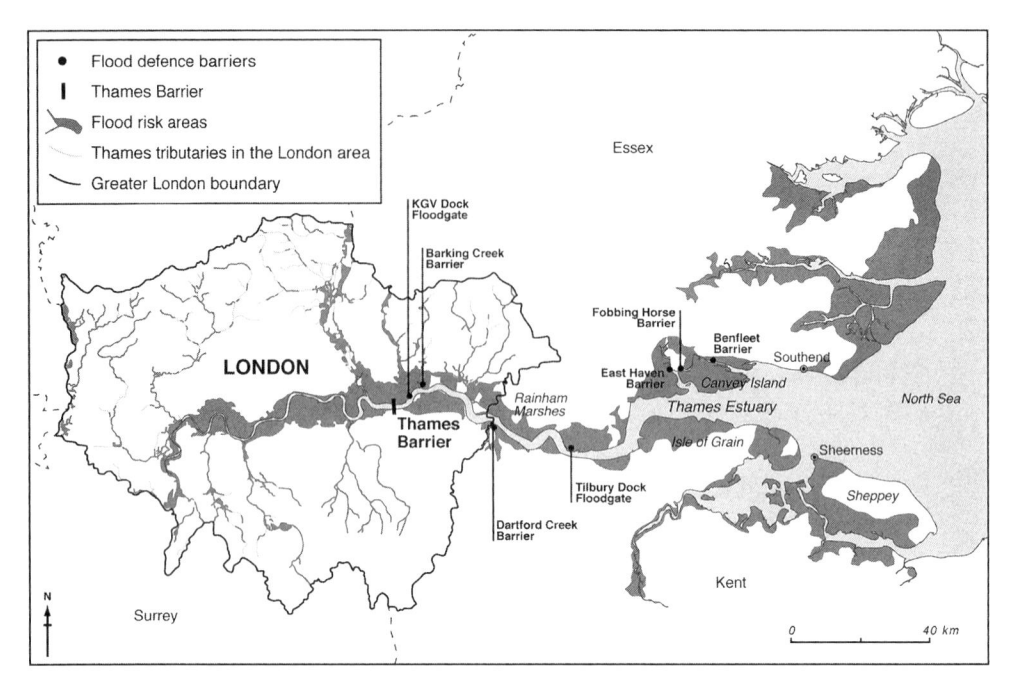

6.1 London and Thames estuary flood risk (2008). Data adapted from the Greater London Authority and the UK Environment Agency.

people died, added a sense of urgency to the deliberations, since it showed the scale of disruption that could be caused by even a small-scale inundation of a major European city.[14]

Legislation passed in 1972 paved the way for the building of a major new defensive structure, the Thames Barrier, which was completed in 1982. This apparent inertia between the flood of 1953 and the final completion of the barrier can be attributed to a range of competing theories about how best to mitigate the threat of flooding to London.[15] Indeed, discussions about a barrage or some other kind of protective structure for London date from the eighteenth century.[16] The eventual shift of design emphasis toward a barrier that could be opened or closed as needed stemmed in part from changing patterns of river use. The rise of containerization, along with the latest developments in hydrology and materials science, contributed to a greater flexibility in design than might have been possible a decade earlier. The design process deployed no less than six specially built hydraulic models, along with the first mathematical model ever used for the analysis of silt movement within an estuary environment.[17]

The Thames Barrier is one of the most striking examples of engineering architecture in London, with its six stainless steel-clad gates spanning the river between Charlton on the south bank and Silvertown on the north bank (see figures 6.2 and 6.3). The metal gates, which vary in weight between 750 and 1,500 tonnes, had to be "manoeuvred to within a maximum tolerance of 10mm, fixed by divers working in zero visibility," and rest on a series of specially cast concrete sills weighing up to 10,000 tonnes on the chalk riverbed.[18] "Although surge barriers have been built in other parts of the world," wrote the engineer Mary Kendrick shortly after its completion, "the Thames Barrier has many unique features, being the first rising-sector gate barrier to be constructed in a high-energy tidal environment."[19] The barrier was commissioned by the Greater London Council (GLC) in 1970, which had been given responsibility for flood prevention along with other aspects of emergency planning after its creation in 1965. This new municipal authority, created on the advice of the Herbert

Commission in 1960, which had been set up in 1957 to explore the future of London's local government, replaced the smaller London County Council dating from 1889. From a design perspective the aim was to give the barrier "a visual identity which would express the great significance of the project."[20] Each of the barrier gate roofs has a curvilinear shape that bears a resemblance to other examples of public architecture such as Hans Scharoun's Berlin Philharmonic concert hall completed in 1963 and Jørn Utzon's design for the Sydney Opera House completed in 1973.[21] As an example of municipal architecture, the Thames Barrier exhibits elements of utilitarian functionalism but also evokes continuities with modernist approaches to technological design. It remains one of the most important achievements of the GLC during its troubled twenty-one-year existence, not just in terms of design but also in terms of long-term infrastructure planning.

The completion of the Thames Barrier in 1982 marks the end of a distinctive phase in the intersection between engineering, water, and local government for London, which began with the creation of the Metropolitan Board of Works in 1856, saw the construction of the Thames Embankment from 1862, and oversaw the gradual rationalization of water infrastructure across the city.[22] After 1986 and the disbandment of the GLC, London's government has been marked by an increasingly fragmented approach. Yet the municipal model of infrastructure provision was driven by a technomanagerial paradigm for which democratic legitimacy was neither sought nor required. A model of technocratic rationality evolved that was often unresponsive to the social and cultural complexity of the city.[23] Even if the GLC's eventual successor, the weaker Greater London Authority created in 2000, has produced several reports on flood risk, its power to effect change is extremely limited because most public services have been steadily removed from the democratic arena through privatization, quango formation, and administrative fragmentation.

The Thames Barrier was used for the first time on 1 February 1983, exactly thirty years after the 1953 flood that had presaged its construction. It remains the second largest flood barrier in the world after the

6.2 The Thames Barrier (2014). Photo: Matthew Gandy.

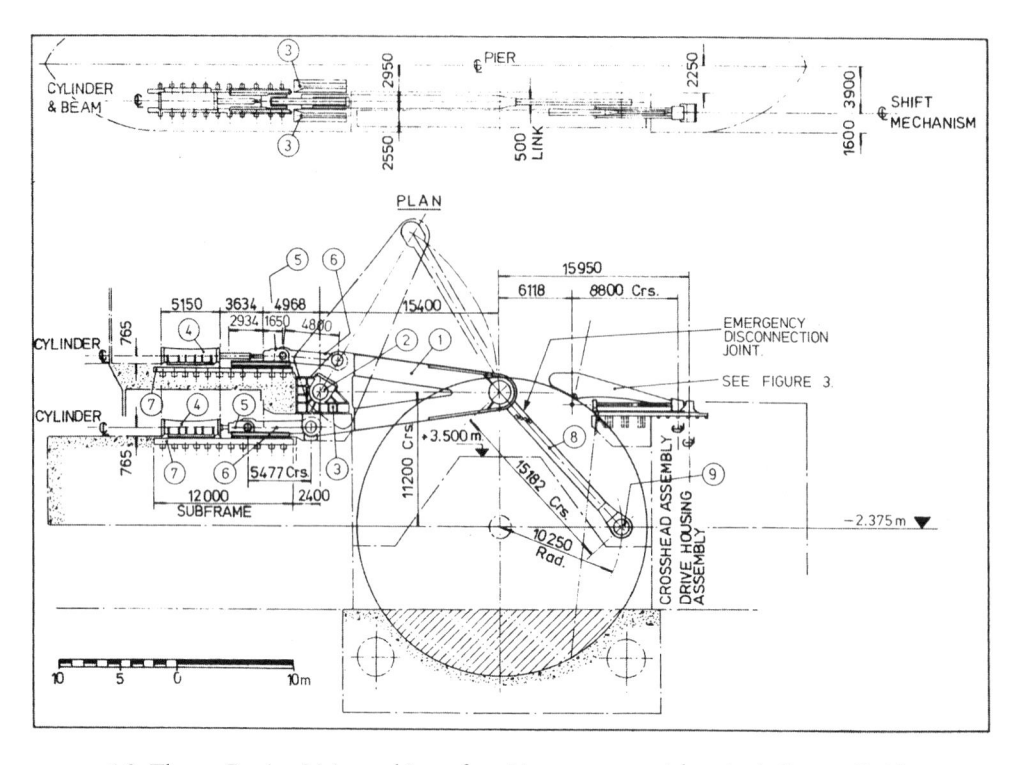

6.3 Thames Barrier. Main machinery for raising sector gate (plan view). Source: D. M. S. Fairweather and R. R. H. Kirton, "Operating Machinery," in *Thames Barrier Design*, proceedings of a conference held at the Institution of Civil Engineers, 5 October 1977 (London: Institution of Civil Engineers, 1978).

nine-kilometer-long Oosterschelderkering in the Netherlands, also constructed in response to the 1953 flood.[24] In addition to the barrier London's current flood protection system comprises eight smaller barriers on tributaries of the Thames, 36 major industrial floodgates, some 400 private structures, and 337 kilometers (209 miles) of tidal walls and embankments to protect the 1.5 million people living and working in the 116 square kilometers (just under 45 square miles) of tidal floodplain.[25]

The difficulty for London is that flood risk is now increasing and climate change is already occurring. Evidence suggests that sea level rise from thermal expansion and melting ice is now under way.[26] The shifting path of the North Atlantic jet stream has been linked with altered patterns of precipitation for the UK and elsewhere in northwest Europe.[27] London's flora and fauna are changing under the influence of climate change, with new arrivals and changing distributions: botanists note that lemons and grapefruits, as well as avocado, *Persea americana*, are now bearing fruit in London for the first time.[28] "Owing to the inertia of the climate system," writes the climatologist Ken Caldeira, "it is already too late to stop any further warming from occurring," though stabilization of emissions could theoretically mitigate the future intensity and frequency of flooding.[29] Evidence suggests, however, that even the aim to stabilize the predicted rise in global temperature is in doubt, which leaves London (and many other cities) facing acute dilemmas in terms of their flood defense infrastructure. Furthermore, feedback loops involving the large-scale release of methane hydrates, a decline in the North Atlantic thermohaline circulation, and especially the disintegration of the West Antarctic Ice Sheet could lead to a global sea level rise of between five and six meters by the end of the century.[30]

The predicted inundation of London is not solely due to climate change. Levels of groundwater in London have also been rising since the 1960s, after the closure of industries such as breweries and paper mills that had previously extracted large volumes of water, combined with glacial isostatic adjustment (also known as "postglacial rebound") whereby southern Britain is slowly sinking.[31] Flooding is now an increasing problem for transport infrastructure: by the early 2000s, for example, London Underground was having to pump 30,000 cubic meters of water out of its tunnels each day. Increased surface water runoff across London can also periodically overwhelm the city's drainage network independent of tide- or surge-related events, and following sudden cloudbursts many Underground stations have to be closed for extended periods, causing widespread disruption.[32] These so-called "flash floods" are being exacerbated by the trend toward more extensive concretization of the city for parking spaces (including the loss of many front gardens) and the growing use of cheaper "hard" landscaping design for new development projects: as London's topography becomes more impervious, its vulnerability to extreme weather events increases.[33]

Apart from a few exceptional non-flood-related circumstances, the trigger for the use of the Thames Barrier is derived from a combination of fluvial and hydrological data readings that include the tide height (typically a spring tide), the surge height (from the North Sea), and levels of fluvial flow at the tidal limit of the river 22 miles (35 kilometers) away at Teddington Lock. As a result of increasing flood risk the Thames Barrier is now being used more frequently: it was closed on just four occasions in the 1980s, 35 times in the 1990s, and 80 times in the 2000s. In the winter floods of early 2014 alone, the barrier was closed a record twenty times over a period of ten days, marking a further intensification in London's reliance on the barrier (figure 6.4).[34] Its increasing use has arisen from a combination of factors: higher-frequency storm surges along the North Sea coast; increased wind-generated surge effects in the English Channel; and higher flood levels and fluvial flows within the River Thames itself.[35]

There is now tacit acknowledgment among many government agencies, including London's own administration, the Greater London Authority, that increased flood risk is inevitable.[36] Surveys predict that in coming decades the number of people at high risk could increase from the current figure of about 1.25 million—equivalent to a sixth of the city's population—to over 3.5 million by the year 2080 if the upper range of predicted

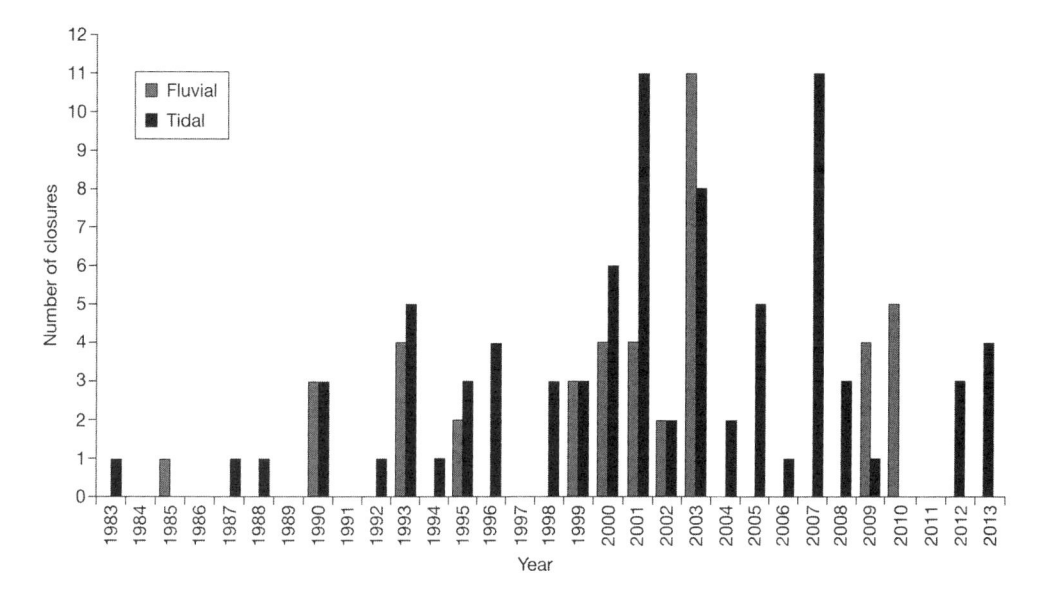

6.4 Thames Barrier closures between 1983 and 2013. Data from the UK Environment Agency.

sea level rise occurs. Yet such projections are fraught with real difficulty, not just in relation to climatic uncertainties but also from the dynamics of urbanization itself and the future patterns of land use, population change, and infrastructure provision in those areas at greatest risk. A report by the British Antarctic Survey suggests that the earlier projections by the Intergovernmental Panel on Climate Change dating from 2007 failed to take full account of the impact of melting glaciers and ice sheets on global changes in sea level. The report contends that following a one-meter rise in sea level, the Thames Barrier, which is currently adequate to protect London from a one-in-one-thousand-year level of flood risk, would be only able to provide a one-in-twelve-year degree of protection.[37] If low-lying parts of the capital were to be flooded now, following a breach of the Thames Barrier, this would force the closure of 143 rail and Underground stations, 16 hospitals, and over 400 schools. Eight power stations would be affected along with 28 museums, 18 major art galleries, and three world heritage sites.[38] The vulnerability of London to flood damage has also increased since the late 1980s because of the growing concentration of new construction activity within the city's floodplain, such as the redevelopment of London Docklands and a plethora of other riverside projects.[39] The UK's 2011 census shows that the highest increases in population for the whole country over the previous decade have been in inner London riverside boroughs such as Newham and Tower Hamlets at greatest risk from flooding.[40]

The Thames Gateway redevelopment program, launched in 2003, marks perhaps the most intense phase of urban boosterism that London has ever seen, of which the 2012 Olympics forms just one element.[41] After decades of demographic decline London has experienced a decisive reversal in its growth since the late 1980s, with a population of over eight million revealed in the 2011 census, approaching its previous peak population of nearly nine million in 1939. If further floodplain development were to be effectively abandoned or scaled back, then strategic planning for the future of London would have to be fundamentally recast in areas such as airport capacity, housing, and infrastructure provision.[42] The axial dynamics of future development would be shifted away from the east, with intensified pressures elsewhere in the metropolitan region.

The Thames Barrier itself will require modifications by 2030—less than twenty years away—but what happens with further sea level rise by the end of the century remains a topic of speculation. It is anticipated that the barrier's design limit of 50 closures a year will be regularly exceeded from around 2035 onward and that after 2070 it will no longer be an adequate source of protection.[43] London will face a choice toward the end of the twenty-first century between a vast program of public works or some kind of accommodation with the risk of regular inundation. The main options are either a much larger flood defense structure nearer to the mouth of the Thames estuary that would dwarf current infrastructure provision, or the designation of a large area of regular flooding where development would be radically curtailed.[44] At a national level there has been a shift of emphasis from flood prevention to acknowledgment of greater flood risk: the earlier technomanagerial logic has been supplanted by a new risk calculus framed by fiscal limits on public expenditure and the increasing incidence of flooding.[45] The restoration of London's floodplain would mark a dramatic reversal in the impetus toward an engineering-led solution, not least because the estuary is now the focus of intense development pressures. The emerging emphasis on floodplain restoration is complicated, however, by its lineage from an ecological imaginary that is drawn from the past yet projected into the future. Indeed, the apparent tension between technomodernity and growing interest in riparian simulacra is not as wide as it may seem.

## 6.2   From Speculative Dreams to Ecological Utopias

> We've got mile after mile and acre after acre of land for our future prosperity. No other city in the world has got right at its centre such an opportunity for profitable progress. So it's important that the right people mastermind the new London.

*Harold Shand*[46]

The Thames estuary is a speculative frontier for development, a zone of cultural exploration, and an area of exceptional biodiversity. Following the controversial redevelopment of the London Docklands in the 1980s, attention has moved progressively east, toward the Lea Valley (the site of the 2012 Olympics), further riverside sites in Deptford, Woolwich, and elsewhere, and since the mid-1990s beyond the metropolitan boundaries of London toward coastal areas of south Essex and north Kent. This interconnected landscape of canals, rivers, marshes, and former industrial sites has provoked a range of responses. The architectural critic Jonathan Glancey has dismissed the Thames estuary as a "cockney Siberia" comprising some of the most isolated places in southern Britain.[47] Glancey appears to delight in the sense of remoteness and disconnection lying east of London, like a slice of neoromanticist architectural noir:

> East of the Thames Barrier, between the south shores of Essex and the north coast of Kent, lies an all but foreign land. A tract of marshes, mists, rip tides, rubbish tips, freighters, sinking buildings, the world's largest sugarcane refinery, reeking sewage and rotting hulks left over from the Napoleonic wars and D-Day, interspersed, like some hastily edited film, with sporadic tower blocks, remote pubs, and hippy encampments. Here, if anywhere on the map of England, be monsters.[48]

These isolated parts of the Thames estuary that meet with Glancey's mix of aesthetic fascination and metropolitan disdain lie beyond the protected zone of the Thames Barrier. The depiction of the estuary by Glancey and others as a kind of inimical terra incognita or zone of emptiness has helped to transform the eastern edge of London into a speculative frontier. As the architectural historian Ben Campkin shows in relation to recent regeneration discourse in London, the stigmatization of specific neighborhoods, through their associations with dirt, danger, or spatial disorder, has served as a powerful pretext for their erasure and redevelopment.[49]

The distinctive tidal landscapes of the Thames estuary have served as the setting for literary explorations of London that emphasize poverty and marginality. Charles Dickens, for example, who grew up in Chatham on the north Kent coast, uses the estuary for the opening of *Great Expectations* (1860), where Pip realizes that "the dark, flat, wilderness beyond the churchyard, intersected with dykes and mounds and gates, with scattered cattle feeding on it, was the marshes; and that the low leaden line beyond, was the river; and that the distant savage lair from which the wind was rushing, was the sea."[50] Some forty years later these landscapes are evoked by Joseph Conrad in the opening of *Heart of Darkness* (1902), where he describes the approach to London along the north Kent coast:

> The sea-reach of the Thames stretched before us like the beginning of an interminable waterway. … A haze rested on the low shores that ran out to sea in a vanishing flatness. The air was dark above Gravesend, and farther back still seemed condensed into a mournful gloom, brooding motionless over the biggest, and the greatest, town on earth.[51]

While Dickens uses these landscapes to denote spaces of abandonment and social malaise, including the presence of prison ships, for Conrad the estuary serves as a metaphor for the unknown, represented by the power of the sea but also a looming distance from modernity. The Manichean cosmology of *Heart of Darkness* begins to take shape even before the real journey has properly begun. Conrad's depiction of the estuary can be read as a precursor to the architectonic dichotomies of the present with their intense juxtaposition of different spatial imaginaries.

Since the 1990s, development proposals for the estuary have invariably rested on the depiction of parts of east London, and the landscapes beyond, as essentially devoid of social, cultural, or ecological value. A narrative of remediation is frequently deployed in which black and white photographs of these spaces, evoking a distinctive ruin aesthetic, are routinely juxtaposed with computer-generated imagery of high-end speculative housing developments or a quasi-public realm of atria and esplanades.[52] The more far-fetched proposals for the Thames estuary include a "floating airport" to replace Heathrow, although the pilots would have to be trained to cope with sudden engine failure caused by flocks of birds (the estuary is one of the most important sites for breeding birds in Europe). Some developers have called for a series of "water cities" in the Thames estuary, combined with a huge "moveable flood barrage" that would also serve as a new road bridge across the mouth of the estuary.[53] Similarly, the "Thames hub" proposal by architect Norman Foster for the Hoo peninsula would involve a four-runway airport, a new combined flood barrier and tidal power generator, and high-speed infrastructure networks for transport and communications.[54]

Ranged against this logic of speculative development, however, are those who wish to protect the floodplain landscape, and the insurance industry which fears that ill-judged development proposals could lead them to incur huge losses. Recognition that the proposed development zones in the Thames estuary are at risk from flooding has led to a flurry of design proposals such as homes on stilts, but most conventional development has proceeded in spite of warnings.[55] Between 1989 and 2010, for example, a fifth of new dwellings constructed in London were in areas at high risk from flooding: a figure significantly higher than for other English regions. The UK government's emphasis since the late 1990s on building new homes on so-called brownfield sites has also intensified development pressures to build on postindustrial locations near the river.[56] The situation is complicated, however, by higher land values for scenic riverside settings in areas at greatest risk and the reluctance by planning authorities to pursue higher-density housing schemes on less vulnerable sites.[57] In general, the insurance industry has been skeptical about further development in London's floodplain, not least because a new wave of construction will itself intensify flood risk.[58] Indeed, until comparatively recently, the insurance industry had better flood risk data than the UK government, a weakness that has become manifest in tensions between the UK's Environment Agency, established in 1996, as the statutory body responsible for managing flood risk in England, and local planning authorities responsible for determining what development can take place.[59]

The Environment Agency has sought to restrict development in areas at risk by producing a series of flood maps since 1999 that are intended to guide local planning decisions. A study of these flood maps by the geographers James Porter and David Demeritt uncovers a "clash of institutional priorities and professional cultures" between the Agency and local planning authorities.[60] They argue that the Environment Agency has used these maps to deflect responsibility for managing future risk onto local planning authorities, yet the lack of sufficient cartographic precision has merely served to produce "sterilized areas" demarcated by flood-related planning blight. The disjuncture between different sources of knowledge and expertise has produced a landscape of hydrological uncertainty. The intersections between environmental risk and different tiers of state authority have produced a circulatory dynamics of governmentality that cannot be effectively mediated by the use of maps as a "technology of indirect rule."[61] This tension is exacerbated by the fact that only the local planning authorities are democratically accountable, yet they have lost much of their own technical expertise in flood control as a result of fiscal retrenchment and outsourcing of public services. As the strategic planning role of the Environment Agency has been weakened, partly in response to the methodological limitations of its own attempt to standardize cartographic practice and the growing resistance of planning authorities to its authority, the Agency has shifted the emphasis of flood maps toward public education, marking an attempt to individualize responsibility for managing flood risk.[62]

From an ecological perspective the Thames estuary not only mediates flood risk but also protects biodiversity. The conversion of Rainham Marshes from a series of military firing ranges and ordnance depots into a nature reserve in 2006 attests to the ecological significance of the estuary but represents only a small fragment of the riverside landscapes (figure

6.5).⁶³ "Despite appearances," reflects the architectural critic Ken Worpole, "the landscape at Rainham has been carefully orchestrated, both architecturally and ecologically," to produce "a subtle open-air theatre of memory."⁶⁴ Part of the marshes on the edge of London were given a statutory designation in 1986 as a Site of Special Scientific Interest by the then Nature Conservancy Council, as "the largest remaining expanse of wetland bordering the upper reaches of the Thames Estuary," combining high levels of biodiversity (especially for birds, plants, and insects) with archaeological interest as a cultural landscape little changed since the Middle Ages.⁶⁵ In the year 2000 nearly 5,600 hectares (13,800 acres) in the Thames estuary were designated as a Special Protection Area under the Ramsar Convention of 1971 for wetlands of international scientific importance. Described in the Ramsar submission as "a complex of brackish, floodplain grazing marsh ditches, saline lagoons and intertidal saltmarsh and mudflat," the area hosts fifteen scarce or endangered species of plants, over twenty noteworthy invertebrate species, and tens of thousands of wintering wildfowl.⁶⁶

The Thames estuary forms part of London's natural flood defenses, since these low-lying areas would help to reduce the impact of a tidal surge before it reached the city. A recent report by the Royal Institute of British Architects suggests the creation of a series of new tidal lagoons to help reduce flood risk, thereby combining landscape architecture with flood control.⁶⁷ The role of ecosystems in mitigating environmental threats has sometimes been referred to as "ecological infrastructure," a term that has its origins in the early 1980s and now encompasses aims such as the control of surface runoff, the lessening of the urban heat island effect, and new approaches to waste water treatment.⁶⁸ The Greater London Authority's *London Plan*, published in 2011, has emphasized a shift toward regarding green space as a form of "green infrastructure" linked to what has been termed the All London Green Grid, within which rivers and waterways are expected to play a distinctive role.⁶⁹ The increasing prominence and malleability of the term "infrastructure" in planning discourse for London marks a movement away from its earlier predominantly technical usage to

6.5  Rainham Marshes, Essex (2010). Photo: Matthew Gandy.

encompass a much wider range of developments. Yet the utilitarian impulse behind the idea of ecological or green infrastructure does not capture the full range of social, cultural, or scientific discourses associated with urban nature. The estuary has long been a source of inspiration for artists and photographers, who emphasize its "wildness" and comparative tranquillity, and these spaces have also acquired significance for scientific interest and environmental education.[70]

An enhanced role for the Thames estuary in mediating flood risk marks a shift of emphasis away from more elaborate physical defense structures toward developing new types of interaction with water.[71] At a national level this change of approach toward greater accommodation of flood risk was signaled by the UK government's policy document *Making Space for Water*, published in 2005. The document, based on a wide-ranging consultation exercise, introduces a more holistic approach to flood control that acknowledges the fiscal and practical limits to reliance on physical defenses. It emphasizes instead a greater range of land use options to control floods, including the creation of "washlands" and wetlands and the strategic abandonment of some sites susceptible to flooding or coastal erosion (a process referred to as "managed realignment").[72] The consultation has elicited a range of responses including many from organizations not hitherto closely engaged in flood control, as part of a more broadly defined understanding of the relationship between water and public policy.[73]

The complexity of an ecologically oriented approach to flood risk management contrasts with the relative simplicity of the technomanagerial paradigm. The development of a floodplain-oriented approach implies a different kind of relationship between scientific expertise and society since a much greater range of factors are involved, including questions of land use, landscape design, and intersections with other social and environmental objectives.[74] Within the London area there have already been some attempts to extend floodplain ecosystems for tributaries of the Thames. In the case of the river Ravensbourne in southeast London, however, an innovative scheme to recreate water meadows was very nearly abandoned in the wake of localized flooding and demands to install conventional flood barriers. The recent flood events have elicited memories of floods in 1968 when many homes near the river were inundated with water after prolonged heavy rainfall within the river's catchment area. A precarious relationship exists between ecological design and anxious communities most at risk from flooding, since barriers can themselves produce a sense of security.[75] The legitimacy of any new kind of ecological approach to flood control will ultimately rest on its technical efficacy rather than its aesthetic allure. Historically, public faith in flood barriers, seawalls, and other kinds of coastal defenses has contributed toward the clustering of developments in areas at greatest risk. With less reliance on obvious types of defensive structures, the relationship between urban development and perceived environmental security may become harder to discern.

The shift of emphasis toward the role of an enhanced natural floodplain connects with the idea of "ecological restoration" in landscape design, as we saw in the case of Los Angeles. The term ecological restoration spans a range of practices including the repair of damaged ecosystems, the control of soil erosion, and the reintroduction of native species.[76] Marshes and wetlands are the habitats that have been most relentlessly eliminated under modernity—principally for agriculture but also for public health reasons (as in the case of Lagos) and to reclaim land for development. In the Thames estuary the building of better flood defenses, especially after the disaster of 1953, saw an accelerated rate of further drainage of remaining marshlands in areas such as north Kent.[77]

Pivotal to many ecological restoration schemes is the conception of an idealized ecosystem from the past. Yet if an idealized Thames floodplain were to be restored, which moment in environmental history would be invoked? Ecological restoration offers a spectrum of possibilities, from the attempted reconstruction of lost ecosystems—exemplified by "re-wilding" and the reintroduction of large mammals—to more pragmatic concerns with the enhancement or modification of contemporary ecosystems. The attempted simulacrum of a premodern or even postglacial landscape before extensive human modification would make little sense, since these landscapes would have largely disappeared anyway as a result of sea level rise and nonanthropogenic sources of climate change. The Thames floodplain of the late Neolithic era (c. 6000 BC), before significant human alteration, would have been a "dark and intimidating landscape" including trees such as yew, *Taxus baccata*, a poisonous species that casts a deep shade.[78] As postglacial sea levels began to rise, a more mixed and open canopy would have formed, with other trees such as alder, *Alnus glutinosa*, becoming more dominant. By the Bronze Age, from around 3400 BC onward, there is clear evidence of alder trackways constructed in response to increased waterlogging and also the use of clearings for livestock grazing.[79] Despite painstaking ecological, palynological, and archaeological reconstructions, however, we can never know for sure what these former landscapes really looked like.[80] Even without human influence, what might constitute a "natural" floodplain now is of necessity very different from that of the past. Rather than the reconstruction of an imaginary past landscape, a new floodplain might take account of the innate hybridity of urban nature as a global synthesis or *brassage planétaire*, so that the illusory quest for authenticity would be replaced with explicit recognition of the mix of biophysical and cultural influences that shape urban space.[81] Archaeological evidence shows that the London floodplain has had a global flora since at least Roman times: traces of many adventitious and introduced plants can be found, including coriander, olive, sour cherry, and many other species, drawn not just from the Mediterranean but also from Africa.[82]

The connection between London and what might be considered a "natural" landscape is therefore complex, contentious, and historically specific. The use of nature as an ideological pivot has been significant in delineating cultural responses to cities and modernity. Yet the ecological critique of modernity harbors disparate cultural and political strands ranging from the neoromanticist veneration of an imaginary past to complex critiques of the relationship between science and society that gathered momentum from the late 1940s onward.[83] During the second half of the nineteenth century

London became the setting for radical critiques of the social and environmental effects of industrial capitalism. We find a marked divergence between the largely aesthetic concerns advanced by John Ruskin and his followers, with their retreat from the cities, and the nascent "green socialism" articulated by William Morris with its roots in a critique of the British class system.[84] Rather than a rejection of modernity, Morris called for cities themselves to be radically changed as part of a new society. In *News from Nowhere* (1890) he develops the utopian idea of a future London immersed in nature. The central London landmark of Trafalgar Square, for example, which had been the scene of extensive police brutality in November 1887, is transformed into a "space of whispering trees and odorous blossoms."[85] For Morris, the ideal city would be "countrified" with "sunny meadows" and "teeming gardens."[86] While some of his contemporaries envisioned garden cities set apart from existing urban centers, and readers of Edward Bellamy conceived centralized mechanistic utopias which anticipate aspects of state socialism in the twentieth century, for Morris it is London itself that has been transformed into an intricate garden.[87] Drawing on a mix of Marx's political economy and Fourier's social radicalism, Morris evokes a precapitalist synthesis of nature and culture in which slums have been replaced with meadows.

## 6.3   Avant-Garde Scenarios and Tranquil Adaptation

In a recent symposium held in London on cities and climate change, the architect Manfred Wolff-Plottegg derided the protection of low-lying urban centers from flooding. "Sustainability is a traditional tool," argued Wolff-Plottegg, "which is designed to prolong elements which instead should be deleted." In response, Kate Macintosh, representing Architects and Engineers for Social Responsibility, reminded the audience that "if we have the sea level rises that are predicted, a very substantial amount of the cities of the world will simply disappear," to which Peter Weibel, of the ZKM Centre for Art and Media, Karlsruhe, responded, "Yes. So what?"[88] At one level this apparent impasse between architectural provocation and engineering practicality reflects a tension between avant-garde urbanism and the day-to-day challenge of enabling modern cities to function effectively. This exchange also touches on the ambiguity of the sustainability concept in terms of what elements should be protected. Should the focus be on safeguarding specific aspects of the environment, or on the wider socioeconomic dynamics that have produced particular manifestations of nature? And to what extent is the current emphasis on sustainability about preventing the environmental disruption of global capitalism of which urbanization forms an integral part?

In an exhibition held at the Museum of London in 2010 entitled *Postcards from the Future*, the digital artists Robert Graves and Didier Madoc-Jones presented a series of future scenarios for London, endorsed by the influential environmentalist James Lovelock. In one striking image Parliament Square has been turned over to rice paddies to combat growing food shortages, while another shows an aerial view of central London under water in one of its regular spring tide inundations. The work of Graves and Madoc-Jones tends to replicate the more dystopian readings of London's future, featuring not just extreme weather events but also mass movements of environmental refugees and the growth of vast shantytowns in the city's parks.

Water has played a distinctive role in science fiction urbanism and the emergence of spatial anxieties about abandoned or dilapidated cities. The architectural historian David Gissen notes how the "dry ruins" of the past have been gradually supplanted since the 1960s by the "undrainable city," exemplified by "crumbling puddle-ridden photographs" and growing anxieties about the metropolis's unseen and uncontrollable entrails.[89] The threat of water, in the case of London from the subterranean realm as well as fluvial and tidal sources, has become a recurring dimension of dystopian future projections of social and economic chaos.

These gloomy visions stand in stark contrast with other projections that emphasize new forms of social and cultural adaptation. In *Flooded London*, for example, a project devised by the media production studio

Squint/Opera 2, we find a series of striking images of London in the year 2090. The 3D images, which were shown as part of the London Architecture Festival of 2008, reflect a placid postcatastrophe scenario in which London has adapted to its now permanently flooded state. In one image a man is about to take a late afternoon dive from the ivy-clad interior of the dome of St. Paul's Cathedral. In another, a familiar-looking Victorian terrace in the southeast London neighborhood of Honor Oak is renamed "suburban bucolia." Luxuriant vegetation spreads across the street toward a lake, beyond which the London skyline is shrouded in mist (figure 6.6). According to the Squint/Opera 2 studio, the "images are optimistic and reveal that far from being a tragedy, the floods have brought about a much-improved way of life in the capital."[90]

These counterdystopian architectonic projections present an avant-garde response to climatic uncertainty as a space for reinventing relations between society and nature. The depiction of what the architectural critic Marcus Fairs refers to as a "tranquil utopia" by Squint/Opera 2 lies in sharp contrast to the neo-Hobbesian vision evoked by J. G. Ballard and many other science fiction writers.[91] To find further examples of utopian flooding scenarios we need to explore other cultural and historical contexts, such as the Soviet-era fiction of Ivan Yefremov. In his novel *Andromeda* (1958), for example, Yefremov presents a peaceful interconnected world after the ice-caps have melted. Imaginary scientific advances include an atomic fuel that allows travel at the speed of light, thought transmission without speech, and new materials made from silicon compounds with extraordinary properties. A doubling in breadth of the subtropical belt has increased food production, and the melting of the Antarctic ice sheet has revealed vast new reserves of minerals:

> In the Age of Realignment artificial suns were made and "hung" over the north and south polar regions. These greatly reduced the size of the polar ice-caps that had been formed during the ice ages of the Quaternary Period and brought about

CHAPTER 6

6.6 Honor Oak, London, imagined in the year 2090. Courtesy of Squint/Opera 2 (2009).

extensive climatic changes. The level of the oceans was raised by seven metres, the cold fronts receded sharply and the ring of trade winds that had dried up the deserts on the outskirts of the tropic zone became much weaker.[92]

The prosperous warmer world evoked by Yefremov may have no basis in climatology, but it is suggestive of an optimism almost entirely lacking in much contemporary environmental discourse about human capacity to create a better future. Or as the literary critic Fredric Jameson suggests, "if Soviet images of Utopia are ideological, our own characteristically Western images of dystopia are no less so, and fraught with equally virulent contradictions."[93] This is not to suggest that the current environmental crisis, with climate change at its epicenter, does not pose grave threats, but rather to highlight that if we do not imagine a better kind of human society, the outcome of current trends will intensify existing patterns of conflict and inequality.

## 6.4   Future Imperfect

London's water infrastructure encompasses the nineteenth-century engineered metropolis, large-scale dimensions of twentieth-century techno-modernity, and new socioecological formations, including alternative approaches to flood control. Despite the neoliberal impetus toward divestment in the public realm and the inadequacies of strategic planning, there is vibrant debate over the city's environmental future. Rather than creating a technological bulwark against environmental change, which may not work anyway, radical alternatives posit a transformation in the socioecological dynamics of contemporary cities. If the decision is taken to withstand rather than accommodate sea level rise, then a dilemma facing London and other low-lying cities is how long-term investment in additional flood defenses would be paid for. Historically, the physical transformation of urban infrastructure involved a combination of technical, institutional, and financial innovations. If London were to embark on the construction of a new, much larger flood barrier, the cost would have to be covered through infrastructure bonds or some other form of long-term finance. Yet the political impasse over taxation, high levels of debt, and wider processes of restructuring within global banking and finance pose major difficulties in securing investment for drainage and flood control infrastructure. Unlike premium infrastructure projects such as airports, toll roads, or high-speed telecommunication nodes, the large-scale physical reconstruction of cities to improve flood defenses cannot attract the same levels of private investment, since the benefits are spread much more widely and accrue over significantly longer time periods.

The narrowing of options to adaptation or resilience obscures the degree to which climate change could be contained if there was the political will to tackle levels of greenhouse gas emissions.[94] In the case of London, however, the increased threat of inundation is not only carbon-related but also geological in origin: even if the projected sea level rise could be stabilized, there is still the longer-term impact of isostatic adjustment to contend with. The concept of resilience is also complicated by its connection with wider security agendas that impinge on urban policymaking across a range of fields, merging with various antistatist, postwelfare, and neoliberal discourses.[95]

The idea of resilience, through its strategic and organizational connections with state formation, raises questions about the autonomy of technical expertise and the role of politics in environmental discourse. The idea of urban resilience is inherently ambiguous, since it is the state that has enabled the advance of neoliberalism (through legislative and regulatory change) and it is the state that must ultimately intervene in the face of environmental threats or other disasters.[96] Since the early 1990s, in the post-Rio context, there has been pressure to develop new scientific models for sustainable urban policymaking that can effectively incorporate all aspects to urbanization. Yet the epistemological basis for these new socio-ecological models does not extend to the historical and political aspects of urban change.[97] We already know how a "zero carbon" city might intersect with fields such as architecture, ecology, or engineering, but the sociopolitical dimensions are much more ill-defined.

The historian Roy Porter has characterized London as "a muddle that worked," not through any concerted strategy but rather as the precarious outcome of a series of disparate developments.[98] London's administrative fragmentation has posed immense difficulties in tackling social and environmental problems. In essence, the political contradictions behind the growth of modern London remain unresolved, exemplified by the continuing political and economic dominance of finance capital, stark and growing inequalities in wealth, and a nondemocratic culture of institutional sclerosis rooted in the city's imperial past. Under what governmental and ideological constraints, therefore, will the future of London be anticipated, imagined, or influenced?

Epilogue

It is a damp and overcast morning in Lewisham, South London, in early January 2013. There is a striking white bird sitting in a tree toward the edge of Ladywell Fields, a park alongside the River Ravensbourne, a flood-prone tributary of the Thames. The bird is a little egret, *Egretta garzetta*, a scarce migrant from Africa, attracted to the park by the newly created riparian ecosystem, which is part of a river restoration project to help protect low-lying communities from flooding, improve public space, and also enhance urban biodiversity. The local river, which passes along the edge of the site largely unnoticed and encased in metal railings, has been partly "re-meandered" so that a secondary channel, diverted away from the main channel, now flows through the middle of the park. What was once a neglected or even avoided space has now become the focus of intense public interest. In this part of London a shift from a narrowly technomanagerial response to flood risk, exemplified by concrete defense structures dating from the 1960s, toward a different kind of approach is under way, spurred not just by local curiosity about London's "lost rivers" but also by an increasingly international dimension to the redesign of urban floodplains.[1]

This localized production of metropolitan nature encompasses a spectrum of ecological imaginaries, ranging from interest in the behavioral and psychological effects of contact with nature to tensions over the role of managed or spontaneous ecological dynamics for newly created riparian landscapes. Even if the hegemony of certain types of technical expertise has been partially displaced by new initiatives in urban design, there remain uncertainties over the role of local knowledge, the symbolic significance of new cultural appropriations of nature, and the wider economic implications of river revitalization in low-income neighborhoods. Furthermore, the realization of experimental projects such as the new waterscapes of Ladywell Fields in South London relies on diverse and potentially transitory sources of funding that do not necessarily mark a prolonged or systemic shift in the socioecological dimensions to contemporary urbanization.[2] The possibilities for new forms of "urban ecological citizenship," to use Jennifer Wolch's formulation for Los Angeles, rest on different constellations of knowledge but also require reflection on the different modalities of power within the urban arena.[3]

The reintroduction of water into the surface topography of the city has multiple implications for the socioecological dynamics of urban space. Many cities have begun to explore ways of allowing more naturalistic watercourses to play a more prominent role in both landscape design and flood control. The award-wining Bishan-Ang Mo Kio Park in Singapore, for example, designed by Atelier Dreiseitl, has overcome potential problems with mosquitoes by ensuring that there are plentiful fish, dragonflies (Odonata), and other natural predators for the larvae. The role of water in landscape design is being reconfigured in response to changing cultural and ecological sensibilities.[4] Elsewhere, however, the presence of standing water, the building of new waterways, or the extension of urban wetlands has produced landscapes of epidemiological uncertainty.

Water is ostensibly universal, as a metabolic component of urban life, yet it remains highly differentiated in its cultural and material appropriations. It connects between disparate locales as part of the precarious "corporeal unity" of urban space, yet reveals significant historical disjunctures and geographical variations. Mughal-era Delhi, for instance, is described by the historian Vijay Prashad as "famous for its canals which blessed the city with gardens, fresh drinking water and water to flush drains." The city's underground sewer system was comprised of masonry conduits "as functional as anything at the time," before the chaotic upheaval of the colonial era and the introduction of an inferior surface sewer network.[5] Similarly, the development of distinctive infrastructure networks for desert cities, such as the underground *qanat* technologies used widely in North Africa and Central Asia, reveals a range of sociotechnological innovations that are routinely marginalized or overlooked.[6] The increasing interest in rainwater harvesting, the closer differentiation of water by use, and other conservation measures illuminate the arbitrary dimensions to technomodernity and the technological path dependencies that have influenced the design and organization of urban space.

The fickle materialities of water serve to elude both technomodern attempts to control nature as well as increasingly sophisticated attempts to model socioecological systems. The severe floods across central Europe in 2002, for example, form part of a trend toward greater frequency and severity of events that leaves the existing scope of defensive structures and insurance coverage in doubt.[7] In the wake of floods affecting Dresden, Meissen, and other German cities, the inadequacy of private insurance coverage provoked calls for comprehensive public insurance, but a workable model has proved elusive.[8] Mandatory public insurance for flood risk already exists in the Netherlands, but this arrangement is rooted in the specific institutional and political legacy of the devastating inundation of 1953.[9] At an international level, the steady growth of "catastrophe bonds," reinsurance products, and other kinds of financial derivatives shows that the threat of climate change is now taken very seriously, even if the political response remains muted or fragmentary. Yet the increasingly pervasive translation of environmental risk into financial products moves the emphasis away from tackling the causes of environmental problems.[10]

When Ulrich Beck introduced the concept of "risk society" in the 1980s, he envisaged ubiquitous forms of environmental risk rather than the uneven landscapes of risk that have emerged in the twenty-first century. He emphasized the "leveling pressure" (*gleichmachenden Druck*) of global risk as an organizational focus for a postclass political paradigm that would transcend the existing system of nation states.[11] Yet the establishment of a new relationship between risk and what Beck terms the "subpolitical realm" has proved highly elusive in practice. There is now a widening divide between the insured and the uninsured whereby variations in material vulnerability intersect with poverty, inequality, and unequal access to knowledge. We have moved a long way from Beck's original conception of "reflexive modernity," exemplified by nuclear risk, toward the biopolitical realm, where to live outside the market is to become mere flesh and blood. The establishment of a new *bíos* implies the ever tighter imposition of boundaries, borders, and distinctions.[12]

If we consider the relationship between cities and strategic policy-making, there is a significant time lag between catastrophic events and any kind of coordinated long-term response. A mix of technical disagreements, political expediency, administrative inertia, and economic uncertainty produces a common pattern of extended delay. The process of physical reconstruction is invariably also one of institutional reorganization, producing different constellations of power, finance, and technical expertise. An accelerated rather than incremental pattern of environmental change, impelled by the "abrupt" climate change scenario, could overwhelm the capacity of many cities to respond.[13] At this point forecasts or models, as conventionally understood, begin to lose their practical utility, and we are left to contend with the ideological parameters of a science fiction imaginary in which worlds are turned upside down but remain strangely familiar.

The relationship between water and modernity is marked by tensions between different forms of expert knowledge and the establishment of a functional public realm. The idea of a dedicated cadre of experts—the "orchestra of specialists" evoked by Walter Gropius and Martin Wagner—working on behalf of a clearly defined public interest is superficially appealing yet increasingly anachronistic.[14] The acknowledgment of more dispersed and polyvalent sources of power, not necessarily residing within the apparatus of the modern state, offers a different angle on the rationality and scope of technical expertise.[15] Even if the representative reach of this array of experts could be widened, along with the inclusion of other forms of knowledge, the decline of this distinctive phase of strategic planning, rooted in progressive political agendas, has rendered the contemporary city increasingly opaque as a focus for democratic intervention. In terms of large-scale infrastructure projects, especially in the fast-growing cities of the global South, different conjunctions of capital and state power can be discerned. The presence of contrasting modernities, exemplified by the rise of neo-Haussmannite approaches to urban planning that owe more to the imperial reconstruction projects of the nineteenth century than to the technological modernism of the twentieth, shows that the relationship between cities and urban infrastructure cannot be restricted to narrowly normative or teleological readings of urban change. The legacy of Haussmann, as we saw, was not so straightforward in any case, marking a transitional phase before the rise of the "bacteriological city," scientific urbanism, and various forms of municipal managerialism.

Why has the engineering of modernity gone awry? Why has technical expertise failed to provide universal access to potable water or better protection from floods? In the case of water and sanitation the science is not mysterious, so that persistent inequalities must remain political in origin.[16] The framing of the "hydro subject" is of necessity related to the scope of urban citizenship, even if only obliquely, and to the different forms of "pressure," to use the anthropologist Nikhil Anand's metaphor, that can be deployed to secure the basic stuff of life. Water both constitutes and delimits the public realm, not necessarily as a stable or coherent social formation, but as a set of spheres of contestation and negotiation.[17] Where inadequate access to water and sanitation is the majority experience, the distinction between idealized urban models becomes an increasingly ideological exercise in enforcing sociospatial distinctions. In some cases the upgrading of infrastructure networks, especially if linked to urban redevelopment projects or other forms of land speculation, can itself become a tool of governmentality that serves to sharpen and reinforce social inequalities. Equally, distinctions between formal and informal, and other binary categorizations, can disguise the complexities of access to water and sanitation.[18] Dual or even multiple modes of governmentality now characterize the postcolonial metropolis, in a clear perpetuation of the uneven patterns of the past.

New forms of middle-class mobilization, especially within the cities of the global South, serve to foreclose options for wider access to basic infrastructure amid widening landscapes of poverty and neglect. The global diffusion of elite lifestyles is marked by the increasing ubiquity of nonagricultural irrigation for lawns, golf courses, and other opulent manifestations of semiarid urbanization. The extending ecological frontier is becoming an arena for voracious patterns of resource extraction that link water with other raw materials. Though water's imprint on the political, economic, and institutional characteristics of contemporary societies is arguably less than that of oil, there are nonetheless analytical commonalities revealed by tracing the flow of water, the hydrosocial dynamics of urban space, and the uneven geographies of risk.[19]

The rhetorical allure of terms such as "ecology," "sustainability," or "resilience" marks a corollary of the shifting ideological contours of mainstream environmental discourse and its entanglement with what has been referred to somewhat ironically as the "postpolitical condition." This fast-changing theoretical terrain, initially introduced by authors such as Chantal Mouffe, Jacques Rancière, and Slavoj Žižek and recently elaborated in relation to the urban process by Mustafa Dikeç, Erik Swyngedouw, and others, points to a pervasive evisceration of the language of critical environmental discourse.[20] The restriction of policy deliberation to an array of narrowly technical or administrative concerns, exemplified by the lexical mutation of vested interests and other manifestations of power into "stakeholders" and other obfuscatory euphemisms, is emblematic of an etiolated and highly restrictive political domain. The urban process is conceptualized as a manageable system comprising discrete and measurable components rather than a politically contested outcome of historical change. The city is reified as a hierarchically structured "living organism" or a "complex adaptive system" which can be understood in terms of its organizational capacity to accommodate environmental change.[21] An emphasis on indices derived from the ecological footprint of cities, commodities, or individual consumers supplants a consideration of the wider structural relationships between resource depletion, global capital, and emerging patterns of urbanization.

Urban space comprises an architectonic palimpsest of artifacts such as old water pumps, disused canals, or ceremonial features that are ostensibly unrelated to modern infrastructure systems. These archaeological traces, both recent and more distant, signal the apparent incongruity of many material landscapes. The nineteenth-century fascination with ruins resurfaced in the last quarter of the twentieth century with an interest in those spaces that appear to be disconnected from the visual grammar of late modernity. The historian Antoine Picon has used the term "anxious landscapes" to denote the "entanglements of buildings and infrastructures" that elude conventional categorizations.[22] But how do these anomalous or uncategorizable spaces relate to the historiography of modernity itself? The intersections between landscape and infrastructure have generated an array of utilitarian spaces that blur distinctions between those structures that are operational and those that are redundant. The recognition of "decomposing modernity," to use the anthropologist James Ferguson's term, signals the presence of multiple modernities that unsettle teleological assumptions about progress and identity.[23] Water, in its multiple cultural and material instantiations, poses a question of spatial legibility, not in a navigational mode à la Kevin Lynch but rather in the direction of the cognitive limits to late modernity and the complexities of material culture. The embodied dimensions of modernity remain entangled in the vagaries of "fictitious capital" and the radical disjuncture between the roles of water as metabolic necessity and as speculative chimera. Indeed, the tensions related to "the scale and durability" of fixed forms of capital and the cultural, political, and technological peculiarities of water infrastructure are enduring threads that underpin the fractured and uneven experiences of modernity.[24]

Water remains vital and elusive; in its oscillation between different cultural and material realms, it underpins both the limits and the possibilities of modernity. Over the last decade the study of water has itself evolved, connecting with new fields, while the political and environmental dimensions to relations between water and society, including the impact of climate change, have become more pressing. By adopting an interdisciplinary and historically grounded approach we can begin to unravel some of the complexities behind landscape, infrastructure, and modernity.

# Notes

## Introduction

1. Gaston Bachelard, *Water and dreams: an essay on the imagination of matter*, trans. Edith R. Farrell (Dallas: Pegasus Foundation, 1983), 6. The original quote reads: "L'eau est vraiment l'élément transitoire. Il est la métamorphose ontologique essentielle entre le feu et la terre." Gaston Bachelard, *L'eau et les rêves: essai sur l'imagination de la matière* (Paris: José Corti, 1947 [1942]), 8–9.

2. See Wendy O'Shea-Meddour, "Gaston Bachelard's *L'eau et les rêves*: conquering the feminine elements," *French Cultural Studies* 14 (1) (2003): 81–99.

3. Recent figures compiled by the UN and the WHO suggest that over 780 million people lack access to safe drinking water and some 2.5 billion people have no basic sanitation.

4. See Andrew Barry and Georgina Born, introduction to *Interdisciplinarity: reconfigurations of the social and natural sciences* (London: Routledge, 2013), 1–56. See also Sara B. Pritchard, "Joining environmental history with science and technology studies: promises, challenges, and contributions," in Dolly Jørgensen, Finn Arne Jørgensen, and Sara B. Pritchard, eds., *New natures: joining environmental history with science and technology studies* (Pittsburgh: University of Pittsburgh Press, 2013), 1–17.

5. The modern usage of the term "infrastructure" in an Anglo-American context is generally traced either to the impetus toward improving flood defenses in the United States in the 1920s or to earlier military concerns with organizational logistics. The English usage is derived from the French word *infrastructure*, used in relation to engineering works from the mid-1870s onward, particularly for railway construction. See

Émile Littré, *Dictionnaire de la langue française, supplément* (Paris: Librairie Hachette, 1886), 200; and Dirk van Laak, "Der Begriff 'Infrastruktur' und was er vor seiner Erfindung besagte," *Archiv für Begriffsgeschichte* 41 (1999): 280–299. On the broader meanings of "infrastructure" see, for example, Paul N. Edwards, "Infrastructure and modernity: force, time, and social organization in the history of sociotechnical systems," in Thomas J. Misa, Philip Brey, and Andrew Feenberg, eds., *Modernity and technology* (Cambridge, MA: MIT Press, 2003), 185–225; Thomas P. Hughes, "The evolution of large technical systems," in Wiebe E. Bijker; Thomas P. Hughes, and Trevor J. Pinch, eds., *The social construction of technological systems: new directions in the sociology and history of technology* (Cambridge, MA: MIT Press, 1987), 51–82; Colin McFarlane and Jonathan Rutherford, "Political infrastructures: governing and experiencing the fabric of the city," *International Journal of Urban and Regional Research* 32 (2) (2008): 363–374; and Susan Leigh Star, "The ethnography of infrastructure," *American Behavioral Scientist* 43 (3) (1999): 377–391. See also Reyner Banham, "Tennessee Valley Authority: l'ingegneria dell'utopia / Tennesse Valley Authority: the engineering of utopia," *Casabella: International Architectural Review* 542/543 (1988): 74–81, 125. More recently, the concepts of landscape and infrastructure have been brought into closer articulation. See Pierre Bélanger, "Landscape as infrastructure," *Landscape Journal* 28 (1) (2009): 79–95; Matthew Gandy, "Landscape and infrastructure in the late-modern metropolis," in Sophie Watson and Gary Bridge, eds., *The new Blackwell companion to the city* (Oxford: Blackwell), 57–65. The concept has also acquired an extended corporeal definition in the context of its absence: see AbdouMaliq Simone, "People as infrastructure: intersecting fragments in Johannesburg," *Public Culture* 16 (3) (2004): 407–429. On ecological dimensions to urban infrastructure see also Jochen Monstadt, "Conceptualizing the political ecology of urban infrastructures: insights from technology and urban studies," *Environment and Planning A* 41 (2009): 1924–1942; Kazys Varnelis, ed., *The infrastructural city: networked ecologies in Los Angeles* (Barcelona: Actar, 2008), 6–16. More recently, the term "infrastructure" has itself been thrown into doubt as part of an alternative theorization of urban materiality: see, for example, Swati Chattopadhyay, *Unlearning the city: infrastructure in a new optical field* (Minneapolis: University of Minnesota Press, 2012).

6. See Timothy Mitchell, *Rule of experts: Egypt, techno-politics, modernity* (Berkeley: University of California Press, 2002).

7. Simon Schama, *Landscape and memory* (New York: Alfred A. Knopf, 1995), 10. On the history of water and landscape see also David Blackbourn, *The conquest of nature: water, landscape and the making of modern Germany* (London: Jonathan Cape, 2006); Stephanie C. Kane, *Where rivers meet the sea: the political ecology of water* (Philadelphia: Temple University Press, 2012); Ari Kelman, *A river and its city: the nature of landscape in New Orleans* (Berkeley: University of California Press, 2003); Chandra

Mukerji, *Impossible engineering: technology and territoriality on the Canal du Midi* (Princeton: Princeton University Press, 2009); Sara B. Pritchard, *Confluence: the nature of technology and the remaking of the Rhône* (Cambridge, MA: Harvard University Press, 2011); Robert Stolz, *Bad water: nature, pollution, and politics in Japan, 1870–1950* (Durham: Duke University Press, 2014); Terje Tvedt, *The river Nile in the age of the British: political ecology and the quest for economic power* (London: I. B. Taurus, 2003); Richard White, *The organic machine: the remaking of the Columbia River* (New York: Hill and Wang, 1995); Donald Worster, *Rivers of empire: water, aridity, and the growth of the American west* (New York: Pantheon Books, 1985).

8. The word "modernity," derived from the earlier word "modern," denoting a self-conscious break with the past, becomes more widely used from the seventeenth and eighteenth centuries onward but often in a negative way. During the nineteenth century, and especially into the twentieth, the meaning begins to shift toward a generally positive connotation and becomes associated with advances in science, technology, medicine, and other fields. See Raymond Williams, *Keywords*, 2nd ed. (London: Fontana, 1983 [1976]). See also Martin Daunton and Bernhard Rieger, eds., *Meanings of modernity: Britain from the late-Victorian era to World War II* (Oxford: Berg, 2001); Mikael Hård and Andrew Jamison, eds., *The intellectual appropriation of technology: discourses on modernity, 1900–1939* (Cambridge, MA: MIT Press, 1998); Anson Rabinbach, *The human motor: energy, fatigue, and the origins of modernity* (Berkeley: University of California Press, 1992 [1990]).

9. See, for example, Nancy Leys Stepan, *Picturing tropical nature* (London: Reaktion, 2001); Ola Uduku, "Modernist architecture and the 'tropical' in West Africa: the tropical architecture movement in West Africa, 1948–1970," *Habitat International* 30 (3) (2006): 396–411.

10. See Fredric Jameson, *A singular modernity: essay on the ontology of the present* (London: Verso, 2002), and Bernard Yack, *The fetishism of modernities: epochal self-consciousness in contemporary social and political thought* (Notre Dame, IN: University of Notre Dame Press, 1997).

11. Ultimately, however, the question is not so much one of refining semantic distinctions as of reflecting on how these terms have actually been deployed in different contexts. See, for example, Paul Rabinow, *French modern: norms and forms of the social environment* (Cambridge, MA: MIT Press, 1989).

12. Examples of the early technological sophistication of urban water supply systems include Rome's Aqua Appia, commissioned in 312 BC, along with the elaborate networks constructed by Renaissance city-states such as Milan, Naples, and Venice. See, for example, Donald Hill, *A history of engineering in classical and mediaeval times* (London: Routledge, 1996); Jean-Pierre Leguay, *L'eau dans la ville au Moyen Âge*

(Rennes: Presses Universitaires de Rennes, 2002); Roberta J. Magnusson, *Water technology in the Middle Ages: cities, monasteries, and waterworks after the Roman Empire* (Baltimore: Johns Hopkins University Press, 2001); and Dietrich Werner, *Wasser für antike Rom* (Berlin: VEB Verlag für Bauwesen, 1986).

13. See, for example, Carlo F. Antonelli, "Acque sporche: Londra e il 'Metropolitan Board of Works', 1855–1865," *Storia Urbana* 16 (61) (1992): 61–81; Denis Bocquet, Konstantinos Chatzis, and Agnès Sander, "From free good to commodity: universalizing the provision of water in Paris (1830–1930)," *Geoforum* 39 (2008): 1821–1832; John Broich, *London: water and the making of the modern city* (Pittsburgh: University of Pittsburgh Press, 2013); Richard J. Evans, *Death in Hamburg: society and politics in the cholera years, 1830–1910* (Oxford: Clarendon Press, 1987); Laura Ferrari, *L'acqua nel paesaggio urbano: letture esplorazioni ricerche scenari* (Florence: Florence University Press, 2005); Jean-Pierre Goubert, *The conquest of water: the advent of health in the industrial age*, trans. Andrew Wilson (Cambridge: Polity, 1989 [1986]); Christopher Hamlin, *Public health and social justice in the age of Chadwick: Britain, 1800–1854* (Cambridge: Cambridge University Press, 1988); Paola Penzo, "L'urbanistica e l'amministrazione socialista a Bologna, 1914–1920," *Storia Urbana* 18 (66) (1994): 109–143; Carl Smith, *City water, city life: water and the infrastructure of ideas in urbanizing Philadelphia, Boston, and Chicago* (Chicago: University of Chicago Press, 2013); Frank M. Snowden, *Naples in the time of cholera, 1884–1911* (Cambridge: Cambridge University Press, 1995); and Vanessa Taylor and Frank Trentmann, "Liquid politics: water and the politics of everyday life in the modern city," *Past and Present* 211 (2011): 199–241.

14. See, for example, Karen Bakker, *Privatizing water: governance failure and the world's urban water crisis* (Ithaca: Cornell University Press, 2010); Jessica Budds, "Water, power, and the production of neoliberalism in Chile, 1973–2005," *Environment and Planning D: Society and Space* 31(2) (2013): 301–318; Maria Kaika, *City of flows: modernity, nature, and the city* (New York: Routledge, 2005); Alex Loftus, *Everyday environmentalism: creating an urban political ecology* (Minneapolis: University of Minnesota Press, 2012); Erik Swyngedouw, "Power, nature, and the city: the conquest of water and the political ecology of urbanization in Guayaquil, Ecuador: 1880–1990," *Environment and Planning A* 29 (2) (1997): 311–332.

15. See, for example, Federico Caprotti and Joanna Romanowicz, "Thermal eco-cities: green building and thermal urban metabolism," *International Journal of Urban and Regional Research* 37 (6) (2013): 1949–1967; Nate Gabriel, "Urban political ecology: environmental imaginary, governance, and the non-human," *Geography Compass* 8 (1) (2014): 38–48; David Gissen, *Subnature: architecture's other environments* (New York: Princeton Architectural Press, 2009).

16. See Ian Graham Ronald Shaw, Paul F. Robbins, and John Paul Jones III, "A bug's life: the spatial ontologies of mosquito management," *Annals of the Association of*

*American Geographers* 100 (2) (2010): 373–392; William K. Reisen, Richard M. Takahashi, Brian D. Carroll, and Rob Quiring, "Delinquent mortgages, neglected swimming pools, and West Nile Virus, California," *Emerging Infectious Diseases* 14 (11) (2008): 1747–1749. On the spatial dynamics of foreclosure see also Alex Schafran, "Origins of an urban crisis: the restructuring of the San Francisco Bay area and the geography of foreclosure," *International Journal of Urban and Regional Research* 37 (2) (2013): 663–688.

17. See, for example, Rita Kusriastuti, "Indonesia has most cases of dengue fever in ASEAN," *Jakarta Post* (12 June 2012); José G. Rigau-Pérez, Gary G. Clark, Duane J. Gubler, Paul Reiter, Eduard J. Sanders, and A. Vance Vorndam, "Dengue and dengue haemorrhagic fever," *Lancet* 352 (1998): 971–977; Agus Suwandono, Herman Kosasih, Nurhayati, Rita Kusriastuti, Syahrial Harun, Chairin Ma'roef, Suharyono Wuryadi, Bambang Herianto, Djoko Yuwono, Kevin R. Porter, Charmagne G. Beckett, and Patrick J. Blair, "Four dengue virus serotypes found circulating during an outbreak of dengue fever and dengue haemorrhagic fever in Jakarta, Indonesia, during 2004," *Transactions of the Royal Society of Tropical Medicine and Hygiene* 100 (2006): 855–862; and Erik Swyngedouw and Nik Heynen, "Urban political ecology, justice, and the politics of scale," *Antipode* 35 (5) (2003): 898–918.

18. Duane J. Gubler and Gary G. Clark, "Dengue/dengue hemorrhagic fever: the emergence of a global health problem," *Emerging Infectious Diseases* 1 (2) (April-June 1995): 55–57.

19. Soumyajit Banerjee, Gautam Aditya, and Goutam K. Saha, "Household disposables as breeding habitats of dengue vectors: linking wastes and public health," *Waste Management* 33 (2013): 233–239. See also Alex M. Nading, "Dengue mosquitoes are single mothers: biopolitics meets ecological aesthetics in Nicaraguan community health work," *Cultural Anthropology* 27 (4) (2012): 572–596.

20. See, for example, Steve Hinchliffe and Nick Bingham, "People, animals, and biosecurity in and through cities," in S. Harris Ali and Roger Keil, eds., *Networked disease: emerging infections in the global city* (Oxford: Wiley-Blackwell, 2008), 214–227; Meike Wulf, "Towards an urban epidemiology," in Matthew Gandy, ed., *Urban constellations* (Berlin: jovis, 2011), 75–79.

21. See, for example, Reiner Grundmann, "On control and shifting boundaries: modern society in the web of systems and networks," in Olivier Coutard, ed., *The governance of large technical systems* (London: Routledge, 1999), 239–257.

22. Arjun Appadurai, interview with the author, in the documentary *Liquid city: water, landscape and social formation in 21st century Mumbai* (UK/India, 2007; director Matthew Gandy; available from the UCL Urban Laboratory).

23. See, for example, Jane Bennett, "The agency of assemblages and the North American blackout," *Public Culture* 17 (3) (2005): 445–465; Stephen Graham and Nigel Thrift, "Out of order: understanding repair and maintenance," *Theory, Culture, and Society* 24 (1) (2007): 1–25; Stephen Graham, ed., *Disrupted cities: when infrastructure fails* (London: Routledge, 2010); Jonathan Rutherford, "Infrastructure transition and the flexible city," Flexible City Symposium, University of Oxford, 24–25 October 2013); Wiebe E. Bijker, "Dikes and dams, thick with politics," *Isis* 98 (1) (2007): 109–123.

24. See for example Françoise de Bonneville, *Le livre du bain* (Paris: Flammarion, 1997); Nadir Lahiji and Daniel S. Friedman, eds., *Plumbing: sounding modern architecture* (New York: Princeton Architectural Press, 1997); Barbara Penner, *Bathroom* (London: Reaktion, 2014); Lawrence Wright, *Clean and decent: the history of the bathroom and the W.C.* (London: Routledge and Kegan Paul, 1960). The expression "auto-house-electrical-appliance complex" is taken from Annemieke J. M. Roobeek, "The crisis in Fordism and the rise of a new technological paradigm," *Futures* 19 (1987): 33.

25. See Richard Dennis, *Cities in modernity* (Cambridge: Cambridge University Press, 2008).

26. See Andrew Harris, "Vertical urbanisms: opening up the geographies of the three-dimensional city," *Progress in Human Geography* (in press); Eyal Weizman, *Hollow land: Israel's architecture of occupation* (London: Verso, 2007). The "vertical axis" in particular has been associated with the hidden realms of the modern city, including not just subterranean spaces but other restricted zones of modernity. See, for example, Paul Dobraszczyk, *Into the belly of the beast: exploring London's Victorian sewers* (Reading: Spire, 2009); Bradley Garrett, *Explore everything: place-hacking and the city* (London: Verso, 2013); and Luzar Kunstmann, *La culture en clandestins. L'UX* (Paris: Hazan, 2009).

27. Sigfried Giedion, *Mechanization takes command: a contribution to anonymous history* (Oxford: Oxford University Press, 1948), 628. See also William W. Braham, "Siegried Giedion and the fascination of the tub," in Lahiji and Friedman, *Plumbing*, 200–224.

28. Stephen Graham and Simon Marvin, *Splintering urbanism: networked infrastructures, technological mobilities and the urban condition* (London: Routledge, 2001). See also Martin Melosi, *The sanitary city: urban infrastructure in America from colonial times to the present* (Baltimore: Johns Hopkins University Press, 2000).

29. See, for example, Bakker, *Privatizing water*; Kathryn Furlong, "The dialectics of equity: consumer citizenship and the extension of water supply in Medellín, Colombia," *Annals of the Association of American Geographers* 103 (5) (2013): 1176–1192.

30. See Matthew Gandy, "The bacteriological city and its discontents," *Historical Geography* 34 (2006): 14–25.

31. Mark Mazower, *Governing the world: the history of an idea* (London: Penguin, 2012), xi.

32. Wright, *Clean and decent*.

33. Alf Bremer, Gabriele Klahr, Christian Porst, and Michael Stein, *Kreuzberg Chamissoplatz. Geschichte eines Berliner Sanierungsgebietes* (Berlin: Propolis, 2007).

34. See, for example, Amita Baviskar, "Between violence and desire: space, power, and identity in the making of metropolitan Delhi," *International Social Science Journal* 55 (175) (2003): 89–98; D. Asher Ghertner, "Nuisance talk and the propriety of property: middle class discourse of a slum free Delhi," *Antipode* 44 (4) (2012): 1161–1187.

35. See, for example, Matthew Gandy, "Rethinking urban metabolism: water, space and the modern city," *City* 8 (3) (2004): 371–387; Maria Kaïka and Erik Swyngedouw, "Fetishising the modern city: the phantasmagoria of urban technological networks," *International Journal of Urban and Regional Research* 24 (1) (2000): 120–138; Joel A. Tarr and Gabriel Dupuy, eds., *Technology and the rise of the networked city in Europe and America* (Philadelphia: Temple University Press, 1988); and Joel A. Tarr, "The metabolism of the industrial city: the case of Pittsburgh," *Journal of Urban History* 28 (2002): 511–545.

36. See Michael Hau, *The cult of health and beauty in Germany: a social history, 1890–1930* (Chicago: University of Chicago Press, 2003). On Weimar-era conceptions of the body as machine see also Matthew Biro, *The dada cyborg: visions of the new human in Weimar Berlin* (Minneapolis: University of Minnesota Press, 2009).

37. On the political dimensions of dams see, for example, Heather Hoag and May-Brit Ohman, "Turning water into power: debates over the development of Tanzania's Rufiji basin, 1945–1985," *Technology and Culture* 49 (2008): 624–651; Maria Kaïka, "Dams as symbols of modernization: the urbanization of nature between geographical imagination and materiality," *Annals of the Association of American Geographers* 96 (2) (2006): 276–301.

38. Mitchell, *Rule of experts*, 38.

39. Matthew Gandy, "Cyborg urbanization: complexity and monstrosity in the contemporary city," *International Journal of Urban and Regional Research* 29 (1) (2005): 26–49.

40. The more restricted readings of urban metabolism include the flow-based models associated with "industrial ecology" and "materials flow analysis." See, for example,

Marina Fischer-Kowalski and Walter Hüttler, "Society's metabolism: the intellectual history of materials flow analysis, Part II. 1970–1998," *Journal of Industrial Ecology* 2 (4) (1999): 107–136.

41. See Neil Brenner, ed., *Implosions/explosions: toward a theory of planetary urbanization* (Berlin: jovis, 2014).

42. Andrew Barry, "Technological zones," *European Journal of Social Theory* 9 (2) (2006): 239–253.

43. Timothy Mitchell, "Carbon democracy," *Economy and Society* 38 (3) (2009): 422.

44. On desalination see Menachem Elimelech and William A. Phillip, "The future of seawater desalination: energy, technology, and the environment," *Science* 333 (2011): 712–717; Erik Swyngedouw, "Into the sea: desalination and the hydro-fix in Spain," *Annals of the Association of American Geographers* 103 (2) (2013): 261–270.

45. See Timothy Mitchell, *Carbon democracy: political power in the age of oil* (London: Verso, 2013 [2011]), 267.

46. Norbert Elias, *Über den Prozeß der Zivilisation* (Frankfurt: Suhrkamp, 1997).

47. Elisabeth Heidenreich, *Fliessräume. Die Vernetzung von Natur, Raum und Gesellschaft seit dem 19. Jahrhundert* (Frankfurt: Campus, 2004). See also Alain Corbin, *The foul and the fragrant: odor and the French social imagination*, trans. Miriam L. Kochan, Roy Porter, and Christopher Prendergast (Cambridge, MA: Harvard University Press, 1988 [1986]).

48. See Hans Peter Duerr, "Die mittelalterlichen Badstuben," in *Nacktheit und Scham*, vol. 1 of *Der Mythos vom Zivilisationsprozeß* (Frankfurt: Suhrkamp, 1988), 38–58.

49. Georges Vigarello, *Concepts of cleanliness: changing attitudes in France since the middle ages*, trans. Jean Birrell (Cambridge: Cambridge University Press, 1988 [1985]).

50. See, for example, Christina Trupat, "'Bade zu Hause!' Zur Geschichte des Badezimmers in Deutschland seit der Mitte des 19. Jahrhunderts," *Technikgeschichte* 63 (3) (1996): 219–236; Wright, *Clean and decent*.

51. See Peter Joyce, *The rule of freedom: liberalism and the modern city* (London: Verso, 2003); Michel Foucault, "Governmentality," in Graham Burchell, Colin Gordon, and Peter Miller, eds., *The Foucault effect: studies in governmentality* (Chicago: University of Chicago Press, 1991), 87–104. See also Taylor and Trentmann, "Liquid politics."

52. Simon Gunn, "From hegemony to governmentality: changing conceptions of power," *Journal of Social History* 39 (3) (Spring 2006): 705–720.

53. Vigarello, *Concepts of cleanliness*, 181. On changing relations between architecture and water engineering see, for example, Alberto Abriani, "Dal Sifone alla città / From the siphon to the city," *Casabella: International Architectural Review* 542/543 (1988): 24–29, 117; Edoardo Benvenuto, "Architetti, ingegneri e matematici / Architects, engineers and mathematicians," *Casabella* 542/543 (1988): 16–19, 115–116; Bertrand Gille, *Les ingénieurs de la Renaissance* (Paris: Hermann, 1967); Thomas Kluge, "Umgang mit Wasser als hydraulischer Maschinerie. Kollektive Hintergrundvorstellungen zur Mechanisierung des Wassers," in Bernd Busch and Larissa Förster, eds., *Wasser* (Cologne: Wienand, 2000), 312–323; Heinrich Tepasse, *Stadttechnik im Städtebau Berlins*, vol. 1, *19. Jahrhundert* (Berlin: Gebr. Mann, 2001).

54. Ross Beveridge and Matthias Naumann, "The Berlin Water Company: from 'inevitable' privatization to 'impossible' remunicipalization," in Matthias Bernt, Britta Grell, and Andrej Holm, eds., *The Berlin Reader* (Bielefeld: transcript, 2013), 192. See also Ross Beveridge, *A politics of inevitability: the privatisation of the Berlin Water Company, the global city discourse, and governance in 1990s Berlin* (Wiesbaden: VS Research, 2012).

55. See Karen Bakker, "Neoliberal versus postneoliberal water: geographies of privatization and resistance," *Annals of the Association of American Geographers* 103 (2) (2013): 253–260.

56. Beveridge and Naumann, "The Berlin Water Company."

57. Arwen Colell, "(Re-)constructing urban infrastructure: civil society movements are claiming the grid," paper given to the Flexible City Symposium, St. Anne's College, University of Oxford, 24–25 October 2013. See also Kurt Berlo and Oliver Wagner, *Stadtwerke-Neugründungen und Rekommunalisierungen. Energieversorgung in kommunaler Verantwortung* (Wuppertal: Wuppertal Institut für Klima, Umwelt, Energie, 2013); and Jeevan Vasager, "Berliners to vote on public takeover of power supply," *Financial Times* (2 November 2013). Although the Berlin energy referendum narrowly failed, other German cities such as Hamburg have seen successful campaigns to bring energy supply under public ownership.

58. Andrew Feenberg, *Between reason and experience: essays in technology and modernity* (Cambridge, MA: MIT Press, 2010), 155.

59. See, for example, Pushpa Arabindoo, "Mobilising for water: hydro-politics of rainwater harvesting in Chennai," *International Journal of Urban Sustainable Development* 3 (1) (2011): 106–126; Alexander Brash, Jamie Hand, and Kate Orff, eds., *Gateway: visions for an urban national park* (New York: Princeton Architectural Press, 2011); Anuradha Mathur and Dilip da Cunha, *Mississippi floods: designing a shifting landscape* (New Haven: Yale University Press, 2001); Tse-Hui Teh, "Hydro-urbanism:

reconfiguring the urban water cycle in the lower Lea river basin, London," PhD dissertation, University College London, 2011; and Jane Wolff, *Delta primer: a field guide to the California delta* (Richmond, CA: William Stout, 2003).

60. See Matthew Gandy, *Concrete and clay: reworking nature in New York City* (Cambridge, MA: MIT Press, 2002); David Soll, *Empire of water: an environmental and political history of New York City* (Ithaca: Cornell University Press, 2013).

61. Amita Baviskar, "Between micro-politics and administrative imperatives: decentralization and the watershed mission in Madhya Pradesh, India," *European Journal of Development Research* 16 (1) (2004): 26–40.

62. See, for example, Norbert Kühn, "Intentions for the unintentional: spontaneous vegetation as the basis for innovative planting design in urban areas," *Journal of Landscape Architecture* (Autumn 2006): 46–53.

63. Myung-Rae Cho, "The politics of urban nature restoration: the case of Cheonggyecheon restoration in Seoul, Korea," *International Development Planning Review* 32 (2) (2010): 146–165.

64. Émile Zola, *L'Assommoir* (Paris: Flammarion, 1997 [1876]), 56.

65. James Joyce, *Ulysses* (London: Penguin, 1992 [1922]), 783.

66. J. G. Ballard, *Quotes*, ed. V. Vale and Mike Ryan (San Francisco: RE/Search, 2004), 211–212.

67. See Reyner Banham, "The Becher vision," in Bernd and Hilla Becher, *Water Towers* (Cambridge, MA: MIT Press, 1988), 7–8; Susanne Lange, *Bernd and Hilla Becher: life and work* (Cambridge, MA: MIT Press, 2002); Weston Naef, "The art of Bernd and Hilla Becher," in Becher and Becher, *Water Towers*, 9–12. See also Matthew Biro, "From analogue to digital photography: Bernd and Hilla Becher and Andreas Gursky," *History of Photography* 36 (3) (2012): 353–366. Earlier precedents include a range of works included in William Jenkins, *New topographics: photographs of a man-altered landscape* (Rochester, NY: International Museum of Photography, 1975).

68. See, for example, Nate Millington, "Post-industrial imaginaries: nature, representation and ruin in Detroit, Michigan," *International Journal of Urban and Regional Research* 37 (1) (2013): 279–296.

69. Julian Stallabrass, "Sebastião Salgado and fine art journalism," *New Left Review* 157 (1997): 131–162. See also Sebastião Salgado, *L'homme et l'eau* (Paris: terre bleue, 2005).

70. Richard Rorty, *Contingency, irony, and solidarity* (Cambridge: Cambridge University Press, 1989), 83.

71. See Richard Rorty, *Philosophy and the mirror of nature* (Princeton: Princeton University Press, 1979).

72. See Jürgen Habermas, "Excursus on leveling the genre distinction between philosophy and literature," in *The philosophical discourse of modernity* (Cambridge: Polity, 1987 [1985]), 185–210; Axel Honneth, "The other of justice: Habermas and the ethical challenge of postmodernism," in Stephen K. White, ed., *The Cambridge companion to Habermas* (Cambridge: Cambridge University Press, 1995), 289–323.

73. Richard Rorty interviewed by Robert Harrison for *Entitled Opinions* (2005), copy held in the radio archive, Stanford University, cited in William Storage, "A matter for the engineers" (13 September 2012), http://themultidisciplinarian.com/2012/09/13/richard-rorty-a-matter-for-the-engineers/ (accessed 24 October 2013).

74. Feenberg, *Between reason and experience*, 136.

75. See Hilmar Bärthel, *Wasser für Berlin* (Berlin: Verlag für Bauwesen, 1997); Mark Anthony O'Malley, "Plumbing the body politic: a political ecology of water and waste in Berlin, 1850–1880," PhD thesis, Department of Geography, University of California, Berkeley, 1997.

76. On the limits and possibilities of comparative insights in a global context see, for example, Jennifer Robinson, "Cities in a world of cities: the comparative gesture," *International Journal of Urban and Regional Research* 35 (1) (2011): 1–23.

## Chapter 1

1. "A deformed tangle of filth and entrails beyond the imagination of Piranesi" (author's translation), in Félix Nadar, *Le Paris souterrain de Félix Nadar 1861: des os et des eaux* (Paris: Caisse Nationale des Monuments Historiques et des Sites, 1982 [1861]), 41.

2. "The magnificent sewers of Paris have always been a source of public fascination and have been honored with illustrious visits. Not a single foreign monarch or distinguished person has left Paris without visiting the sewers" (author's translation), in Eugène Belgrand, *Les travaux souterrains de Paris V: les égouts et les vidanges* (Paris: Dunod, 1887), 174.

3. Frederick Hiorns, *Town-building in history: an outline review of conditions, influences, ideas, and methods affecting "planned" towns through five thousand years* (London: George G. Harrap, 1956), 247.

4. Edmund N. Bacon, *Design of cities* (London: Thames and Hudson, London, 1967), 179. See also Marshall Berman, *All that is solid melts into air: the experience of modernity* (London: Verso, 1982), 131–172 ; Fabienne Chevallier, *Le Paris moderne: histoire des politiques d'hygiène (1855–1898)* (Rennes: Presses Universitaires de Rennes, 2010); T. J. Clark, *The painting of modern life: Paris in the art of Manet and his followers* (London:

Thames and Hudson, 1985); Jeanne Gaillard, *Paris la Ville, 1852–1870* (Paris: Champion, 1977); David Garrioch, *The making of revolutionary Paris* (Berkeley: University of California Press, 2002); David Harvey, *Consciousness and the urban experience: studies in the history and theory of capitalist urbanization 1* (Oxford: Blackwell, 1985); André Morizet, *Du vieux Paris au Paris moderne, Haussmann et ses prédécesseurs* (Paris: Hachette, 1932); Jean-Luc Pinol, ed., *Atlas historiques de villes de France* (Barcelona: Centre de Cultura Contemporània de Barcelona, 1996); Donald J. Olsen, *The city as a work of art: London, Paris, Vienna* (New Haven: Yale University Press, 1986), 35–57; Marcel Roncayolo and Louis Bergeron, "D'Haussmann à nos jours," in Louis Bergeron, ed., *Paris: genèse d'un paysage* (Paris: Picard, 1989), 217–298; Aldo Rossi, *The architecture of the city*, trans. Diane Ghirardo and Joan Ockman (Cambridge, MA: MIT Press, 1982 [1966]); Howard Saalman, *Haussmann: Paris transformed* (New York: George Braziller, 1971); Anthony Sutcliffe, *The autumn of central Paris* (London, 1970); Anthony Sutcliffe, *Toward the planned city: Germany, Britain, the United States and France 1780–1914* (Oxford: Basil Blackwell, 1981); Anthony Sutcliffe, *Paris: an architectural history* (New Haven: Yale University Press, 1993).

5. An exaggerated emphasis on the significance of the Haussmann era is to be found in Leonardo Benevolo, *The European city*, trans. Carl Ipsen (Oxford: Blackwell, 1993); J. M. Chapman and Brian Chapman, *The life and times of Baron Haussmann: Paris in the Second Empire* (London: Weidenfeld and Nicholson, 1957); David P. Jordan, *Transforming Paris: the life and labors of Baron Haussmann* (New York: Free Press, 1995). Benevolo suggests that in Second Empire Paris there was a decisive shift from private to public interests, yet as this chapter shows in the case of sewers and urban sanitation, public interests only began to significantly prevail in the 1890s.

6. The most comprehensive piece of recent scholarship on the Paris sewers is provided by the historian Donald Reid, but his study does not explore in any detail the interrelationships between modernity and urban space or the cultural ramifications of new urban infrastructures. Reid's primary interest is in labor history over a very long time frame, from the premodern era until the late twentieth century. See Donald Reid, *Paris sewers and sewermen: realities and representations* (Cambridge, MA: Harvard University Press, 1991). Barrie Ratcliffe, in a paper devoted to the city's nineteenth-century sanitation arrangements, explores the sewage system before Haussmann and emphasizes how urban historians have tended to overlook the significance of this earlier era. See Barrie M. Ratcliffe, "Cities and environmental decline: elites and the sewage problem in Paris from the mid 18th to the mid 19th century," *Planning Perspectives* 5 (1990): 189–222. In contrast, this chapter suggests that it is the last decade of the nineteenth century that presents a crucial transformation from the continuity between the Haussmann period and earlier times.

7. In cinema alone we can draw on many examples such as Carol Reed's *The Third Man* (1948), Andrzej Wajda's *Kanal* (1956), and Stephen King's *It* (1987). As for literature, examples include Émile Zola's *Germinal* (1885), Gaston Leroux's *The Phantom of the Opera* (1911), and Victor Hugo's *Les Misérables* (1862). More recently the theme of the urban labyrinth has been developed in Thomas Pynchon's *V: A Novel* (1963) and Harlan Ellison's *Strange Wine* (1978).

8. Victor Hugo, *Les Misérables*, trans. Norman Denny (Harmondsworth: Penguin, 1982 [1862]), 1075. Hugo based his vivid description of the pre-Haussmann Paris sewers largely on the accounts given by the early nineteenth-century public health activist Parent-Duchâtelet. For an interesting analysis of Hugo's conception of the subterranean city see Louis Chevalier, *Laboring classes and dangerous classes in Paris during the first half of the nineteenth century*, trans. Frank Jellinek (Princeton: Princeton University Press, 1973 [1958]). See also Wendy Lesser, *The life below the ground: a study of the subterranean in literature* (Boston: Faber and Faber, 1987); David L. Pike, *Subterranean cities: the world beneath Paris and London, 1800–1945* (Ithaca: Cornell University Press, 2005); Rosalind H. Williams, *Notes on the underground: an essay on technology, society and the imagination*, 2nd ed. (Cambridge, MA: MIT Press, 2008).

9. Anthony Vidler, *The architectural uncanny: essays in the modern unhomely* (Cambridge, MA: MIT Press, 1992), 192. See also Samuel C. Florman, *The existential pleasures of engineering* (New York: St. Martins Press, 1976).

10. "Les galeries souterraines, organes de la grande cité, fonctionneraient comme ceux du corps humain, sans se montrer au jour; l'eau pure et fraîche, la lumière et la chaleur y circuleraient comme les fluides divers dont le mouvement et l'entretien servent à la vie. Les sécrétions s'y exécuteraient mystérieusement, et maintiendraient la santé publique sans troubler la bonne ordonnance de la ville et sans gâter sa beauté extérieure." Baron Haussmann, *Mémoire sur les eaux de Paris, présenté à la commission municipale par m. le préfét de la Seine* (Paris: Vinchon, 1854), 53.

11. A variety of interdisciplinary approaches to the understanding of water, modernity, and architectural form are to be found in Nadi Lahiji and Daniel S. Friedman, eds., *Plumbing: sounding modern architecture* (New York: Princeton Architectural Press, 1997). See also Erik Swyngedouw, "The city as a hybrid: on nature, society and cyborg urbanization," *Capitalism, Nature, Socialism* 7 (1996): 65–80.

12. Maria Morris Hambourg, "A portrait of Nadar," in Maria Morris Hambourg, Françoise Heilbrun, and Philippe Néagu, eds., *Nadar* (New York: Metropolitan Museum of Modern Art, 1995), 2–35. Nadar's family had strong republican and internationalist sentiments: his father had taken part in the Lyonnais uprising during the French revolution and had also published Félicité de Lamennais's radical tract *Essai sur l'indifférence en matière de religion* in 1817. On Nadar's role in the history of

photography see Alison Gernsheim and Helmut Gernsheim, *L. J. M. Daguerre: the history of the diorama and the daguerreotype* (London: Secker and Warburg, 1965); Alison Gernsheim and Helmut Gernsheim, *The history of photography from the camera obscura to the beginning of the modern era* (New York: McGraw-Hill, 1969); Beaumont Newhall, *The history of photography* (New York: Museum of Modern Art, 1982); Aaron Scharf, *Art and photography* (London: Allen Lane, 1968); John Szarkowski, *Photography until now* (New York: Museum of Modern Art, 1989).

13. Harvey, *Consciousness and the urban experience*, 65.

14. See Anne d'Eugny, *Au temps de Baudelaire, Guys et Nadar* (Paris: Éditions du Chêne, 1945); Nigel Gosling, *Nadar* (London: Secker and Warburg, 1976); Jean Prinet and Antoinette Dilasser, *Nadar* (Paris: Colin, 1966); Shelley Rice, *Parisian views* (Cambridge, MA: MIT Press, 1997).

15. Sylvie Aubenas, "Beyond the portrait, beyond the artist," in Hambourg, Heilbrun, and Néagu, *Nadar*, 95–107.

16. Hambourg, "A portrait of Nadar"; Françoise Heilbrun, "Nadar and the art of portrait photography," in Hambourg, Heilbrun, and Néagu, *Nadar*, 35–59; Rosalind Krauss, "Tracing Nadar," *October* 5 (1978): 29–47; Philippe Néagu, "Nadar and the artistic life of his time," in Hambourg, Heilbrun, and Néagu, *Nadar*, 59–77; André Rouillé, "When I was a photographer: the anatomy of a myth," in Hambourg, Heilbrun, and Néagu, *Nadar*, 107–114.

17. The debate surrounding modernity and aesthetic realism is a pivotal theme in relation to critical reception of the emergence of impressionist and postimpressionist art. See Anthony Sutcliffe, "The impressionists and Haussmann's Paris," *French Cultural Studies* 6 (2) (1995): 197–219.

18. Félix Nadar, *Quand j'étais photographe* (Paris: Seuil, 1994 [1900]); Nadar, *Le Paris souterrain de Félix Nadar 1861*. See also Charles Dollfus, *Balloons,* trans. C. Mason (London: Prentice-Hall International, 1962 [1960]); Paul Rey, "Du ballon de Nadar à Apollo XIII: la détection à distance des ressources naturelles," *Mémoires de l'Académie des Sciences, Inscriptions et Belles-Lettres de Toulouse* 132 (1970): 97–103.

19. Walter Benjamin, "Paris, the capital of the nineteenth century," trans. Howard Eiland, in Walter Benjamin, *The writer of modern life: essays on Charles Baudelaire*, ed. Michael W. Jennings (Cambridge, MA: Belknap Press of Harvard University Press, 2006 [1935]), 34.

20. Hambourg, "A portrait of Nadar," 28.

21. Hugo, *Les Misérables*, 1072. The word "cloaca" is described in the *Oxford English Dictionary* as meaning "Sewer; excrementory cavity in birds, reptiles etc.; gathering place of moral evil." See Reid, *Paris sewers and sewermen*, 15.

22. See Wolfgang Schivelbusch, *Disenchanted night: the industrialization of light in the nineteenth century*, trans. Angela Davies (Oxford: Berg, 1988 [1983]). See also Rice, *Parisian views*.

23. David Pinkney, *Napoleon III and the rebuilding of Paris* (Princeton: Princeton University Press, 1958).

24. See Reid, *Paris sewers and sewermen*. Demand easily outstripped available places on allocated visiting days, and the sewers were listed as a major attraction in popular tourist guides of the late nineteenth century. See, for example, Karl Baedeker, *Paris and its environs* (Leipzig: Karl Baedeker, 1876).

25. Pinkney, *Napoleon III and the rebuilding of Paris*, 143.

26. Raymond Williams, *The country and the city* (Oxford: Oxford University Press, 1973).

27. Quoted in Alain Corbin, *The foul and the fragrant: odor and the French social imagination*, trans. Miriam L. Kochan, Roy Porter, and Christopher Prendergast (Cambridge, MA: Harvard University Press, 1986 [1982]), 115. Paris had been known as the "city of mud" (*ville de boue*) since Roman times. See Reid, *Paris sewers and sewermen*, 10–11.

28. On the consequences of rapid urban growth see also Chevalier, *Laboring classes and dangerous classes in Paris*; François Delaporte, *Disease and civilization: the cholera in Paris, 1832*, trans. Arthur Goldhammer (Cambridge, MA: MIT Press, 1986); Harvey, *Consciousness and the urban experience*, 63; Gerry Kearns, "Biology, class and the urban penalty," in Gerry Kearns and Charles W. J. Withers, eds., *Urbanising Britain: essays on class and community in the nineteenth century* (Cambridge: Cambridge University Press, 1991), 12–30.

29. Chapman and Chapman, *The life and times of Baron Haussmann*; Jordan, *Transforming Paris*; Henri Malet, *Le Baron Haussmann et la rénovation de Paris* (Paris: Éditions municipales, 1973); Elbert Peets, "Famous town planners: Baron Haussmann," *Town Planning Review* 12 (1926): 181–190.

30. Pinkney, *Napoleon III and the rebuilding of Paris*.

31. Saint-Simon (1760–1825) played a significant role in the development of new political and economic ideas which reflected the growing influence of industrialists and scientists in nineteenth-century France. Like Comte, he was an early advocate of technologically based positivist solutions to political problems and proved highly influential within the French engineering profession. See Harvey, *Consciousness and the urban experience*; Antoine Picon, *Les saint-simoniens: raison, imaginaire et utopie* (Paris: Belin, 2002).

32. Pinkney, *Napoleon III and the rebuilding of Paris*.

33. Clark, *The painting of modern life*, 37. On the scale of public works under Haussmann, see also Gaillard, *Paris la Ville*.

34. Harvey, *Consciousness and the urban experience*, 87.

35. Saalman, *Haussmann: Paris transformed*.

36. The first covered sewer in Paris was the *égout de ceinture* (the "beltway sewer") built by the *prévôt* [city father] Hughes Aubriot in 1370. By 1636 a report revealed that the city had built a network of only twenty-four sewers, most of which were either seriously dilapidated or choked with rotting refuse. No more than a quarter of these sewers were covered, and the inability of city authorities to improve the sanitary conditions of the city became symbolic of the ineptitude and iniquity of pre-Revolutionary France. By 1826 the major Amelot, La Roquette, and Chemin Vert sewers had become completely blocked with refuse, and with the obstruction of the sewers the city faced a growing crisis of undrained stagnant water and overflowing refuse dumps. See Laure Beaumont-Maillet, *L'eau à Paris* (Paris: Hazan, 1991); Philippe Cebron de Lisle, "L'eau a Paris au dix-neuvieme siècle," PhD dissertation, Université de Lille III, 1991; Konstantinos Chatzis and Olivier Coutard, "Water and gas: early developments in the utility networks of Paris," *Journal of Urban Technology* 12 (3) (2005): 1–17; A. Chevallier, "Notice historique sur la police et la distribution des eaux dans Paris depuis 360 jusqu'à l'époque actuelle, pour servir à l'histoire de la salubrité et de l'hygiène publique des grandes villes," *Annales d'hygiène publique et de médecine légale* 45 (1851): 5–71; P. Diffre, "Historique de l'alimentation en eau de Paris," *Bulletin de Bureau de Recherches Géologiques et Minières* 4 (1967): 3–22; J. Dupuit, *Traité théorique et pratique de la conduite et de la distribution des eaux*, 2nd ed. (Paris: Dunod, éditeur, libraire des corps impériaux des ponts et chaussées et des mines, 1865); M. Emmery, *Statistique des égouts de la ville de Paris* (Paris: Chez Carillan-Goeury, 1837); Haussmann, *Mémoire sur les eaux de Paris*; G.-E. Haussmann, "De l'égout des eaux," in *Documents relatifs aux eaux de Paris* (Paris: Paul Dupont, 1861 [1854]), 54–83; Pierre-Yves Mauguen, "Les galeries souterraines d'Haussmann. Le système des égouts parisiens, prototype ou exception?," *Annales de la recherche urbaine* 44–45 (1989): 163–175; Reid, *Paris sewers and sewermen*, 12–15; Daniel Roche, "Le temps de l'eau rare, du moyen âge à l'époque moderne," *Annales: économies, sociétés, civilisations* 39 (1984): 383–399; Clément Vautel, *Le service des eaux à Paris* (Paris: F. Juven, 1904).

37. Eugène Belgrand, *Recherches statistiques sur les sources du bassin de la Seine qu'il est possible de conduire à Paris* (Paris: Vinchon, 1854); Belgrand, *Les travaux souterrains de Paris V*; L. Figuier, *Les eaux de Paris* (Paris: Michel Lévy Frères, 1862); Baron Haussmann, *Second mémoire sur les eaux de Paris présenté par Le Préfet de la Seine au Conseil Municipal* (Paris: Typographie de Charles de Mourgues Frères, 1858). See also

Pierre-Denis Boudriot, "Les égouts de Paris aux XVIIe et XVIIIe siècles: les humeurs de la ville préindustrielle," *Histoire, économie et société* 9 (2) (1990): 197–211; Chevalier, *Laboring classes and dangerous classes in Paris.*

38. Haussmann, *Second mémoire sur les eaux de Paris.* See also Pinkney, *Napoleon III and the rebuilding of Paris.*

39. Eugène Belgrand, *Historique du Service des Eaux, depuis l'année 1854 jusqu'à l'année 1874* (Paris: Dunod, 1875); Belgrand, *Les travaux souterrains de Paris V.* Belgrand stressed the importance of better tools and ventilation for sewer workers as part of his modernization program; A. Dumont, *Les eaux de Lyon et de Paris* (Paris: Dunod, éditeur, libraire des corps impériaux des ponts et chaussées et des mines, 1862); V. L'Olivier, F. Maxwell Lyte, and L. M. Stoffel, *Utilisation des eaux des égoûtes de la ville de Paris et assainissement du cours de la Seine* (Paris: Imprimerie Polyglotte de L. Hugonis, 1873). See also F. Caron, *Paris et ses réseaux: naissance d'un mode de vie urbain XIX$^e$–XX$^e$ siècles* (Paris: Bibliothèque historique de la ville de Paris, 1990); Pinkney, *Napoleon III and the rebuilding of Paris,* 35.

40. The term "night soil" refers to the contents of various types of cesspools and cesspits and the fact that the unpleasant and often dangerous activity of emptying these structures was usually carried out at night. Some of the earliest public health ordinances in France were directed at the problem of human waste: in 1533, for example, a royal decree ordered that every new property must have a cesspool constructed, yet there were no attempts to impose design specifications on these underground structures until the early nineteenth century. See M. Deligny, *Les projets de loi et de règlement relatifs à l'envoi direct des vidanges à l'égout* (Paris: Chaix, Paris, 1883).

41. Haussmann, *Mémoire sur les eaux de Paris,* 48. The only vestige of a combined sewer system Haussmann permitted was the construction of public urinals [*pissoirs*] along the new boulevards. See also Pinkney, *Napoleon III and the rebuilding of Paris,* 136.

42. In 1854 Haussmann wrote disparagingly of the English emphasis on the use of combined sewers for storm water and human excrement: "Même, ainsi corrigé, ce système aurait encore pour résultat inévitable l'infection des galeries d'égout, dont la pente ne peut être que très-faible, d'après le relief du sol de Paris, et qu'aucune chasse d'eau, si forte qu'elle soit, ne lave et n'assainit jamais complètement." Haussmann, *Mémoire sur les eaux de Paris,* 48. See also A. Dumont and G. Dumont, *Les eaux de Nîmes, de Paris et de Londres* (Paris: Dunod, éditeur, libraire des corps impériaux des ponts et chaussées et des mines, 1874); H.-B. Hederstedt, "An account of the drainage of Paris," *Minutes of the Proceedings of the Institution of Civil Engineers* 24 (1865): 257–279; Pinkney, *Napoleon III and the rebuilding of Paris,* 144.

43. A. Chevallier, *Essai sur la possibilité de recueillir les matières fécales, les eaux vannes, les urines de Paris, avec utilité pour la salubrité, et advantage pour la Ville et pour l'agriculture* (Paris: Baillière, 1860); Maxime Paulet, *Les vidanges-engrais de Paris* (Paris: Smith, 1858). See also André E. Guillerme, *The age of water: the urban environment of the north of France AD 300–1800* (College Station: Texas A&M University Press, 1988 [1983]); Mikulas Teich, "Circulation, transformation, conservation of matter and the balancing of the biological world in the eighteenth century," *Ambix* 29 (1982): 17–28.

44. F. Ducuing, *Des eaux d'égout et des vidanges: leur utilisation par irrigation dans leur parcours jusqu'à la mer* (Paris: Société des études, 1875). See also Ratcliffe, "Cities and environmental decline."

45. At the close of the Haussmann era there were still wide disparities in the number of homes connected to the new water distribution system: some 82 percent of homes in the affluent inner-city *arrondissements* and only 48 percent in the predominantly working-class outer *arrondissements*. No legal sanctions existed to force existing property owners to link up to the new system, and new buildings were only required to drain storm water. See Konstantinos Chatzis and Olivier Coutard, "'Eau et gaz à tous les étages': Compared patterns of network development in Paris," paper presented to the conference "Urban Futures—Technological Futures," Durham, UK, 23–25 April 1998.

46. In the post-Haussmann era the idea of using sewer water for irrigation emerged as a temporary solution to the desire for organic continuity in the circulation of urban water. Examples include M. Aubrey-Vitet, *L'épuration et l'utilisation des eaux d'égouts* (Paris: A. Quantin, 1880); Eugène Belgrand, *Transformation de la vidange et suppression de la voirie de Bondy: achèvement des égouts et emploi de leurs eaux dans l'agriculture* (Paris: de Mourgues, 1871); V. Châtel, *Assainissement de la Seine: épuration des eaux d'égouts avant leur jet dans le fleuve* (Asnières: Trouttet, 1876); M. A. Durand-Claye, *Réponse à l'article dans La Revue des Deux-Mondes, par Aubrey-Vitet sur la question des égouts de Paris* (Paris: Choix, 1882); G.-E. Haussmann, *Mémoires du Baron Haussmann*, 2nd ed., vol. 3 (Paris: Victor-Havard, 1893), 319; C. Joly, *La question des eaux d'égouts en France et en Angleterre* (Paris: A. Michels, 1877); C. Joly, *Les eaux d'égout à Paris* (Paris: Chaix, 1885); Ann F. La Berge, "Edwin Chadwick and the French connection," *Bulletin of the History of Medicine* 62 (1988): 23–42; Mémoire de l'inspecteur general des Ponts et Chaussées, *Transformation de la vidange et suppression de la voirie de Bondy: achèvement des égouts et emploi de leurs eaux dans l'agriculture* (Paris: Mourgues, 1875); E. Miotat, *Suppression complète de la vidange: assainissement des égouts et des habitations* (Paris: Ducher, 1881). Joly's writings are particularly interesting for their explicit recognition that the modernization of the Paris sewers remained incomplete: "Il est triste de penser, qu'en 1877, on en est encore à hésiter pour envoyer aux égouts la totalité des vidanges solides et liquides, et que les maisons les plus favorisées n'ont encore que le système

diviseur proposé par Gourlier en 1788. Sur 300,000 mètres cubes d'eaux vannes qu'on envoie quotidiennement à la Seine, que peuvent faire 1,000 mètres cubes de matières solides à l'état frais?" (Joly, *La question des eaux d'égouts*, 2.)

47. M. A. Durand-Claye, *Le système de Liernur* (Paris: G. Masson, 1880). See also Ratcliffe, "Cities and environmental decline." Some commentators emphasized ever more elaborate technical modifications to the design and operation of cesspits, including the implementation of ideas developed by Gourlier, Giraud, and others in the late eighteenth century for the separation of urine and feces within the home. See J. B. Berlier, *Projet de vidange pneumatique pour la Ville de Paris* (Paris: Grande Imprimerie, 1883); F. Liger, *Fosses d'aisances: latrines, urinoirs et vidanges* (Paris: Baudry, 1875). For Suilliot the problem of what to do with human waste was "la question éternelle des Vidanges," leaving only two real options, "assainissement" (disposal) or "utilisation" (use). See H. Suilliot, *La question des vidanges* (Paris: Vauthrin, 1880), 3. Interestingly, Haussmann had also reflected on the technical possibilities for a dual sewer system in his submission to the city's municipal council in 1854 as part of the "canalisation complète de Paris." See Haussmann, *Mémoire sur les eaux de Paris*, 49–51.

48. Jordan, *Transforming Paris*, 271. It would be difficult to argue that existing sanitation arrangements were clearly backward in public health terms, since Parisian mortality rates compared favorably with other European cities at the turn of the century. See Gerry Kearns, "Zivilis or Hygaeia: urban public health and the epidemiological transition," in Richard Lawton, ed., *The rise and fall of great cities: aspects of urbanization in the western world* (London: Belhaven, 1989), 96–124.

49. As early as 1883 the City Council urged that the connection and subscription to water and sewer services should be compulsory in order to extend adequate sanitation to poorer parts of city, but this initial attempt to ensure integration into the city's new water and sewer system was to be subjected to a successful legal challenge by private property interests. See also Chatzis and Coutard, "'Eau et gaz à tous les étages'"; Sabine Barles, *La ville délétère. Médecins et ingénieurs dans l'espace urbain XVIIIe–XIXe siècle* (Seyssel: Champ Vallon, 1999).

50. See Chatzis and Coutard, "'Eau et gaz à tous les étages.'"

51. Désiré Magloire Bourneville, *Le tout-à-l'égout et l'assainissement de la Seine* (Paris: Bureaux du Progrès Médical, 1892). Bourneville scorned opposition to the *tout-à-l'égout* solution, pointing out that human feces had already been entering the sewer system throughout the nineteenth century at Place Maubert, Rue Saint-Jacques, and a number of other locations across the city. See also Roger-Henri Guerrand, "La bataille du tout-à-l'égout," *Histoire* 53 (1983): 66–74; Gérard Jacquemet, "Urbanisme parisien: la bataille du tout-à-l'égout à la fin du XIXe siècle," *Revue d'Histoire Moderne et Contemporaine* 26 (1979): 505–548.

52. V. Amondruz, *Assainissment de Paris: le tout à l'égout rationnel obtenu par la vidange hydraulique* (Paris: Ducher, 1883); E. Bert, *Loi du février 1888 concernant la répression des fraudes dans la commerce des engrais* (Paris: Chevalier-Maresq, no date); Bourneville, *Le tout-à-l'égout et l'assainissement de la Seine*; L. Gauthier, *Le tout à l'égout, considérations sur les différents systèmes de vidange à Paris* (Paris: Chamerot, 1888); E. Gérards, *Paris souterrain* (Paris: Garnier Frères, 1908); J.-J. Poulet-Allamagny, "Note historique," in Nadar, *Le Paris souterrain de Félix Nadar 1861*, 61; G. Ville, *État de l'opinion du monde agricole en 1890 sur la doctrine des engrais chimiques* (Paris: Libraire Agricole, no date).

53. Corbin, *The foul and the fragrant*, 72–73; Georges Vigarello, *Concepts of cleanliness: changing attitudes in France since the Middle Ages*, trans. Jean Birrell (Cambridge: Cambridge University Press, 1988 [1985]).

54. See William W. Braham, "Sigfried Giedion and the fascination of the tub," in Lahiji and Friedman, *Plumbing*, 200–224; Michelle Facos, "A sound mind in a sound body: Bathing in Sweden," in Susan C. Anderson and Bruce H. Tabb, eds., *Water, leisure and culture: European historical perspectives* (Oxford: Berg, 2003), 105–117; Sigfried Giedion, *Mechanization takes command: a contribution to anonymous history* (New York: W. W. Norton, 1948); Ivan Illich, *H2O and the waters of forgetfulness: reflections on the historicity of "stuff"* (Berkeley: Heyday Books, 1985).

55. Corbin, *The foul and the fragrant*, 175. In the seventeenth century bathing had been considered doubly dangerous: the rendering of skin moist and soft was considered "feminine" while simultaneously exposing the body to the threat of unhealthy air and disease. Even Francis Bacon, the pioneer of the empirical sciences, cautioned against allowing "the liquor's watery part" from penetrating the body. See Constance Classen, David Howes, and Anthony Synnott, *Aroma: the cultural history of smell* (London: Routledge, 1994), 70.

56. The late eighteenth century and early decades of the nineteenth century saw increasing attempts to regulate the use of cesspits more effectively. See, for example, M. Dubois, M. Huzard, and M. Héricart de Thury, *Rapport sur les fosses mobiles inodores* (Paris: Bureau de l'Entreprise Générale, 1819); P. Giraud, *Fosses d'aisances et leurs inconvénients* (Paris: Cailleau, 1786); M. Goulet, *Inconvéniens des fosses d'aisance, possibilité de les supprimer* (Paris: Auteur, 1787); M. Labarraque, A. Chevallier, and A. J. B. Parent-Duchâtelet, *Rapport sur les améliorations à introduire dans les fosse d'aisances, leur mode de vidange, et les voiries de la Ville de Paris* (Paris: Renouard, c. 1840) (reprinted from *Annales d'Hygiène Publique* 14 [2]: 1–76); M. Laborie, *Observations sur les fosses d'aisances et moyens de prévenir les inconvéniens de leur vidange* (Paris: College de Pharmacie, 1778); J. P. Lucquel, *De l'assainissement de la vidange et de la supression des voiries de la Ville de Paris* (Paris: Lacquin, 1840); Municipalité de Paris, *Arrêté concernant la salubrité et la vidange des fosses d'aisance, puits et puisards* (2 janvier) (Paris: Lottin, 1792);

Préfecture de police, conseil de salubrité, *Rapport sur les améliorations à introduire dans les voiries, les modes à vidange et les fosses d'aisances de la Ville de Paris* (Paris: Lottin, 1835); J. P. Sucquet and L. Krafft, *L'assainissement de la vidange et de la supression des voieries de la Ville de Paris* (Paris: Felix Locquin, 1840). Sucquet and Krafft (p. 6) provide a vivid depiction of the stench and inconvenience of night soil collection in the early decades of the nineteenth century: "Comme par le passé, chaque maison deviendra tour à tour un foyer d'émanations infectes, et la ville sera sillonnée tous les soirs par ces charettes qu'on devine à leur odeur et qu'on ne peut éviter à une certaine heure dans Paris."

57. Corbin, *The foul and the fragrant*, passim. Norbert Elias develops a similar theme in his exploration of the changing "frontier of embarrassment" in Norbert Elias, *The civilizing process*, vol. 1, *The history of manners*, trans. Edmund Jephcott (Oxford: Basil Blackwell, 1978 [1939]). See also Alain Corbin, *Time, desire and horror: toward a history of the senses*, trans. Jean Birrell (Cambridge: Polity, 1995 [1991]).

58. Reid, *Paris sewers and sewermen*, 27; Corbin, *The foul and the fragrant*, 172–173; J.-P. Goubert, *The conquest of water: the advent of health in the industrial age*, trans. A. Wilson (Cambridge: Polity Press, 1989 [1986]); R. Magraw, "Producing, retailing, consuming: France 1830–70," in Brian Rigby, ed., *French literature, thought and culture in the nineteenth century: a material world* (London: Macmillan, 1993), 59–85. For the late nineteenth-century architect and essayist Adolf Loos, for example, the increasing use of water and the technological sophistication of plumbing was a vital indicator of cultural advancement. Innovative changes in the design of housing were particularly advanced in England, where pressurized water faucets in kitchens and water closets were combined with the disposal of sewage through drains, but their adoption in France proved much slower. See Adolf Loos, "Plumbers," trans. Harry Francis Mallgrave, in Lahiji and Friedman, *Plumbing*, 15–19.

59. J. Chrétien, *Les odeurs de Paris: étude analytique des causes qui concourent a l'insalubrité de la ville et des moyens de les combattre* (Paris: Baudry, 1881).

60. See Helen Molesworth, "Bathrooms and kitchens: cleaning house with Duchamp," in Lahiji and Friedman, *Plumbing*, 74–92.

61. Paul Hayes Tucker, *Monet at Argenteuil* (New Haven: Yale University Press, 1982), quoted in Francis Frascina, Nigel Blake, Briony Fer, Tamar Garb, and Charles Harrison, *Modernity and modernism: French painting in the nineteenth century* (New Haven: Yale University Press, 1993), 121. Historical evidence of poor water quality at Asnières is provided in E. Belgrand, *Direction des eaux et des égôuts: rapport présenté à M. Le Préfet sur le projet de dérivation des eaux d'égout, depuis l'embouchure du collecteur général d'Asnières, jusqu'à l'extrémité nord-ouest de la forêt de Saint-Germain* (Paris: Gauthier-Villars, Paris 1875); E. Belgrand, *La situation du service des eaux et égouts* (Paris: Chaix, 1878); Bourneville, *Le tout-à-l'égout et l'assainissement de la Seine*; L. Cabanon, *Dépotoir et fabrique*

*d'ammoniaque: l'infection de la Seine* (Neuilly: L'abeille, 1886); Châtel, *Assainissement de la Seine*.

62. Ernst Bloch, *The principle of hope*, trans. N. Plaice, S. Plaice, and P. Knight (Cambridge, MA: MIT Press, 1986 [1954]); Peter Collier and Robert Lethbridge, eds., *Artistic relations: literature and the visual arts in nineteenth-century France* (New Haven: Yale University Press, 1994); C. Crossley, "Romanticism and the material world: mind, nature and analogy," in Rigby, *French literature, thought and culture in the nineteenth century*, 9–22; Joan Ungersma Halperin, *Félix Fénéon: aesthete and anarchist in fin-de-siècle Paris* (New Haven: Yale University Press, 1988); M. Kelly, "Materialism in nineteenth-century France," in Rigby, *French literature, thought and culture in the nineteenth century*, 23–41; Linda Nochlin, "Seurat's *La Grande Jatte*: an anti-utopian allegory," in *The politics of vision: essays on nineteenth-century art and society* (London: Thames and Hudson, 1989), 170–193; Meyer Schapiro, "Seurat and La Grande Jatte," *Columbia Review* 17 (1935): 14–15.

63. See Paul Smith, *Seurat and the avant-garde* (New Haven: Yale University Press, 1996).

64. See Nicholas Green, *The spectacle of nature: landscape and bourgeois culture in nineteenth-century France* (Manchester: Manchester University Press, 1990). An important exception to the absence of "work" from these riparian landscapes can be found in Monet's *Men unloading coal* (1875), which is reproduced in John Leighton and R. Thomson, *Seurat and the bathers* (London: National Gallery, 1997), 112.

65. A. J. B. Parent-Duchâtelet, *Essai sur les cloaques ou égouts de la Ville de Paris* (Paris: Crevot, 1824). Other early explorations of the sewers include those by Bruneseau, Inspector-General of Health, between 1805 and 1812. H. Horeau, *Nouveaux égouts proposés à la Ville de Paris* (Paris: Didot, 1831). See also Chevalier, *Laboring classes and dangerous classes in Paris*; B. P. Lécuyer, "L'hygiène en France avant Pasteur, 1759–1850," in Claire Salomon-Bayet, ed., *Pasteur et la révolution pastorienne* (Paris: Payot 1986); Lion Murard and Patrick Zylberman, *L'hygiène dans la République: la santé publique en France, ou l'utopie contrariée 1870–1918* (Paris: Fayard, 1996).

66. Quoted in Charles Bernheimer, "Of whores and sewers: Parent-Duchâtelet, engineer of abjection," *Raritan: A Quarterly Review* 6 (1987): 78. The word "putrid," so often used by Parent-Duchâtelet and others in reference to the Paris sewers, is derived from the Latin word *puta* meaning whore. Parent-Duchâtelet was a critical early figure in the emergence of what one might term "medical hygienics" and was a founder of the influential periodical *Annales d'Hygiène Publique et de la Médecine Légale* in 1829. The term *Cloaca Maxima*, which also has an explicitly gendered etymology, is used by Baron Haussmann in his *Second mémoire sur les eaux de Paris*, 107.

67. A classic example is Jules Michelet, *La Femme* (Paris: Flammarion, 1981 [1859]). See also A. Callen, "Immaterial views? Science, intransigence and the female spectator of modern French art in 1879," in Rigby, *French literature, thought and culture in the nineteenth century*, 184–197; Alain Corbin, *Women for hire: prostitution and sexuality in France after 1850*, trans. Alan Sheridan (Cambridge, MA: Harvard University Press, 1990 [1978]); Green, *The spectacle of nature*; Laura S. Strumingher, "L'ange de la maison: mothers and daughters in nineteenth-century France," *International Journal of Women's Studies* 2 (1979).

68. Leighton and Thomson, *Seurat and the bathers*.

69. Examples of water-based statues of the female nude completed during the Second Empire include the reconstruction of Lescot's Fountain of the Nymphs in the Square des Innocents and the Square Louvois and the Fountain for the Four Parts of the World (1867–1874) designed by Gabriel Davioud. See George F. Chadwick, *The park and the town: public landscape in the nineteenth and twentieth centuries* (London: Architectural Press, 1966), 152–162; C. Daly, "La caisse des travaux de Paris," *Revue Générale de l'Architecture* (1858); Henry-Russell Hitchcock, *Architecture: nineteenth and twentieth centuries* (New Haven: Yale University Press, 1977); François Loyer, *Paris nineteenth century: architecture and urbanism*, trans. C. L. Clark (New York: Abbeville Press, 1988); Françoise Masson and Marc Gaillard, *Les fontaines de Paris* (Amiens: Martelle, 1995); Charles W. Moore and Jane Lidz, *Water and architecture* (London: Thames and Hudson, 1994); Morizet, *Du vieux Paris au Paris moderne*; Penelope Woolf, "Symbol of the Second Empire: cultural politics and the Paris Opera House," in Denis Cosgrove and Stephen Daniels, eds., *The iconography of landscape: essays on the symbolic representation, design and use of past environments* (Cambridge: Cambridge University Press, 1988), 214–236; David Van Zanten, *Building Paris: architectural institutions and the transformation of the French capital 1830–1870* (Cambridge: Cambridge University Press, 1994).

70. Green, *The spectacle of nature*, 147. Late nineteenth-century Europe saw a sharpening of gender inequalities and differentiations within which nature-based ideologies played an important role. See, for example, Bonnie S. Anderson and Judith P. Zinsser, *A history of their own: women in Europe from prehistory to the present* (New York: Harper and Row, 1988); Betty Roszak and Theodore Roszak, eds., *Masculine/feminine: readings in the sexual mythology of women* (New York: Harper and Row, 1969).

71. Sigmund Freud, "The 'uncanny'," in *The standard edition of the complete psychological works of Sigmund Freud*, trans. James Strachey (London: Hogarth Press, 1955 [1919]), vol. 17, 217–256.

72. For useful overviews of the emergence of the "uncanny" see Adam Bresnick, "Prosopoetic compulsion: reading the uncanny in Freud and Hoffmann," *Germanic*

*Review* 71 (1996): 114–132; Terry Castle, *The female thermometer: eighteenth-century culture and the invention of the uncanny* (New York: Oxford University Press, 1995); Maria M. Tatar, "The houses of fiction: toward a definition of the uncanny," *Comparative Literature* 33 (1981): 167–182. For Martin Jay, a spatial reading of the uncanny is inherently gendered through "desire for a womb-like state of ontological security," stemming from a polarity between the female body as nurturing and protective and as a threatening disturbance to a masculinist spatial order. Jay suggests that the most intense sense of the uncanny is rooted in a reunion with the mother's body, and especially through the spatial disorientation of the "watery womb" of the city. See Martin Jay, "The uncanny nineties," *Salmagundi* 108 (1995): 20–29.

73. Teresa Brennan, "Psychoanalytic feminism," in A. M. Jaggar and I. M. Young, eds., *A companion to feminist philosophy* (Oxford: Blackwell, 1988); Moira Gatens, *Imaginary bodies: ethics, power and corporeality* (London: Routledge, 1996); Luce Irigaray, "The gesture in psychoanalysis," in Teresa Brennan, ed., *Between feminism and psychoanalysis* (London: Routledge, 1989), 135; Kaja Silverman, *The acoustic mirror: the female voice in psychoanalysis and cinema* (Bloomington: Indiana University Press, 1988); Donn Welton, ed., *Body and flesh: a philosophical reader* (Oxford: Blackwell, 1998); Sue Vice, "Psychoanalytic feminist theory," in Stevi Jackson and Jackie Jones, eds., *Contemporary feminist theories* (Edinburgh: Edinburgh University Press, 1998), 162–176; Vidler, *The architectural uncanny*.

74. Julia Kristeva explores the cultural significance of excrement as the ultimate abject object "expelled from the body," a part of identity that becomes a defiling otherness with the intensified differentiation of boundaries and identity under modernity. See Julia Kristeva, *Powers of horror: an essay on abjection*, trans. L. S. Roudiez (New York: Columbia University Press, 1982 [1980]); Julia Kristeva, "Psychoanalysis and the polis," trans. M. Waller, in *The Kristeva Reader*, ed. Toril Moi (Oxford: Blackwell [1982] 1986), 301–320. See also Judith Butler, *Gender trouble: feminism and the subversion of identity* (London: Routledge, 1990). Butler describes how "the boundary between the inner and outer is confounded by those excremental passages in which the inner effectively becomes outer, and this excreting function becomes, as it were, the model by which other forms of identity-differentiation are accomplished" (134). Under modernity we find a radical indeterminacy of bodies and nature, through the increasingly abstract and complex interactions in urban space and the dissolving of traditional conceptions of the organic cycle linking the body to social space. On the gendered dimensions of the experience of modernity (and their downplaying within the existing literature) see Rosalyn Deutsche's essay "Men in space," in her *Evictions: art and spatial politics* (Cambridge, MA: MIT Press, 1996), 195–202; Griselda Pollock, *Vision and difference: femininity, feminism and the histories of art* (London: Routledge, 1988); Janet Wolff, "The invisible *flâneuse*: women and the literature of modernity," *Theory, Culture and Society* 2 (1985): 37–46.

75. See Richard Sennett, *Flesh and stone: the body and the city in Western civilization* (London: Faber and Faber, 1994).

76. Edmond Texier, *Tableau de Paris. Ouvrage illustré de quinze cents gravures* (Paris: Paulin et le Chevallier, 1852–1853).

77. Corbin, *The foul and the fragrant*, 135; Felicity Edholm, "The view from below: Paris in the 1880s," in Barbara Bender, ed., *Landscape: politics and perspectives* (Oxford: Berg, 1993), 139–169; Harvey, *Consciousness and the urban experience*, 94.

78. Clark, *The painting of modern life*, 29 and 45. See also John M. Merriman, *The margins of city life: explorations on the French urban frontier, 1815–1851* (Oxford: Oxford University Press, 1991).

79. Josef W. Konvitz, Mark H. Rose, and Joel A. Tarr, "Technology and the city," *Technology and Culture* 31 (1990): 284–294.

80. On the evolution of public health ideas see William Coleman, *Death is a social disease: public health and political economy in early industrial France* (Madison: University of Wisconsin Press, 1982); Delaporte, *Disease and civilization*; Marcel Gillet, ed., *L'homme, la vie et la mort dans le nord au XIXe siècle* (Lille: Presses universitaires, 1972); Goubert, *The conquest of water*; Martha L. Hildreth, *Doctors, bureaucrats and public health in France, 1888–1902* (New York: Garland, 1987); Sue Jones, "Public hygiene and hygienists in Rouen (France) 1880–1930," PhD thesis, Cambridge University, 1992; Ange-Pierre Leca, *Et le choléra s'abattit sur Paris 1832* (Paris: Albin Michel, 1982); Murard and Zylberman, *L'hygiène dans la République*; Salomon-Bayet, *Pasteur et la révolution pastorienne*.

81. Haussmann's urban geometry owed more to the imaginary cities of the Italian Renaissance than to the technical impetus of new advances in engineering science. Examine, for example, Filarete's water-based design for the ideal city of Sforzinda (c. 1457–1464) in Moore and Lidz, *Water and architecture*, 38.

82. Tautological conceptions of the relationship between modernity and the process of "Haussmannization" are criticized in Clark, *The painting of modern life*, 14.

83. Jerry B. Gough, "Lavoisier's memoirs on the nature of water and their place in the chemical revolution," *Ambix* 30 (1983): 89–106; A.-L. Lavoisier, "Calculs et observations sur le projet d'établissement d'une pompe à feu pour fournir de l'eau à la ville de Paris," *Histoire de l'Académie Royale des Sciences, année 1771* (1774): 17–44; A. N. Meldrum, "Lavoisier's work on the nature of water and the supposed transmutation of water into earth (1768–1773)," *Archeion* 14 (1932): 246–247. See also Christopher Hamlin, *A science of impurity: water analysis in nineteenth-century Britain* (Berkeley: University of California Press, 1990).

84. Eugène Belgrand, *Les travaux souterrains de Paris II: les aqueducs romains* (Paris: Dunod, éditeur, libraire des corps des ponts et chaussées et des mines, 1875); Haussmann, *Mémoires*, vol. 3, 351–352.

85. See Pierre Lavedan, "L'influence de Haussmann: l'haussmannisation," *Urbanisme et Habitation* 34 (1953): 302–317; Charlene Marie Leonard, *Lyon transformed: public works of the Second Empire 1853–1864* (Berkeley: University of California Press, 1961).

86. Pinkney, *Napoleon III and the rebuilding of Paris*, 140.

87. Quoted in Elizabeth Wilson, *The sphinx in the city: urban life, the control of disorder, and women* (London: Virago, 1991), 37.

88. Peter Stallybrass and Allon White, *The politics and poetics of transgression* (London: Methuen, 1986). See also Mary Douglas, *Purity and danger: an analysis of the concepts of pollution and taboo* (London: Routledge and Kegan Paul, 1978); David Sibley, *Geographies of exclusion: society and difference in the West* (London: Routledge, 1995).

89. The Haussmann era put in place a dual water system comprising a "noble network" supplied by spring water for wealthy subscribers and a second inferior network (using sources such as the dirty Ourcq canal) for street cleaning, fountains, and water services for the poor. See Chatzis and Coutard, "'Eau et gaz à tous les étages.'"

## Chapter 2

1. Andor Kraszna-Krausz, "Menschen am Sonntag (People on Sunday)," *Close Up* 6 (4) (April 1930): 317.

2. "The perfection of a city lies either in the provision of fulfilling work or in the provision of ostensibly aimless calm and relaxation" (author's translation). Martin Wagner, "Berlin am Wasser," *Deutsche Bauzeitung* 65 (1–2) (1931): 20.

3. See, for example, Peter E. Gordon and John P. McCormick, eds., *Weimar thought: a contested legacy* (Princeton: Princeton University Press, 2013); Anton Kaes, Martin Jay, and Edward Dimendberg, eds., *The Weimar Republic sourcebook* (Berkeley: University of California Press, 1994).

4. Though antecedents to the cultural and intellectual ebullience of this period can be traced earlier, the most significant changes begin to appear during the 1920s. See, for example, Peter Gay, *Weimar culture: the outsider as insider* (New York: W. W. Norton, 1968); Ludwig Hilberseimer, *Berliner Architektur der 20er Jahre* (Berlin: Gebr. Mann Verlag, 1967); Andreas Killen, *Berlin electropolis: shock, nerves, and German modernity* (Berkeley: University of California Press, 2006); Detlev J. K. Peukert, *The Weimar Republic: the crisis of classical modernity* (New York: Hill and Wang, 1987); Eberhard Roters, ed., *Berlin 1910–1933* (Fribourg: Office du Livre, 1982); and Eric

D. Weitz, *Weimar Germany: promise and tragedy* (Princeton: Princeton University Press, 2007).

5. William Gaunt, "A modern utopia? Berlin—the New Germany—the New Movement," *The Studio* 98 (441) (1929): 859.

6. See Beth Irwin Lewis, *George Grosz: art and politics in the Weimar Republic* (Princeton: Princeton University Press, 1991); Maria M. Tatar, *Lustmord: sexual murder in Weimar Germany* (Princeton: Princeton University Press, 1995).

7. See Lotte Eisner, *The haunted screen: expressionism in the German cinema and the influence of Max Reinhardt*, trans. Roger Greaves (Berkeley: University of California Press, 1969 [1965/1952]).

8. Hilberseimer, *Berliner Architektur der 20er Jahre*. See also Timothy O. Benson, "Fantasy and functionality: the fate of utopia," in Timothy O. Benson, ed., *Expressionist utopias: paradise, metropolis, architectural fantasy* (Berkeley: University of California Press, 2001), 13–55. The desire to reconnect the industrial city with nature has been a long-standing countercurrent to utopian abstraction within architectonic thought that can be traced through the writings of Peter Kropotkin, Leberecht Migge, and others. See David H. Haney, *When modern was green: life and work of landscape architect Leberecht Migge* (London: Routledge, 2010).

9. See Achim Wenschuh, ed., *Hans Scharoun: Zeichnunge, Aquarelle, Texte* (Berlin: Akademie der Künste, 1993).

10. Friedrich Leyden, *Groß-Berlin. Geographie der Weltstadt* (Berlin: Mann Verlag, 1933). Leyden shows that Berlin attained its accelerated rate of growth later than other European "world cities" (*Weltstädten*) such as London, Paris, Rome, and Amsterdam, particularly after the end of the Franco-Prussian War and German unification in 1871. Across Europe as a whole the "late" urbanization pattern is discernable in Copenhagen, Stockholm, Saint Petersburg, and other cities toward the north and east, as well as some of the southern cities such as Milan and Naples (in the wake of Italian unification). See Peter Clark, *European cities and towns: 400–2000* (Oxford: Oxford University Press, 2009).

11. On Weimar-era art see, for example, Maud Lavin, *Cut with the kitchen knife: the Weimar photomontages of Hannah Höch* (New Haven: Yale University Press, 1993); Bärbel Schrader, *The golden twenties: art and literature in the Weimar Republic* (New Haven: Yale University Press, 1988); Janet Ward, *Weimar surfaces: urban visual culture in 1920s Germany* (Berkeley: University of California Press, 2001).

12. Contemporary observers detected a distinctive "modern consciousness" that flowed from new connections between the visual arts and wider society. See Gaunt, "A modern utopia?," 865. See also Manfredo Tafuri, *The sphere and the labyrinth:*

*avant-gardes and architecture from Piranesi to the 1970s*, trans. Pellegrino d'Acierno and Robert Connolly (Cambridge, MA: MIT Press, 1987 [1980]), 220; David Frisby, *Fragments of modernity: theories of modernity in the work of Simmel, Kracauer and Benjamin* (Cambridge, MA: MIT Press, 1986), 136.

13. See, for example, Susanne Frank, *Stadtplanung im Geschlechterkampf: Stadt und Geschlecht in der Großstadtentwicklung des 19. und 20. Jahrhunderts* (Opladen: Leske und Budrich, 2003); Dagmar Herzog, *Sexuality in Europe: a twentieth-century history* (Cambridge: Cambridge University Press, 2011); Katharina von Ankum, ed., *Women in the metropolis: gender and modernity in Weimar culture* (Berkeley: University of California Press, 1992).

14. Frisby, *Fragments of modernity*, 267; David Frisby, "Social theory, the metropolis, and expressionism," in Benson, *Expressionist utopias*, 88–111. See also Kurt H. Wolff, ed., *The sociology of Georg Simmel* (New York: Free Press, 1950).

15. Key works include Walter Benjamin, *Einbahnstrasse* (Berlin: Rowohlt, 1928); Werner Hegemann, *Das steinerne Berlin: Geschichte der größten Mietskasernenstadt der Welt* (Braunschweig and Wiesbaden: Vieweg, 1988 [1930]); and Siegfried Kracauer, *Straßen in Berlin und anderswo* (Berlin: Das Arsenal, 1987 [1964]).

16. Hegemann, *Das steinerne Berlin*.

17. See Aldo Rossi, *The architecture of the city*, trans. Diane Ghirardo and Joan Ockman (Cambridge, MA: MIT Press, 1982 [1966]).

18. See, for example, Helmut Lethen, *Neue Sachlichkeit 1924–1932. Studien zur Literatur des "Weissen Sozialismus"* (Stuttgart: J. B. Metzlersche Verlagsbuchhandlung, 1970); Helmut Lethen, *Cool conduct: the culture of distance in Weimar Germany*, trans. Don Reneau (Berkeley: University of California Press, 2002); Richard W. McCormick, *Gender and sexuality in Weimar modernity: film, literature, and the "new objectivity"* (Basingstoke: Palgrave, 2001); Karl Prümm, "Neue Sachlichkeit. Anmerkungen zum Gebrauch des Begriffs in neueren literaturwissenschaftlichen Publikationen," *Zeitschrift für deutsche Philologie* 91 (1972): 606–616.

19. See Günter Agde and Karin Herbst-Meßlinger, eds., *Die rote Traumfabrik. Meschrabpom-Film und Prometheus (1921–1936)* (Berlin: Bertz and Fischer, 2012); Bruce Arthur Murray, *Film and the German left in the Weimar Republic: from Caligari to Kuhle Wampe* (Austin: University of Texas Press, 1990).

20. The Bauhaus, established in 1919, was first led by Walter Gropius until 1928, then by Hannes Meyer between 1928 and 1930, and finally by Mies van der Rohe, who oversaw its relocation to Berlin before its demise in 1933. See, for example, Walter Gropius, *The new architecture and the Bauhaus*, trans. P. Morton Shand (London: Faber, 1935); Frank Whitford, *Bauhaus* (London: Thames and Hudson, 1984).

21. Gustav Böß and Willy Müller-Wieland, *Berlin von Heute: Stadtverwaltung und Wirtschaft* (Berlin: Gsellius, 1929), 128.

22. See, for example, Regine Auster, "Schutz den Wäldern und Seen! Die Anfänge des sozialpolitischen Naturschutzes in Berlin und Brandenburg," in Gert Gröning and Joachim Wolschke-Bulmahn, eds., *Naturschutz und Demokratie!?* (Munich: Martin Meidenbauer, 2006), 155–167; Fritz Herrgeist, *Die Wasser-, Boden- und Deichverbände in Ost- und Westpreußen 1868 bis 1938* (Cologne and Berlin: Grote, 1983); Rainer Stürmer, *Freiflächenpolitik in Berlin in der Weimarer Republik: ein Beitrag zur Sozial- und Umweltschutzpolitik einer modernen Industriestadt* (Berlin: Arno Spitz, 1991); Jürgen von Reuß, "Freiflächenpolitik als Sozialpolitik," in Klaus Homann, Martin Kieren, and Ludovica Scarpa, eds., *Martin Wagner 1885–1975. Wohnungsbau und Weltstadtplanung. Die Rationalisierung des Glücks* (Berlin: Akademie der Künste, 1987), 49–65.

23. Auster, "Schutz den Wäldern und Seen!"

24. See, for example, Hartmut Böhme, *Kulturgeschichte des Wassers* (Frankfurt am Main: Suhrkamp, 1988).

25. Both Böß and Wagner were members of the SPD (Sozialdemokratische Partei Deutschlands), which played a key role in the Weimar Republic, its different factions having radically diverged in the early twentieth century. See Carl E. Schorske, *German social democracy, 1905–1917: the development of the great schism* (Cambridge, MA: Harvard University Press, 1955).

26. Some 135,000 new housing units were completed between 1925 and 1929 alone. See Hubert Hoffmann, "Martin Wagner 1885–1957," *Bauwelt* 76 (41–42): 1676–1679; Barbara Miller Lane, *Architecture and politics in Germany, 1918–1945* (Cambridge, MA: Harvard University Press, 1985 [1968]); Horst Matzerath and Ingrid Thienel, "Stadtentwicklung, Stadtplanung, Stadtentwicklungsplanung: Probleme im 19. und im 20. Jahrhundert am Beispiel der Stadt Berlin," *Die Verwaltung: Zeitschrift für Verwaltungswissenschaft* 2 (1977): 173–196; Frank Werner, *Stadtplanung, Berlin 1900–1950* (Berlin: Kiepert, 1969).

27. See, for example, Claus Bernet, "The 'Hobrecht Plan' (1862) and Berlin's urban structure," *Urban History* 31 (3) (2004): 400–419; Hoffmann, "Martin Wagner 1885–1957."

28. Other key figures who worked with Wagner include Richard Ermisch and Ludwig Hoffmann.

29. Of the 135,000 new housing units built in Berlin between 1924 and 1929, for example, only 14,000 were located in the model housing estates constructed by Wagner and his colleagues. See Lane, *Architecture and politics in Germany*.

30. See Tafuri, *The sphere and the labyrinth*.

31. Martin Wagner, "Das neue Berlin," 1929. Martin Wagner papers, Frances Loeb Library, Graduate School of Design, Harvard University.

32. Ludovica Scarpa, *Martin Wagner und Berlin: Architektur und städtebau in der Weimarer Republik*, trans. Heinz-Dieter Held (Braunschweig and Wiesbaden: Vieweg, 1986 [1983]); Ludovica Scarpa, "Martin Wagner oder die Rationalisierung des Glücks," in Homann, Kieren, and Scarpa, *Martin Wagner 1885–1957*, 8–23.

33. Hoffmann, "Martin Wagner 1885–1957." Wagner's interest in the rationalization of space and time can be interpreted as a citywide counterpart to the so-called "Frankfurt kitchen" designed by the Austrian architect Margarete Schütte-Lihotzky in 1926 as part of Ernst May's Römerstadt housing estate in Frankfurt. Although the Frankfurt kitchen was later criticized for its perpetuation of gendered stereotypes within the home, Schütte-Lihotzky sought to rationalize domestic labor through the application of design and technology in order to make the most efficient use of space and time.

34. Wagner, "Das neue Berlin," 109.

35. Ibid.

36. Von Reuß, "Freiflächenpolitik als Sozialpolitik."

37. Martin Wagner, "Das sanitäre Grün," *Öffentliche Gesundheitspflege* 1 (1) (1916): 9–15. See also Helmut Engel, "Das Strandbad ein Baudenkmal!," in Anja Wiese, ed., *Das Strandbad am Großen Wannsee* (Berlin: Stiftung Denkmalschutz Berlin, 2003), 5–18.

38. See Joachim Wolschke-Bulmahn, "The ideological, aesthetic and ecological significance of water in twentieth-century German garden design," in Susan C. Anderson and Bruce H. Tabb, eds., *Water, leisure and culture: European historical perspectives* (Oxford: Berg, 2003), 119–139. See also Gert Gröning and Joachim Wolschke-Bulmahn, "Zur Entwicklung und Unterdrückung freiraumplanerischer Ansätze der Weimarer Republik," *Das Gartenamt* 34 (6) (1985): 443–458; and Stürmer, *Freiflächenpolitik in Berlin in der Weimarer Republik*.

39. See, for example, Diego Claudio Armus, "O discurso da regeneração: espaço urbano, utopias e tuberculose em Buenos Aires, 1870–1930," *Estudos Historicos* 8 (1995): 235–250; Linda Bryder, *Below the magic mountain: a social history of tuberculosis in twentieth-century Britain* (Oxford: Oxford University Press, 1988); Sylvelyn Hähner-Rombach, *Sozialgeschichte der Tuberkulose. Vom Kaiserreich bis zum Ende des 2. Weltkriegs unter besonderer Berücksichtigung Württembergs* (Stuttgart: Franz Steiner, 2000).

40. Wagner, "Berlin am Wasser," 23. Berlin and its hinterland have some 3,000 lakes, 48,000 kilometers of streams, and around 600 kilometers of urban beaches.

41. An example of the "decorative" approach is the Hermann Mächtig design for the Viktoria Park in Kreuzberg discussed by Wolschke-Bulmahn, "The ideological, aesthetic and ecological significance of water."

42. Jennifer Reed Dillon, "Modernity, sanitation and the public bath: Berlin, 1896–1930, as archetype," PhD dissertation, Duke University, 2007. See also Hilmar Bärthel, *Wasser für Berlin* (Berlin: Verlag für Bauwesen, 1997); Brian K. Ladd, "Public baths and civic improvement in nineteenth-century German cities," *Journal of Urban History* 14 (3) (1988): 372–393. On the emergence of modernist public architecture see also David Watkins, *Morality and architecture: the development of a theme in architectural history and theory from the Gothic revival to the modern movement* (Oxford: Clarendon Press, 1977).

43. Wagner, "Berlin am Wasser."

44. Stürmer, *Freiflächenpolitik in Berlin in der Weimarer Republik*.

45. Auster, "Schutz den Wäldern und Seen!"

46. Hegemann, *Steinerne Berlin*.

47. Wagner, "Berlin am Wasser."

48. Tafuri, *The sphere and the labyrinth*. In terms of access to open space, Wagner oversaw an increase in the number of playing fields across the city from 152 in 1921 to 950 in 1929. See Wagner, "Das neue Berlin."

49. Richard Ermisch, "Das neue Strandbad Wannsee," *Das neue Berlin* 6 (1929): 111. In 1924 the city passed a law permitting the redevelopment of Wannsee as part of a wider program of modernization that recognized opportunities for outdoor bathing as integral to a healthy city. In the wake of the 1925 municipal elections, which strengthened the position of the Social Democrats, a range of public works projects envisaged as part of the reshaping of Berlin as a "world city" could now begin in earnest. Engel, "Das Strandbad ein Baudenkmal!" See also Joachim G. Jacobs and Petra Hübner, "Weltstadtbad in Preußens Arkadien. Das Berliner Strandbad Wannsee und seine Außenanlagen," *Die GARTENkunst* 16 (2) (2004): 383–393. The Wagner-Ermisch project for Wannsee shared similarities with the utopian vision for Berlin developed by Hans Poelzig as a "postromantic" synthesis of nature and culture appropriate for a world city. See von Reuß, "Freiflächenpolitik als Sozialpolitik." Other ambitious schemes for Berlin during this period include J. Goldmerstein and Karl Stodieck's proposal for an indoor water complex that could accommodate thirty thousand visitors a day, but this "thermal palace" that appeared to combine both hygienist ideas and the more recent emphasis on leisure was never completed. See Gert Gröning and Joachim Wolschke-Bulmahn, "The *Thermenpalast* (Thermal Palace): an outstanding German water leisure project from the 1920s," in Anderson and Tabb, *Water, leisure and culture*, 141–147.

50. Scarpa, *Martin Wagner and Berlin*.

51. Von Reuß, "Freiflächenpolitik als Sozialpolitik"; Hoffmann, "Martin Wagner 1885–1957."

52. See, for example, Fred J. Gray, "The railway system of Berlin," *Railway Magazine* 18 (January-June 1906): 406–412; Walter-Dieter Zach and Klaus Evers, "Die Berliner Schnellbahnnetze—ihre Entwicklung und ihre Funktion," *Monatsschrift* 7 (2003): 266–294. Contemporary maps depict diminishing time bands (*Isochronen*) for Berlin, so that Wannsee and other bathing beaches were now just minutes away from the city center.

53. See, for example, Laurenz Demps, "Berlin am Wasser: ein historischer Ausflug," in Uli Hellweg and Jörn Oltmans, eds., *Wasser in der Stadt: Perspektiven einer neuen Urbanität* (Berlin: Transit, 2000), 13–56; Ute Seiderer, "Die Zeit und die Flüsse. Vom historisch bedingten Blick auf Natur am Beispiel europäischer Kunst und Literatur," in Bernd Busch and Larissa Förster, eds., *Wasser* (Cologne: Wienand, 2000), 194–215.

54. See, for example, Michael Fried, *Menzel's realism: art and embodiment in nineteenth-century Berlin* (New Haven: Yale University Press, 2002); Dorothy Rowe, *Representing Berlin: sexuality and the city in imperial and Weimar Germany* (Aldershot: Ashgate, 2003).

55. Theodor Fontane, *Wanderungen durch die Mark Brandenburg* (Wiesbaden: Emil Vollmer, 1980 [1880]).

56. Theodor Fontane, *Irrungen, Wirrungen* (Stuttgart: Reclam, 1994 [1891]), 111.

57. See Nicholas Teeuwisse, *Vom Salon zur Secession: Berliner Kunstleben zwischen Tradition und Aufbruch der Moderne 1871–1900* (Berlin: Deutscher Verlag für Kunstwissenschaft, 1986).

58. See Gerhard Hard, "Kiefenverbreitung und Kulturlandschaftsgeschichte," *Allgemeine Forst und Jagdzeitung* 234 (January 1963): 24–26; Hansjörg Küster, *Geschichte der Landschaft im Mitteleuropa: von der Eiszeit bis zur Gegenwart* (Munich: C. H. Beck, 1999).

59. Pine, spruce, and other trees were also extensively planted because deer wouldn't eat them. See Werner Köllner and Albrecht Neumeister, *Einblicke in nicht nur 750 Jahre Geschichte des Grunewaldes: eine Ausstellung des Naturschutzzentrum Berlin, Ökowerk, Teufelsee* (Berlin: Naturschutzzentrum, 1987).

60. See, for example, Matthew Biro, *The dada cyborg: visions of the new human in Weimar Berlin* (Minneapolis: University of Minnesota Press, 2009).

61. Wolschke-Bulmahn, "The ideological, aesthetic and ecological significance of water."

62. As a counterresponse in 1909, however, a police ordinance specified in precise detail what types of bathing costumes could be worn, along with demands to provide separate bathing zones for men, women, and families. Engel, "Das Strandbad ein Baudenkmal!" See also Catherine Horwood, "'Girls who arouse dangerous passions': women and bathing, 1900–39," *Women's History Review* 9 (4) (2000): 653–673.

63. See Matthias Oloew, *100 Jahre Strandbad Wannsee* (Berlin: Nicolai, 2007).

64. On changing cultures of bathing in Europe see Jill Steward, "The culture of the water cure in nineteenth-century Austria, 1800–1914," in Anderson and Tabb, *Water, leisure and culture*, 23–35. For North America see also Thomas A. P. van Leeuwen, *The springboard in the pond: an intimate history of the swimming pool* (Cambridge, MA: MIT Press, 1998).

65. We can also trace elements of a "carnivalesque nature" to the emergence of ice fairs, skating, and other pursuits during the "Little Ice Age" of the early modern era. See Jan Hein Furnée, "The thrill of frozen water: class, gender and ice-skating in the Netherlands, 1600–1900," in Anderson and Tabb, *Water, leisure and culture*, 53–69.

66. Karl Toepfer, *Empire of ecstasy: nudity and movement in German body culture* (Berkeley: University of California Press, 1997), 6. See also Michael Hau, *The cult of health and beauty in Germany: a social history, 1890–1930* (Chicago: University of Chicago Press, 2003); and Peukert, *The Weimar Republic*.

67. See, for example, Ernst Gerhard Eder, "Sonnenanbeter und Wasserratten: Körperkultur und Freiluftbadebewegung in Wiens Donaulandschaft," *Archiv für Sozialgeschichte* 33 (1993): 245–274; Martin Faass, "Lichtgestalten in der Landschaft. Die Nacktkultur und ihre Folgen," in Wilhelm Hornbostel and Nils Jockel, eds., *Nackt: Die Ästhetik der Blöße* (Munich: Prestel, 2002), 127–136; Bernd Wedermeyer-Kolwe, *"Der neue Mensch": Körperkultur im Kaiserreich und in der Weimarer Republik* (Würzburg: Königshausen and Neumann, 2004).

68. These ascetic impulses toward nature are also discernable in other European body cultures such as early twentieth-century Sweden's where the appreciation of nature became entwined with conceptions of national identity. See, for example, Michelle Facos, "A sound mind in a sound body: Bathing in Sweden," in Anderson and Tabb, *Water, leisure and culture*, 105–117.

69. Von Reuß, "Freiflächenpolitik als Sozialpolitik."

70. See Tafuri, *The sphere and the labyrinth*, 208.

71. The picnic motif is also explored in other films at the time such as Jean Renoir's *Partie de campagne* (1936) and Jean Vigo's *À propos de Nice* (1930).

72. See Ludwig Bauer, "Zeichen und kulturelles Wissen. Die Rekonstruktion des Bedeutungspotentials visueller Zeichen am Beispiel von Menschen am Sonntag," in Elfriede Ledig, ed., *Der Stummfilm: Konstruktion und Rekonstruktion* (Munich: Schaudig, Bauer, Ledig, 1988), 33–67; Andreas Hutter and Klaus Kamolz, *Billie Wilder: eine europäische Karriere* (Vienna: Böhlau, 1998); Kracauer, *Straßen in Berlin und anderswo*; Karl Prümm, "Melancholie der Großstadt," *Frankfurter Allgemeine Zeitung* (21 July 2000); and Peter Zander, "Die Wahrheit über den Wannsee," *Berliner Morgenpost* (15 August 1988).

73. On the development and history of water infrastructure in Berlin see Heinrich Tepasse, *Stadttechnik im Städtebau Berlins 19. Jahrhundert* (Berlin: Mann Verlag, 2001).

74. The film only existed in edited and censored versions until its recent restoration by the Netherlands Film Museum: some 7 minutes of the original film are still missing.

75. See also Margrit Tröhler, "*Menschen am Sonntag*. Les acteurs non-professionels entre document, fiction et expérimentation cinématographique et comme carrefour des discours sur le cinéma à la fin des années 20," in "L'uomo visibile /the visible man," VIII International Film Studies Conference, Udine, 21–24 March 2001. The use of forest as a setting for sexual encounters also occurs in Robert Siodmak's later films such as *Stürme der Leidenschaft* (*Storms of passion*) (1932) and *Nachts, wenn der Teufel kam* (*The devil strikes at night*) (1957), as well in Rainer Werner Fassbinder's 1980 adaptation of Alfred Döblin's *Berlin Alexanderplatz*.

76. See, for example, Dietmar Elger, *Expressionism: a revolution in German art* (Cologne: Benedikte Taschen, 1989); Hal Foster, *Prosthetic gods* (Cambridge, MA: MIT Press, 2004); Reinhold Heller, "Bridge to utopia: The Brücke as utopian experiment," in Benson, *Expressionist utopias*, 62–83.

77. Linda Nochlin, *Bathers, bodies, beauty: the visceral eye* (Cambridge, MA: Harvard University Press, 2006), 9.

78. Hans Feld, "Menschen am Sonntag," *Film-Kurier* (5 February 1930).

79. Prümm, "Melancholie der Großstadt." See also Günther Elbin, *Am Sonntag in die Matinee: Moritz Seeler und die Junge Bühne* (Mannheim: Persona Verlag, 1998).

80. Bauer, "Zeichen und kulturelles Wissen"; Richard Combs, "Menschen am Sonntag," *Monthly Film Bulletin* 45 (533) (June 1978): 121–122; Hutter and Kamolz, *Billie Wilder*; Siegfried Kracauer, *From Caligari to Hitler: a psychological history of German film* (Princeton: Princeton University Press, 1947). Nonprofessional actors are also used in other contemporary films such as Alberto Cavalcanti's *Rien que les heures* (1926), Georges Lacombe's *La Zone* (1929), and Ilya Kopalin's *Moscow* (1927).

81. See Eithne O'Neil, "Les Hommes, le dimanche 'Toutes choses tachetées'," *Positif* 265 (November 1999): 100–102.

82. In an interview in 1968 Robert Siodmak makes the claim that *Menschen am Sonntag* is the first neorealist film, some fifteen years before Roberto Rossellini's *Roma città aperta* (*Rome Open City*), but its actual influence is difficult to gauge since very few prints were in circulation in the immediate postwar era. Robert Siodmak, "Wandere zwischen den Film-Welten," interview transcript held in the library of the Deutsche Kinemathek, Museum für Film und Fernsehen, Berlin, no interviewer's name given, dated November 1968. See also Kerstin Decker, "Namenlose, unsterblich," *Tagesspiegel* (11 July 2000).

83. See Combs, "Menschen am Sontag."

84. For the influential German film critic Lotte Eisner, *Menschen am Sonntag*, along with Slaton Dudow's adaptation of Bertolt Brecht's *Kuhle Wampe*, effectively mark the end of the silent era. See Eisner, *The haunted screen*. See also Bauer, "Zeichen und kulturelles Wissen"; Jacques Mandelbaum, "Les traces parlantes d'un film muet allemand," *Le Monde* (13 October 2000); Hans-Jörg Rother, "Vom nutzen der guten Laune," *Frankfurter Allgemeine Zeitung* (15 September 2001).

85. See Sabine Hake, *The cinema's third machine: writing on film in Germany, 1907–1933* (Lincoln: University of Nebraska Press, 1993); Helmut Korte, ed., *Film und realität in der Weimarer Republik* (Munich: Carl Hanser, 1978).

86. Combs, "Menschen am Sonntag"; Hutter and Kamolz, *Billie Wilder*.

87. Kraszna-Krausz, "Menschen am Sonntag (People on Sunday)," 316.

88. Brigitte Borchert, interview with C. Z. Klotzel, *Berliner Zeitung am Mittag* (13 May 1931). On the complexities of public culture and early cinema in Germany see Anke Gleber, "Female flanerie and the *Symphony of the city*," in von Ankum, *Women in the metropolis*, 67–88; Anke Gleber, *The art of taking a walk: flanerie, literature, and film in Weimar culture* (Princeton: Princeton University Press, 1999); Miriam Hansen, "Early silent cinema: whose public sphere?," *New German Critique* 29 (Spring/Summer 1983): 147–184.

89. Alfred Durus [Alfréd Keményi], "Ein interessanter Filmversuch: *Menschen am Sonntag*," *Die Rote Fahne 32* (7 February 1930).

90. Feld, "Menschen am Sonntag."

91. Herbert Jhering, "Menschen am Sonntag," *Berliner Börsen-Courier* (5 February 1930). See also Herbert Jhering, *Von Reinhardt bis Brecht III* (Berlin: Aufbau-Verlag, 1960).

92. Cited in Hutter and Kamolz, *Billie Wilder* (author's translation).

93. Kracauer, *From Caligari to Hitler*.

94. See Frisby, *Fragments of modernity*, 166–167.

95. Prümm, "Melancholie der Großstadt."

96. Bauer, "Zeichen und kulturelles Wissen"; Karl Prümm, "Brüchig wie das Leben," *Tagesspiegel* (18 November 2000).

97. See Jürgen Dettbarn-Reggentin, *Strandbad Wannsee: Badegeschichte aus achtzig Jahren* (Berlin: Nishen, 1987).

98. After leaving Germany Wagner took up the post of planner for Istanbul in 1935, before finally settling in the United States after 1938 as associate professor of planning at Harvard University. The SPD was founded as a socialist party in 1875, and reaffirmed its commitment to social forms of ownership with the Heidelberg Program of 1925. It was banned in 1933 after the Nazis came to power.

99. See, for example, William Sheridan Allen, *The Nazi seizure of power: the experience of one German town 1922–1945*, rev. ed. (Harmondsworth: Penguin, 1989).

100. See Saul Friedländer, *Nazi Germany and the Jews: the years of persecution, 1933–1939* (New York: HarperCollins, 1997); Marion A. Kaplan, *Between dignity and despair: Jewish life in Nazi Germany* (Oxford: Oxford University Press, 1998); Jürgen Matthäus, "Antisemitic symbolism in early Nazi Germany 1933–1935," *Leo Baeck Institute Yearbook* 45 (1) (2000): 183–203.

101. On the longer-term influence of Weimar-era architecture, planning, and design, see Hilberseimer, *Berliner Architektur der 20er Jahre*. Wagner's integrated concept of urban landscape (*Stadtlandschaft*) would later be adopted by Hans Scharoun as part of the rationale for postwar urban reconstruction in Berlin. Despite this intellectual lineage from the Weimar era, Wagner proved a fierce critic of postwar development in the city. See Hoffmann, "Martin Wagner 1885–1957."

102. Walter Gropius and Martin Wagner, "The new city pattern for the people and by the people," *Andean Quarterly* (Winter 1943): 4–12.

103. See David Frisby, *Cityscapes of modernity: critical explorations* (Cambridge: Polity, 2001).

104. See Helen Callaghan and Martin Höpner, "Changing ideas: organized capitalism and the German Left," *West European Politics* 35 (3) (2012): 551–573.

105. See Schorske, *German social democracy, 1905–1917*.

106. On modernism, politics, and architecture see, for example, Eve Blau, *The architecture of Red Vienna, 1919–1934* (Cambridge, MA: MIT Press, 1999); Sabine Hake, *Topographies of class: modern architecture and mass society in Weimar Berlin* (Ann Arbor: University of Michigan Press, 2008).

107. See von Reuß, "Freiflächenpolitik als Sozialpolitik." See also Bernhard Citron, "Die Tragödie Berlin," *Die Weltbühne* 27 (22) (2 June 1931): 792–794.

108. See Scarpa, *Martin Wagner and Berlin*, 13.

## Chapter 3

1. This expression was used by Jorg Sieweke at the Woltz Symposium on "Adaptive infrastructure," held at the School of Architecture, University of Virginia, Charlottesville (8–10 October 2009).

2. Amby Njoku, *Swinging Lagos* (Lagos: Ambrose Editions, 1985), 67.

3. See, for example, Laurent Fourchard, "Lagos, Koolhaas and partisan politics in Nigeria," *International Journal of Urban and Regional Research* 35 (1) (2011): 40–56; Matthew Gandy, "Learning from Lagos," *New Left Review* 33 (2005): 37–52; and Onookome Okome, "Nollywood: spectatorship, audience and sites of consumption," *Postcolonial Text* 3 (2) (2007): 1–21. See also the important collection, Okwui Enwezor, Carlos Basualdo, Ute Meta Bauer, Susanne Ghez, Sarat Maharaj, Mark Nash, and Octavio Zaya, eds., *Under siege: four African cities. Freetown, Johannesburg, Kinshasa, Lagos. Documenta11_Platform4* (Ostfildern-Ruit: Hatje Cantz, 2003). The most comprehensive account of infrastructure development in the city is provided by Ayodeji Olukoju, *Infrastructure development and urban facilities in Lagos, 1861–2000* (Ibadan: Institut Français de Recherche en Afrique, University of Ibadan, 2003).

4. Jeffrey Sachs and Pia Malaney, "The economic and social burden of malaria," *Nature* 415 (2002): 680–685.

5. This etymology is taken from the *Oxford English Dictionary* (2002).

6. T. E. Rice, "A lecture on malarial fever at the Residency, Ibadan, on Saturday 23rd February, 1901," London School of Hygiene and Tropical Medicine archives (hereafter cited as LSHTM archives), Ross/83/06.

7. See Richard Jones, *Mosquito* (London: Reaktion, 2012); Andrew Spielman and Michael D'Antonio, *Mosquito* (London: Faber and Faber, 2001).

8. Paul S. Sutter, "Nature's agents or agents of empire? Entomological workers and environmental change during the construction of the Panama Canal," *Isis* 98 (2007): 724–754.

9. Kunle Akinsemoyin and Alan Vaughan-Richards, *Building Lagos* (Lagos: F. and A. Services, 1976); Liora Bigon, "The former names of Lagos (Nigeria) in historical perspective," *Names* 59 (4) (2011): 229–240.

10. A. Aderibigbe, "Expansion of the Lagos protectorate, 1863–1900," PhD dissertation, University of London, 1959. On the development of colonial cities see, for example, Janet Abu-Lughod, *Rabat: urban apartheid in Morocco* (Princeton: Princeton University Press, 1980); Anthony D. King, *Colonial urban development: culture, social power and environment* (London: Routledge and Kegan Paul, 1976); and Gwendolyn Wright, *The politics of design in French colonial urbanism* (Chicago: University of Chicago Press, 1991).

11. Spielman and D'Antonio, *Mosquito*; Jones, *Mosquito*.

12. Leonard J. Bruce-Chwatt, "Malaria in African infants and children in southern Nigeria," *Annals of Tropical Medicine and Parasitology* 46 (2) (1952): 173–200; Leonard J. Bruce-Chwatt, "Problems of malaria control in tropical Africa," *British Medical Journal* (23 January 1954): 169–174.

13. Leonard J. Bruce-Chwatt, "Lessons learned from applied field research activities in Africa during the malaria eradication era," *Bulletin of the World Health Organization* 62 (1984): 19–26. See also Geoffrey Gimonneau, Jeremy Bouyer, Serge Morand, Nora J. Besansky, Abdoulaye Diabate, and Frederic Simarda, "A behavioral mechanism underlying ecological divergence in the malaria mosquito *Anopheles gambiae*," *Behavioral Ecology* 21(5) (2010):1087–1092; Ann H. Kelly and Uli Beisel, "Neglected malarias: the frontlines and back alleys of global health," *BioSocieties* 6 (1) (2011): 71–87; Randall M. Packard, *The making of a tropical disease: a short history of malaria* (Baltimore: Johns Hopkins University Press, 2007); Michael R. Reddy, Hans J. Overgaard, Simon Abaga, Vamsi P. Reddy, Adalgisa Caccone, Anthony E. Kiszewski, and Michel A. Slotman, "Outdoor host seeking behaviour of *Anopheles gambiae* mosquitoes following initiation of malaria vector control on Bioko Island, Equatorial Guinea," *Malaria Journal* 10 (2011): 184.

14. Francis Gethin Hopkins, Orisadipe Obasa, and Jacob Henry Samuel, Report on high levels of infant mortality in the town of Lagos and its suburbs, presented to the Governor of Lagos Colony, William MacGregor, February 1901, LSHTM archives, Ross/83/05; William M. MacGregor, paper read at the inaugural meeting of the Lagos Institute, 16 October 1901, LSHTM archives, Ross/83/02.

15. Bruce-Chwatt, "Malaria in African infants and children in southern Nigeria." See also M. A. Barber and M. T. Olinger, "Studies on malaria in southern Nigeria," *Annals of Tropical Medicine and Parasitology* 25 (1931): 461–501.

16. I. Mueller; P. Zimmerman, and J. Reeder, "*Plasmodium malariae* and *Plasmodium ovale*: the 'bashful' malaria parasites," *Trends in Parasitology* 23 (6) (2007): 278–283. In Asia and Latin America, by contrast, the less dangerous strain of malaria, *Plasmodium vivax*, tends to be predominant.

17. Bruce-Chwatt, "Malaria in African infants and children in southern Nigeria."

18. Cited in Bruce-Chwatt, "Malaria in African infants and children in southern Nigeria," 197. Jelliffe, who founded East Africa's first child health and pediatrics department at Makele University, Uganda, would later feature prominently in public health campaigns against the marketing of powdered milk in the global South.

19. Bruce-Chwatt, "Malaria in African infants and children in southern Nigeria," 192. The original reports of the Royal Commission on Population were primarily concerned with debates over the causes and consequences of falling fertility in the UK. For a contemporary response see, for example, F. M. Reddington and R. D. Clarke, "The papers of the Royal Commission on Population," *Journal of the Institute of Actuaries* 77 (1951): 81–97.

20. Ronald Ross, "Prophylaxie de la malaria," Congress at Brussels, August-September 1903, LSHTM archives, Ross/102/17.

21. William M. MacGregor, "Lecture on malaria," 28 November 1902, LSHTM archives, Ross/83/19; Ronald Ross and William M. MacGregor, "The fight against malaria: an industrial necessity for our African colonies," *Journal of the African Society* 2 (6) (1903): 149–160, LSHTM archives, Ross/83/20.

22. W. H. Findlay, "Report of a visit to the Lagos district," c. 1901, LSHTM archives, Ross/83/01. See also Robert Home, *Of planting and planning: the making of British colonial cities* (London: Spon, 1997).

23. The heterogeneous character of Lagos in comparison with other colonial cities is noted, for example, by John Gunther, *Inside Africa* (London: Hamish Hamilton, 1955). On the "racial ecology" of colonial cities see, for example, Myron J. Echenberg, *Black death, white medicine: bubonic plague and the politics of public health in colonial Senegal, 1914–1945* (Portsmouth, NH: Heinemann, 2002); Salvatore Speziale, "Società e malattia: tunisi di fronte al colera del 1885," *Africa* 49 (2) (1994): 275–298; Maynard W. Swanson, "The sanitation syndrome: bubonic plague and urban native policy in the Cape Colony," *Journal of African History* 18 (3) (1977): 387–410.

24. Akinsemoyin and Vaughan-Richards, *Building Lagos*.

25. See Okwui Enwezor, Carlos Basualdo, Ute Meta Bauer, Susanne Ghez, Sarat Maharaj, Mark Nash, and Octavio Zaya, eds., *Créolité and creolization. Documenta11_Platform3* (Ostfildern-Ruit: Hatje Cantz, 2003).

26. Stephen Frenkel and John Western, "Pretext or prophylaxis? Racial segregation and malarial mosquitoes in a British tropical colony: Sierra Leone," *Annals of the Association of American Geographers* 78 (2) (1988): 211–228.

27. Ibid., 217. See also William M. MacGregor to Joseph Chamberlain, 16 June 1903, LSHTM archives Ross/83/14. It is conceivable that Frenkel and Western tend to overstate the differences between MacGregor and Ross, since their own correspondence and a wider range of materials is suggestive of a greater degree of commonality, not least in terms of their support for the British colonial presence in Africa.

28. MacGregor, "Lecture on malaria," 11.

29. Frenkel and Western, "Pretext or prophylaxis?"

30. Ibid., 225.

31. Ross, "Prophylaxie de la malaria," 3.

32. See, for example, Ronald Ross, *The prevention of malaria* (London: John Murray, 1910).

33. See Robert Koch, "Dritter Bericht über die Tätigkeit der Malariaexpedition," *Deutsche Medizinische Wochenschrift* 17–18 (1900): 404–411. See also M. De Castro, Y. Tamagata, D. Mtasiwa, M. Tanner, J. Utzinger, J. Keiser, and B. H. Singer, "Integrated urban malaria control: a case study of Dar es Salaam, Tanzania," *American Journal of Tropical Medicine and Hygiene* 71 (2) (2004): 103–117.

34. Henry Strachan, "Notes on malarial fever for the assistance of the Ladies' League," Lagos, 8 May 1901, LSHTM archives, Ross/83/12; MacGregor, "Lecture on malaria." See also Philip S. Zachernuk, *Colonial subjects: an African intelligentsia and Atlantic ideas* (Charlottesville: University Press of Virginia, 2000).

35. William M. MacGregor, paper read at the inaugural meeting of the Lagos Institute, 11.

36. Ibid.

37. Ibid., 12.

38. Alan B. Gilroy, *Malaria control by coastal swamp drainage in West Africa* (London: Ross Institute of Tropical Hygiene, 1948), v. See also Leonard Bruce-Chwatt, "*A. gambiae melas* control by swamp drainage," *Mosquito News* 9 (1949): 56–68.

39. A similar degree of "strategic urgency" in the control of malaria can also be observed for Freetown, Sierra Leone, another British-controlled port and staging post for West Africa, which suddenly became the focus of epidemiological attention in the early 1940s. See G. A. Walton, "On the control of malaria in Freetown, Sierra Leone. 1. *Plasmodium falciparum* and *Anopheles gambiae* in relation to malaria occurring in infants," *Annals of Tropical Medicine and Parasitology* 41 (1947): 380–407; G. A. Walton, "On the control of malaria in Freetown, Sierra Leone. II. Control methods, and the effects upon the transmission of *Plasmodium falciparum* resulting from the reduced

abundance of *Anopheles gambiae*," *Annals of Tropical Medicine and Parasitology* 43 (1949): 117–139.

40. Alan B. Gilroy and Leonard J. Bruce-Chwatt, "Mosquito-control by swamp drainage in the coastal belt of Nigeria," *Annals of Tropical Medicine and Parasitology* 39 (1) (1945): 20–40.

41. Gilroy, *Malaria control by coastal swamp drainage in West Africa*, 11.

42. Leonard J. Bruce-Chwatt, "Malaria in Nigeria," *Bulletin of the World Health Organization* 4 (1951): 301–327.

43. See, for example, M. Watson, *The prevention of malaria* (London: John Murray, 1916).

44. Gilroy, *Malaria control by coastal swamp drainage in West Africa*, 28, 31.

45. Theo C. Hoskyns-Abrahall, *Report of the Lagos Town Planning Commission with recommendations on the planning and development of Greater Lagos* (Lagos: Government Printer, 1946), 35–36.

46. Ibid., 17, 18, 38.

47. See Matthew Gandy, "Planning, anti-planning and the infrastructure crisis facing metropolitan Lagos," *Urban Studies* 43 (2) (2006): 71–96.

48. Pauline H. Baker, *Urbanization and political change: the politics of Lagos 1917–1967* (Berkeley: University of California Press, 1974), 273.

49. National Archives, Ibadan, COMCOL I 3860, "Central Lagos slum clearance."

50. Gilroy and Chwatt, "Mosquito-control by swamp drainage in the coastal belt of Nigeria," 34, 38.

51. Frenkel and Western, "Pretext or prophylaxis?"

52. Gilroy, *Malaria control by coastal swamp drainage in West Africa*, 24.

53. Bruce-Chwatt, "Malaria in Nigeria."

54. On the African state see, for example, James Ferguson and Akhil Gupta, "Spatializing states: towards an ethnography of neoliberal governmentality," *American Ethnologist* 29 (4) (November 2002): 981–1002; James Ferguson, *Global shadows: Africa in the neoliberal world order* (Durham: Duke University Press, 2006).

55. Leonard J. Bruce-Chwatt, "Relationship between cultural patterns and health services in West Africa" (circa 1949), unpublished manuscript, Wellcome Collection, WT1/LBC/F/1/7. See also Leonard J. Bruce-Chwatt, "Summary of malaria in Nigeria," report by the secretary of the Expert Committee on Malaria to the World Health Organization (8 August 1950).

56. Cited in Gilroy, *Malaria control by coastal swamp drainage in West Africa*, ii.

57. See Mahmood Mamdani, *Citizen and subject: contemporary Africa and the legacy of late colonialism* (Princeton: Princeton University Press, 1996).

58. In the case of Italy see, for example, Frank M. Snowden, "From triumph to disaster: fascism and malaria in the Pontine marshes, 1928–1946," in John Dickie, John Foot, and Frank M. Snowdon, eds., *Disastro! Disasters in Italy since 1860: culture, politics, society* (New York: Pallgrave, 2002), 113–140; Federico Caprotti, "Malaria and technological networks: medical geography in the Pontine Marshes, Italy, in the 1930s," *Geographical Journal* 172 (2) (2006): 145–155. For the history of German approaches see David Blackbourn, *The conquest of nature: water, landscape and the making of modern Germany* (London: Jonathan Cape, 2006).

59. On the connections between African colonialism and European fascism see, for example, Sebastian Conrad, *German colonialism: a short history*, trans. Sorcha O'Hagan (Cambridge: Cambridge University Press, 2012); Mark Mazower, *Dark continent: Europe's twentieth century* (London: Allen Lane, 1998).

60. Raymond Williams, *The country and the city* (Oxford: Oxford University Press, 1973).

61. Gyan Prakash, "Subaltern studies as postcolonial criticism," *American Historical Review* 99 (5) (1994): 1475–1490. See also Dipesh Chakrabarty, *Habitations of modernity: essays in the wake of subaltern studies* (Chicago: University of Chicago Press, 2002).

62. Prakash, "Subaltern studies as postcolonial criticism," 1483.

63. Cyprian Ekwensi, *People of the city* (London: Heinemann, 1963 [1954]) and *Jagua Nana* (London: Hutchinson, 1961). On the depiction of cities in Nigerian literature see Chris Dunton, "Entropy and energy: Lagos as city of words," *Research in African Literatures* 39 (2) (Summer 2008): 68–78; Ojo Olorunleke, *The images of Lagos in Nigerian novels* (Otto/Ijankin: Centre for Lagos Studies, 2002); and Emmanuel Obiechina, *Culture, tradition and society in the West African novel* (Cambridge: Cambridge University Press, 1975).

64. Chinua Achebe, *No longer at ease* (London: Heinemann, 1960).

65. Ibid., 104.

66. Ibid., 105–106.

67. MacGregor, "Lecture on malaria."

68. Maxwell Fry and Jane Drew, *Tropical architecture in the humid zone* (New York: Reinhold, 1956), 65, 137. On housing and planning in colonial cities see, for example, Home, *Of planting and planning*; and James Cobban, "Public housing in colonial Indonesia 1900–1940," *Modern Asian Studies* 27 (4) (1993): 871–896.

69. See, for example, Andrew Apter, *The pan-African nation: oil and the spectacle of culture in Nigeria* (Chicago: University of Chicago Press, 2005); Daniel Immerwahr, "The politics of architecture and urbanism in postcolonial Lagos," *Journal of African Cultural Studies* 19 (2) (2007): 165–186; and Hannah Le Roux, "Modern architecture in post-colonial Ghana and Nigeria," *Architectural History* 47 (2004): 361–392.

70. National Archives, Ibadan, COMCOL I 675, Drainage, Ikoyi. From Chief Federal Land Officer to Chief Administrative Officer (11 July 1955).

71. John Godwin, Professor of Architecture, University of Lagos, private communication with the author, 2 July 2010. See also M. A. Adeleke, C. F. Mafiana, A. B. Idowu, M. F. Adekunle, and S. O. Sam-Wobo, "Mosquito larval habitats and public health implications in Abeokuta, Ogun State, Nigeria," *Tanzania Journal of Health Research* 10 (2) (2008): 103–107; J. C. Anosike, B. E. Nwoke, A. N. Okere, E. E. Oku, J. E. Asor, I. O. Emmy-Egbe, and D. A. Adimike, "Epidemiology of tree-hole breeding mosquitoes in the tropical rainforest of Imo State, south-east Nigeria," *Annals of Agricultural and Environmental Medicine* 14 (1) (2007): 31–38.

72. Akinsemoyin and Vaughan-Richards, *Building Lagos*.

73. See, for example, Lawrence H. Dunn, "Tree-holes and mosquito breeding in West Africa," *Bulletin of Entomological Research* 18 (1927): 139–144; W. M. Graham, "Results obtained from a monthly examination of the native domestic water receptacles at Lagos, southern Nigeria, in 1910–1911," LSHTM archives, Ross/147/53/11.

74. P. Olusola Fasan, "Malaria in the school children of Lagos city and Lagos state," *West African Medical Journal* (October 1969): 176–180.

75. Bruce-Chwatt, "Problems of malaria control in tropical Africa." On the adaptation of *Anopheles* to urban environments see also T. Awolola, A. Oduola, J. Obansa, N. Chukwurar, and J. Unyimadu, "*Anopheles gambiae* s.s. breeding in polluted water bodies in urban Lagos, southwestern Nigeria," *Journal of Vector Borne Diseases* 44 (2007): 241–244; W. A. Chinery, "Impact of rapid urbanization on mosquitoes and their disease transmission potential in Accra and Tema, Ghana," *African Journal of Medical Science* 24 (1995): 179–188; M. Sattler, D. Mtasiwa, M. Kiama, Z. Premji, M. Tanner, G. Killeen, and C. Lengeler, "Habitat characterization and spatial distribution of Anopheles spp. mosquito larvae in Dar es Salaam (Tanzania) during an extended dry period," *Malaria Journal* 4 (4) (2005): 1–15.

76. Timothy Mitchell, *Rule of experts: Egypt, techno-politics, modernity* (Berkeley: University of California Press, 2002).

77. Bruce-Chwatt, "Problems of malaria control in tropical Africa." See also Martin Shipway, *Decolonization and its impact: a comparative approach to the end of the colonial empires* (Malden, MA: Blackwell, 2008).

78. Paul F. Russell, "Discussion of paper by Leonard J. Bruce-Chwatt: malaria eradication at the crossroads," *Bulletin of the New York Academy of Medicine* 45 (10) (1969): 1013–1015; Bruce-Chwatt, "Lessons learned from applied field research activities in Africa." See also P. B. Bloland and M. E. Ettling, "Making malaria-treatment policy in the face of drug resistance," *Annals of Tropical Medicine and Parasitology* 93 (1) (1999): 5–23.

79. Aderounmu A. Omotayo, "The incidence of malaria in Lagos," MSc thesis, College of Medicine, University of Lagos, 2002. Omotayo analyzed data from over 6,500 blood samples taken at the parasitology laboratory of the Lagos University Teaching Hospital during the 1990s, and found an increase in the infection rate from 27.8 percent in 1992 to 68.9 percent in 1999. See also Awolola et al., "*Anopheles gambiae* s.s. breeding in polluted water bodies in urban Lagos."

80. J. E. Ahiamba, K. O. Dimuna, and G. R. A. Okogun, "Built environment decay and urban health in Nigeria," *Journal of Human Ecology* 23 (3) (2008): 259–265; Martin J. Donnelly, P. J. McCall, Christian Lengler, Imelda Bates, Umberto D'Alessandro, Guy Barnish, Flemming Konradsen, Eveline Kilnkenberg, Harold Townson, Jean-François Trape, Ian M. Hastings, and Clifford Mutero, "Malaria and urbanization in sub-Saharan Africa," *Malaria Journal* 4 (12) (2005). On the relationship between malaria and urban agriculture in sub-Saharan Africa see Y. Afrane, E. Klinkenberg, P. Drechsel, K. Owusu-Daaku, R. Garms, and T. Kruppa, "Does irrigated urban agriculture influence the transmission of malaria in the city of Kumasi, Ghana?," *Acta Tropica* 89 (2004): 125–134; J. Dossou-Yovo, J. Doannio, F. Riviere, and J. Duval, "Rice cultivation and malaria transmission in Bouake City," *Acta Tropica* 57 (1994): 91–94; B. Mathys, P. Vounatsou, G. Raso, A. Tschannen, E. Becket, L. Gosoniu, G. Cisse, M. Tanner, E. N'Goran, and J. Utzinger, "Urban farming and malaria risk factors in a medium sized town in Côte d'Ivoire," *American Society of Tropical Medicine and Hygiene* 75 (6) (2006): 1223–1231. On the relationship between slum housing, poverty, and vulnerability to malaria see, for example, J. Coene, "Malaria in urban and rural Kinshasa: the entomological input," *Medical and Veterinary Entomology* 7 (1993): 127–137; S. Yamamoto, V. Louis, A. Sie, and R. Sauerborn, "Household risk factors for clinical malaria in a semi-urban area of Burkina Faso: a case-control study," *Transactions of the Royal Society of Tropical Medicine and Hygiene* 104 (2010): 61–65. On the complexities of malaria control in sub-Saharan Africa see also Uli Beisel, "Who bites back first? Malaria control in Ghana and the politics of co-existence," PhD thesis, Open University, 2010. On the impact of dilapidated or poorly maintained infrastructure on disease epidemiology see, for example, J. Carlson, J. Keating, C. Mbogo, S. Kahindi, and J. Beier, "Ecological limitations on aquatic predator colonization in the urban environment," *Journal of Vector Ecology* 29 (2) (2004): 331–339; J. Keating, K. McIntyre, C. Mbogo, J. Githure, and J. Beier, "Characterization of

potential larval habitats for *Anopheles* mosquitoes in relation to urban land-use in Malindi, Kenya," *International Journal of Health Geographics* 3 (9) (2004): 1–13.

81. The leading US engineer, Joseph Augustin LePrince, had enquired as early as 1927, for example, whether in a North American context "engineers as well as others may in fact have been building malaria into communities." J. A. LePrince, "Mosquito control in relation to impounded water supply," *Journal of the American Water Works Association* 17 (1) (January 1927): 31. See also Pete Kolsky, "Engineers and urban malaria: part of the solution, or part of the problem?," *Environment and Urbanization* 11 (1) (1999): 159–164; Sutter, "Nature's agents or agents of empire?"; and P. H. van der Brug, "Malaria in Batavia in the 18th century," *Tropical Medicine and International Health* 2 (9) (1997): 892–902.

82. On politics, society, and the emergence of the "petro state" in contemporary Nigeria see, for example, Apter, *The pan-African nation*; and Michael Watts, "The shock of modernity: petroleum, protest and fast capitalism in an industrializing society," in Allan Pred and Michael Watts, eds., *Reworking modernity: capitalism and symbolic discontent* (New Brunswick: Rutgers University Press, 1992), 21–63.

83. The term "microecologies" in relation to postcolonial landscapes is taken from Ann Laura Stoler, "Imperial debris: reflections on ruins and ruination," *Cultural Anthropology* 33 (2) (2008): 194.

84. Obinna Onwujekwe, Kara Hanson, and Julia Fox-Rushby, "Inequalities in purchase of mosquito nets and willingness to pay for insecticide-treated nets in Nigeria: challenges for malaria control interventions," *Malaria Journal* 3(6) (2004).

85. Maik Nwosu, *Invisible chapters* (Lagos: Malaika and Hyburn, 2001); and Chris Abani, *Graceland* (New York: Farrar, Straus and Giroux, 2004).

## Chapter 4

1. Arjun Appadurai, "Spectral housing and urban cleansing: notes on millennial Mumbai," *Public Culture* 12 (3) (2000): 627.

2. Kiran Nagarkar, *Ravan & Eddie* (New Delhi: Penguin, 1995), 69–70.

3. Somini Sengupta, "Flood of complaint: Mumbai deluge blamed on political failures," *Guardian* (London) (5 August 2005): 17. On the material and symbolic dimensions to the flood of 2005 see also Vyjayanthi Rao, "Proximate distances: the phenomenology of density in Mumbai," *Built Environment* 33 (2) (2007): 227–248. On broader technical issues posed by the flood see Kapil Gupta, "Urban flood resilience planning and management and lessons for the future: a case study of Mumbai, India," *Urban Water Journal* 4 (3) (2007): 183–194.

4. The proportion of slum dwellers across the Mumbai metropolitan region now accounts for about half of the total population, but this population is concentrated in less than a tenth of the total housing area. See, for example, Shashi Shekhar Jha, *Structure of urban poverty: the case of Bombay slums* (Bombay: Popular Prakashan, 1986); Suketu Mehta, *Maximum city: Bombay lost and found* (New York: Alfred A. Knopf, 2004); and Madhura Swaminathan, "Aspects of poverty and living standards," in Sujata Patel and Jim Masselos, eds., *Bombay and Mumbai: the city in transition* (New Delhi: Oxford University Press, 2003), 81–109.

5. Jan Nijman, "The effects of economic globalization on land use and land values in Mumbai," in Richard Grant and John R. Short, eds., *Globalization and the margins* (Basingtoske: Palgrave Macmillan, 2002), 150–169.

6. Partha Chatterjee, "Are Indian cities becoming bourgeois at last?," in Indira Chandrasekhar and Peter C. Seel, eds., *body.city: siting contemporary culture in India* (Berlin: Haus der Kulturen der Welt; Delhi: Tulika Books, 2003), 178.

7. Amita Baviskar, "The politics of the city," *Seminar 516* (2002), http://www.india-seminar.com.

8. On water and Hindu mythology see, for example, Anne Feldhaus, *Water and womanhood: religious meanings of rivers in Maharashtra* (Oxford: Oxford University Press, 1995); Thomas Hoffmann, ed., *Wasser in Asien: elementare Konflikte* (Osnabrück: Secolo, 1997); Cornelia Mallebrein, "Heiliges Wasser und heilige Gewässer am Beispiel des Ganges," in Bernd Busch and Larissa Förster, eds., *Wasser* (Cologne: Wienand, 2000), 536–549.

9. Nagarkar, *Ravan & Eddie*, 71.

10. Mehta, *Maximum city*, 588. See also Gyan Prakash, *Mumbai fables* (Princeton: Princeton University Press, 2010).

11. The political salience of public health in Bombay was also bolstered in 1848 with the first systematic data compiled on causes of death, though in the prebacteriological era these statistics were deployed in defense of environmental rather than contagionist explanations for differences in the prevalence of disease. For more detail on the Vihar works see Henry Conybeare, "Description of the works, recently executed, for the water supply of Bombay, in the East Indies," *Minutes of Proceedings of the Institution of Civil Engineers* 17 (1858): 555–575; S. M. Edwards, *The gazetteer of Bombay City and Island*, vol. 3 (Mumbai: Cosmo Publications, 2001 [1910]).

12. Vijay Prashad, "The technology of sanitation in colonial Delhi," *Modern Asian Studies* 35 (2001): 116.

13. Mariam Dossal, "Henry Conybeare and the politics of centralised water supply in mid-nineteenth century Bombay," *Indian Economic and Social History Review* 25 (1988): 79–96; Mariam Dossal, *Imperial designs and Indian realities: the planning of Bombay City, 1845–1875* (Bombay: Oxford University Press, 1991).

14. Thomas Blom Hansen, "Sovereigns beyond the state: on legality and authority in urban India," in Thomas Blom Hansen and Finn Stepputat, eds., *Sovereign bodies: citizens, migrants and state in the postcolonial world* (Princeton: Princeton University Press, 2005), 169–191.

15. Dossal, "Henry Conybeare and the politics of centralised water supply in mid-nineteenth century Bombay."

16. The city experienced rapid demographic and industrial expansion in the 1860s, leading to intense overcrowding in districts such as Kumbharwada, Chakla, and Khara Talao. Contributory factors toward the city's rapid growth included the benefit to the Indian cotton trade caused by the global disruption to supplies during the American Civil War, and the opening of the Suez Canal in 1869 that halved journey times to Europe.

17. To Sir Charles Wood, Principal Secretary of State for India, from Robert Rawlinson, Report on the proposed scheme of main sewerage and drainage submitted to the Municipal Commissioners of Bombay, dated April 1863, Maharashtra State Archives, Mumbai. See also D. E. Gostling, *Suggestions for a comprehensive system of high pressure water supply from Tulsi and Vehar Lakes into every part of the city and suburbs of Bombay* (Bombay: Caxton Printing Works, 1875); J. A. Jones, *A manual of hygiene, sanitation and sanitary engineering with special references to Indian conditions* (Madras: Government Press, 1896); and C. Sowerby, "Memorandum on the drainage of Bombay" *Bombay Builder* (5 August 1868).

18. Hector Tulloch, *The water supply of Bombay* (Roorkee: Thomason College Press, 1873), 215.

19. Prashad, "The technology of sanitation in colonial Delhi."

20. W. W. Hunter, *Bombay 1885 to 1890: a study in Indian administration* (London: Henry Frowde; Bombay: B. M. Malabari, 1893), 324.

21. For Ira Klein, the experience of late nineteenth-century Bombay provides a unique insight into the complex relationship between imperialism, modernization, and demography: while waterborne diseases such as cholera declined, other diseases associated with overcrowding such as tuberculosis and bubonic plague increased. Ira Klein, "Urban development and death: Bombay City, 1870–1914," *Modern Asian Studies* 20 (1986): 725–754. See also Mridula Ramanna, *Western medicine and public health in colonial Bombay, 1845–95* (Hyderabad: Sangam, 2002).

22. Shekhar Krishnan, "Bombay's blame game," *DNA: Daily News and Analysis* (4 August 2005), 2.

23. D. T. Lakdawala, V. N. Kothari, P. A. Nair, and J. C. Sandesara, *Report on the economic survey of Greater Bombay* (Bombay: Department of Economics, University of Bombay, 1959), 1131.

24. Government of Bombay, *Report of the study group on Greater Bombay* (Bombay: Government Central Press, 1961); S. N. V. Modak and A. Mayer, *Outline of the Master Plan of Greater Bombay* (Bombay: Municipal Printing Press, 1948).

25. Sudipta Kaviraj, "In search of civil society," in Chandrasekhar and Seel, *body.city*, 150–169; Gyan Prakash, *Another reason: science and the imagination of modern India* (Princeton: Princeton University Press, 1999).

26. The second of Mumbai's World Bank-funded water supply and sewerage projects (BWSSP-II), initiated in 1978 and completed in 1990, was deemed "unsatisfactory" under the Bank's Performance Audit Reports. See World Bank, *India—water supply and waste water services in Bombay: first, second and third Bombay water supply and sewerage projects* (World Bank Group: Independent Evaluation Service, 2001). Shekhar Krishnan notes in the context of the 2005 floods that the planned upgrading of the city's drainage system under the so-called Brihanmumbai Storm Water Drainage Project has been delayed by over a decade because of the "state's inability to satisfy World Bank loan conditions." Krishnan, "Bombay's blame game," 3.

27. See, for example, Lisa Björkman, "Becoming a slum: from municipal colony to illegal settlement in liberalization-era Mumbai," *International Journal of Urban and Regional Research* (2014); Stephen Graham, Renu Desai, and Colin McFarlane, "Water wars in Mumbai," *Public Culture* 25 (1) (2013): 115–141; Swaminathan, "Aspects of poverty and living standards." Although Mumbai faces formidable challenges, it remains in a better position than many other Indian cities with respect to its physical infrastructure because of its relative wealth and its legacy of independent service provision: its electricity generation, for example, has been primarily handled by two private companies since the 1920s without any reliance on the Indian national grid. Marie-Hélène Zérah, Institute of Research for Development and Centre de Sciences Humaines, New Delhi, interview with the author (12 April 2003). See also Marie-Hélène Zérah, "Splintering urbanism in Mumbai: contrasting trends in a multi-layered society," *Geoforum* 39 (6) (2008): 1922–1932.

28. Nirupa Bhangar, rural water rights and education activist, Mumbai, interview with the author (11 December 2002).

29. United Nations Habitat Programme, *The State of the World's Cities 2012–13: The prosperity of cities* (London: Earthscan, 2013).

30. V. R. Pednekar, Executive Engineer, Planning and Research, B-Ward, BMC, interview with the author (12 December 2002).

31. See Janaky Ammal, "Mumbai city aims for round-the-clock water supply," *Asian Water* (June 2008): 10–14.

32. See Amita Baviskar, *In the belly of the river: tribal conflicts over development in the Narmada Valley*, 2nd ed. (Delhi: Oxford University Press, 2005); Parita Mukta, "Wresting riches, marginalizing the poor, criminalizing dissent: the building of the Narmada Dam in Western India," *South Asia Bulletin: Comparative Studies of South Asia, Africa and the Middle East* 15 (2) (1995): 99–109.

33. K. Rathod, "Acute water shortage in Thane," *Mid Day* (Mumbai) (28 November 2002); Marie-Hélène Zérah, interview with the author (12 April 2003).

34. Dev Benegal, Mumbai-based filmmaker, interview with the author (12 June 2003).

35. G. Sovani and K. Lokhandwala, "Water-starved Bhayandar residents take to the streets," *Bombay Times* (4 March 1999).

36. K. Lokhandwala and N. N. Namboodiri, "Tanker lobby is squeezing us dry," *Bombay Times* (9 March 1999); K. Lokhandwala and N. N. Namboodiri, "Population, politicians worsen water problem," *Bombay Times* (9 March 1999); and S. Shrivastava, "Politician-bureaucrat nexus blamed for water problem at Mira-Bhayandar," *Times of India* (22 December 1998).

37. Krishnakumar, "BMC engineers want to hold the city to ransom," *Mid Day* (Mumbai) (19 February 2004).

38. Stuart Corbridge, Glyn Williams, Manoj Srivastava, and René Véron, eds., *Seeing the state: governance and governmentality in India* (Cambridge: Cambridge University Press, 2005).

39. Jennifer Davis, "Corruption in public service delivery: experience from South Asia's water and sanitation sector," *World Development* 32 (1) (2004): 53–71.

40. Sheela Patel, interview with the author (4 December 2002); Rajendra Vora and Suhas Palshikar, "Politics of locality, community, and marginalization," in Patel and Masselos, *Bombay and Mumbai*, 161–182; and Marie-Hélène Zérah, interview with the author (12 April 2003).

41. Meera Bapat and Indu Agarwal, "Our needs, our priorities; women and men from the slums in Mumbai and Pune talk about their needs for water and sanitation," *Environment and Urbanization* 15 (2003), 74.

42. Nirupa Bhangar, interview with the author (11 December 2002).

43. On the dynamics of sectarian violence in Gujarat see, for example, Martha C. Nussbaum, *The clash within: democracy, religious violence, and India's future* (Cambridge, MA: Belknap Press of Harvard University Press, 2007).

44. World Bank, *India—water supply and waste water services in Bombay*.

45. R. B. Hardas, Assistant Engineer, Water Works, K-West Ward, BMC, interview with the author (25 November 2002).

46. Problems with technical management are matched by arcane tariff structures that facilitate profligate wastage of potable water by wealthy residents. Nayan Parekh, personal communication, 4 May 2006.

47. Appadurai, "Spectral housing and urban cleansing," 627–651.

48. Kalpana Sharma, "Mumbai after the rain: piecemeal policies," *The Hindu* (9 September 2005).

49. Nirupa Bhangar, interview with the author (11 December 2002).

50. Colin McFarlane, "Geographical imaginations and spaces of political engagement: examples from the Indian Alliance," *Antipode* (2004): 890–916; Sheela Patel, interview with the author (4 December 2002).

51. Arjun Appadurai, "Deep democracy: urban governmentality and the horizon of politics," *Public Culture* 14 (1) (2002): 36.

52. Sundar Burra, Sheela Patel, and Thomas Kerr, "Community-designed, built and managed toilet blocks in Indian cities," *Environment and Urbanization* 15 (2003): 11–32.

53. Darshini Mahadevia, "State supported segmentation of Mumbai: policy options in the global economy," *Review of Development and Change* 3 (1) (1998): 12–41; Darshini Mahadevia, *Globalisation, urban reforms and metropolitan response: India* (Ahmedabad: School of Planning, Centre for Environmental Planning and Technology, 2003).

54. Compare, for example, Appadurai, "Deep democracy," with Partha Chatterjee, *The politics of the governed* (New York: Columbia University Press, 2004).

55. Vyjayanthi Rao, "Slum as theory: the South/Asian city and globalisation," *International Journal of Urban and Regional Research* 30 (2006): 225–232.

56. John Holston, ed., *Cities and citizenship* (Durham: Duke University Press, 1999); Sudipta Kaviraj, "Filth and the public sphere: concepts and practices about space in Calcutta," *Public Culture* 10 (1) (1997): 83–113.

57. Chatterjee, *The politics of the governed*.

58. Corbridge et al., *Seeing the state*.

59. Elinor Ostrőm, "Crossing the great divide: coproduction, synergy and development," *World Development* 24 (6) (1996): 1073–1097; William Reuben, *Civic engagement, social accountability and governance crisis* (Washington, DC: World Bank, 2002).

60. Marie-Hélène Zérah, interview with the author (12 April 2003); Joel Ruet; V. S. Saravanan, and Marie-Hélène Zérah, *The water and sanitation scenario in Indian metropolitan cities*, Occasional Paper 6 (New Delhi: Centre de Sciences Humaines, 2002).

61. Ramesh Bhatia, Deputy Municipal Commissioner, Special Engineering, BMC, interview with the author (9 April 2003).

62. In 2003, for example, the Reserve Bank of India announced a number of significant changes to their risk assessment of large projects that may facilitate the use of bonds rather than bilateral loans for infrastructure finance. Times News Network, "RBI eases infrastructure funding norms," *Times of India* (5 February 2003). On the history of infrastructure investment in India see also Amiya Kumar Bagchi, *Private investment in India* (Cambridge: Cambridge University Press, 1972), and Tirthankar Roy, "Economic history and modern India: redefining the link," *Journal of Economic Perspectives* 16 (3) (2002): 109–130.

63. Nirupa Bhangar, interview with the author (11 December 2002).

64. Sheela Patel, interview with the author (4 December 2002).

65. V. A. Kendra, "Privatising water supply in Mumbai," World Prout Assembly, http://worldproutassembly.org/archives/2006/06/privatising_wat.html (accessed 29 June 2006).

66. Ramesh Bhatia, interview with the author (9 April 2003).

67. Karen Bakker, "Archipelagos and networks: urbanisation and water privatisation in the South," *Geographical Journal* 169 (4) (2003): 328–341; Rupa Chinai, "Manila water system shows failure of privatisation," *Times of India* (27 November 2002).

68. Susan Chaplin, "Cities, sewers and poverty: India's politics of sanitation," *Environment and Urbanization* 11 (1999): 145–158; T. D'Souza, "The rich get their water cheap" *Times of India* (3 December 1999).

69. S. Chaterjee, "Poor quality of life causes disease in Mumbai," *Bombay Times* (19 June 2002); Kaviraj, "Filth and the public sphere."

70. Ranjit Hoskote, "Heritage as commodity, culture as process," *Art News Magazine of India* 5 (3) (2000): 64–66; Brenda Yeoh, "The global cultural city? Spatial imagineering and politics in the (multi) cultural marketplaces of south-east Asia," *Urban Studies* 42 (2005): 945–958.

71. Suresh Hosbet, "The sword against the poor," *Deccan Herald* (22 December 2000), and Usha Ramanathan, "Illegality and the Urban Poor," *Economic and Political Weekly* 41 (29) (2006): 3193–3197.

72. Dionne Bunsha, "Rain or shine, malaria continues to spread its tentacles in Mumbai," *Times of India* (Mumbai) (21 September 1998); M. Lal, "Another plague in the offing," *Down to Earth* (15 August 2001).

73. Simon Szreter, "Economic growth, disruption, deprivation, disease and death: on the importance of the politics of public health for development," *Population and Development Review* 23 (4) (1997): 693–728.

74. Dipesh Chakrabarty, *Rethinking working-class history: Bengal 1890–1940* (Delhi: Oxford University Press, 1989); Rajnarayan Chandavarkar, *The origins of industrial capitalism in India: business strategies and the working class in Bombay 1900–1940* (Cambridge: Cambridge University Press, 1994); Zoya Hasan, S. N. Jha, and Rasheeduddin Khan, eds., *The state, political processes and identity: reflections on modern India* (New Delhi: Sage, 1999).

75. Erik Swyngedouw, *Social power and the urbanization of water: flows of power* (Oxford: Oxford University Press, 2004).

76. See, for example, Matthew Gandy, "Planning, anti-planning and the infrastructure crisis facing metropolitan Lagos," *Urban Studies* 43 (2006): 71–96; and Matthew Gandy, "Zones of indistinction: bio-political contestations in the urban arena," *Cultural Geographies* 13 (2006): 1–20.

77. Barbara Harriss-White, *India working: essays on society and economy* (Cambridge: Cambridge University Press, 2003), 77.

78. Hansen, "Sovereigns beyond the state," 171.

79. Swaminathan, "Aspects of poverty and living standards."

80. Swapna Banerjee-Guha, "Shifting cities: urban restructuring in Mumbai," *Economic and Political Weekly* (12 January 2002): 121–128; Jan Breman, *Footloose labor: working conditions in India's informal sector* (Cambridge: Cambridge University Press, 1996); Sandeep Pendse, "Toil, sweat and the city," in Sujata Patel and Alice Thorner, eds., *Bombay: metaphor for modern India* (New Delhi: Oxford University Press, 1995), 3–25.

81. Christopher de Bellaigue, "Bombay at war," in R. B. Silvers and B. Epstein, eds., *India: a mosaic* (New York: New York Review Books, 2000 [1999]), 35. On the devastating impact of the 1992 Bombay riots see, for example, Arjun Appadurai, "Burning questions: arson and other public works in Bombay," *ANY: Architecture New York* 18 (1997): 44–47; Pradip Kumar Datta, "*Hindutva* and the new Indian

middle class," in Chandrasekhar and Seel, *body.city*, 186–197; Thomas Blom Hansen, *The saffron wave: democracy and Hindu nationalism in modern India* (Princeton: Princeton University Press, 1999); Thomas Blom Hansen, *Wages of violence: naming and identity in postcolonial Bombay* (Princeton: Princeton University Press, 2001); Jayant Lele, "Saffronization of the Shiv Sena: the political economy of city, state and nation," in Patel and Thorner, *Bombay*, 185–212; Jim Masselos, "Postmodern Bombay: fractured discourses," in Sophie Watson and Katherine Gibson, eds., *Postmodern cities and spaces* (Oxford: Blackwell, 1995), 199–215; Dileep Padgaonkar, ed., *When Bombay burned* (New Delhi: UBSPD, 1993).

82. The decision of the Indian Supreme Court in March 2005, for example, to allow extensive parts of the former mill lands in Lower Parel to be sold off to private developers in a context of extensive slum clearances elsewhere in the city is suggestive of an increasingly market-driven emphasis in urban policy that transcends both the Shiv Sena administration of the mid-1990s and their Congress Party successors.

83. Praful Bidwai, "Hindutva in dire straits," *News International* (3 December 2005); P. K. Das, "Slums: the continuing struggle for housing," in Patel and Masselos, *Bombay and Mumbai*, 207–234.

84. Chatterjee, "Are Indian cities becoming bourgeois at last?," 174.

85. Susan Bayly, *Caste, society and politics in India from the eighteenth century to the modern age* (Cambridge: Cambridge University Press, 1999); Nicholas B. Dirks, *Castes of mind: colonialism and the making of modern India* (Princeton: Princeton University Press, 2001); Kaviraj, "In search of civil society."

86. Marie-Hélène Zérah, "Networks and differentiated spaces in Mumbai: a question of scale?," paper presented to the CNRS conference "Placing splintering urbanism," Autun, France, 22–24 June 2005.

87. Chatterjee, "Are Indian cities becoming bourgeois at last?," 177. Legislative initiatives such as the Urban Land Ceiling Act, which seek to bring idle land into the public domain and ensure the construction of low-income housing, have never been effectively implemented. Dionne Bunsha, "A tale of two Mumbais," *Frontline* 22 (12 March 2005). See also Nikhil Anand, "Pressure: the polytechnics of water supply in Mumbai," *Cultural Anthropology* 26 (94) (2011): 542–564; D. Asher Ghertner, "Calculating without numbers: aesthetic governmentality in Delhi's slums," *Economy and Society* 39 (2) (2010): 185–217; and Liza Weinstein, "'One man-handled': fragmented power and political entrepreneurship in globalizing Mumbai," *International Journal of Urban and Regional Research* 38 (1) (2014): 14–35.

88. Dionne Bunsha, "Slums near plush areas are hell holes," *Times of India* (30 March 1998).

89. Dionne Bunsha, "On the highroad to 'Shanghai'," *Frontline* 22 (30 July 2005); Randeep Ramesh, "Poor squeezed out by Mumbai's dream plan: India's biggest city is razing its shanty towns," *Guardian* (London) (1 March 2005).

90. In the field of water and sanitation this neoliberal emphasis was exemplified by the Indo-Dutch corporate event entitled "Water for the Future," held at the Taj Mahal Hotel in November 2002. The promotional film used to inaugurate the occasion was festooned with the slogan "Globalization, trade liberalization and economic restructuring will benefit everyone, including the poor."

91. On the role of transport infrastructure in "reimagining" Mumbai see Andrew Harris, "Concrete geographies: assembling global Mumbai through transport infrastructure," *City* 17 (3) (2013): 343–360.

92. Sunil Khilnani, *The idea of India* (London: Penguin, 1997); Gyan Prakash, "The urban turn," in *Sarai Reader 02: the cities of everyday life* (Delhi: Sarai / Centre for the Study of Developing Societies, 2002), 2–7.

93. Whether such a strategy may prove successful is by no means certain, as evidenced by the relatively sluggish economic performance of Mumbai since the late 1990s in comparison with many other Indian cities: extreme congestion, high land prices, poor infrastructure, and Byzantine regulatory and tax regimes have all combined to thwart any straightforward transition toward a more dynamic or efficient urban form.

94. Rahul Srivastava, Director of PUKAR, Mumbai, interview with the author (12 April 2003).

95. Anuradha Mathur and Dilip da Cunha, *Soak: Mumbai in an estuary* (New Delhi: Rupa, 2009).

96. Baviskar, "The politics of the city."

## Chapter 5

1. Dick Roraback, "Up a lazy river, seeking the source," *Los Angeles Times* (20 October 1985).

2. Commentary by Reyner Banham for the documentary *Reyner Banham loves Los Angeles* (London: British Broadcasting Corporation, 1972).

3. Luis J. Rodríguez, "The concrete river," in Luis J. Rodríguez, *The concrete river* (Willimantic, CT: Curbstone Books, 1991 [1990]), 39.

4. Dick Roraback, "LA River practices own trickle-down theory," *Los Angeles Times* (27 October 1985).

5. Dick Roraback, "Que serra serra along Friar Crespi's Porciuncula," *Los Angeles Times* (14 November 1985).

6. Ibid.

7. Robert M. Fogelson, *The fragmented metropolis: Los Angeles 1850–1930* (Berkeley: University of California Press, 1993 [1967]). On the development of the city see also William Deverell and Greg Hise, eds., *Land of sunshine: an environmental history of metropolitan Los Angeles* (Pittsburgh: University of Pittsburgh Press, 2005); Greg Hise, *Magnetic Los Angeles* (Baltimore: Johns Hopkins University Press, 1997); Hynda Rudd and Tom Sitton, eds., *The development of Los Angeles city government: an institutional history, 1850–2000* (Los Angeles: Los Angeles City Historical Society, 2007).

8. The popular sense of Los Angeles as an "unplanned" or largely spontaneous urban sprawl, as suggested by Reyner Banham and others, has been dispelled by more recent research on the role of industrial clusters and other patterns of urban development. See, for example, Hise, *Magnetic Los Angeles*; and Allen J. Scott, "High technology industrial development in the San Fernando Valley and Ventura County: observations on economic growth and the evolution of urban form," in Allen J. Scott and Edward W. Soja, eds., *The city: Los Angeles and urban theory at the end of the twentieth century* (Berkeley: University of California Press, 1996), 276–310.

9. Reyner Banham, *Los Angeles: the architecture of four ecologies* (London: Penguin Allen Lane, 1971). The cover of the first edition of Banham's book is adorned by a painting by another British aficionado of the city, David Hockney, showing a swimming pool and a single-story modernist-style apartment set against a cloudless sky. See also Reyner Banham, "LA: the structure behind the scene," *Architectural Design* 41 (April 1971): 227–230; Edward Dimendberg, "The kinetic icon: Reyner Banham on Los Angeles as mobile metropolis," *Urban History* 33 (1) (2006): 106–125; Nigel Whitely, *Reyner Banham: historian of the immediate future* (Cambridge, MA: MIT Press, 2002). Banham's interest in a technologically driven modernism can be traced to various writings including Reyner Banham, *Theory and design in the first machine age* (Oxford: Butterworth-Heinemann, 1960); Reyner Banham, "Introduction," in *Umberto Boccioni: an Arts Council exhibition of his graphic art from the collection of Mr and Mrs Harry Lewis Winston of Birmingham, Michigan* (London: Arts Council, 1964).

10. Anthony Vidler, *Histories of the immediate present: inventing architectural modernism* (Cambridge, MA: MIT Press, 2008), 108.

11. Anton Wagner, *Los Angeles: Werden, Leben und Gestalt der Zweimillionenstadt in Südkalifornien* (Leipzig: Bibliographisches Institut, 1935).

12. Ibid., 217–218. By the early 1930s Los Angeles County already had 43 incorporated cities.

13. See, for example, James Duncan, "The superorganic in American cultural geography," *Annals of the Association of American Geographers* 70 (1980): 181–198; *Land and life: selections from the writings of Carl Ortwin Sauer*, ed. J. B. Leighly (Berkeley: University of California Press, 1963); Oskar Schmieder, *Spuren spanischer Kolonisation in US-amerikanischen Landschaften* (Berlin: Reimer, 1928). On wider developments in German geography at the time see Bruno Schelhaas and Ingrid Hönsch, "History of German geography: worldwide reputation and strategies of nationalisation and institutionalisation," in Gary S. Dunbar, ed., *Geography: discipline, profession and subject since 1870* (Dordrecht: Kluwer, 2001), 9–44.

14. Wagner, *Los Angeles*.

15. Ibid., 205.

16. Ibid., 238.

17. See Mike Davis, "Cannibal city: Los Angeles and the destruction of nature," in Russell Ferguson, ed., *Urban revisions: current projects for the public realm* (Cambridge, MA: MIT Press, 1994), 39–57; Mike Davis, *Ecology of fear: Los Angeles and the imagination of disaster* (New York: Metropolitan Books, 1998).

18. Jennifer Wolch, "Green urban worlds," *Annals of the Association of American Geographers* 97 (2) (2007): 373–384.

19. See, for example, Robert Gottlieb and Margaret FitzSimmons, *Thirst for growth: water agencies as hidden government in California* (Tucson: University of Arizona Press, 1991); William L. Kahrl, *Water and power: the controversy over Los Angeles' water supply in the Owens Valley* (Berkeley: University of California Press, 1982); Jared Orsi, *Hazardous metropolis: flooding and urban ecology in Los Angeles* (Berkeley: University of California Press, 2004); and John Walton, *Western times and water wars: state, culture, and rebellion in California* (Berkeley: University of California Press, 1992).

20. Works that tend to emphasize the unique aspects of Los Angeles (or southern California) include Carey McWilliams, *Southern California country: an island on the land* (New York: Duell, Sloan and Pearce, 1946); Carey McWilliams, *California: the great exception* (Berkeley: University of California Press, 1999 [1949]); Laurence A. Rickels, *The case of California* (Minneapolis: University of Minnesota Press, 2001); and especially Banham, *Los Angeles: the architecture of four ecologies*. While Dimendberg places Mike Davis's *City of quartz* (London: Verso, 1990) in the "exceptionalist" category, this work is perhaps closer to the literature that emphasizes the significance of LA as an "archetype" or portent of the future, which includes Micheal Dear, *The postmodern urban condition* (Oxford: Blackwell, 2000); Roger Keil, *Los Angeles: globalization, urbanization and social struggles* (London: Wiley,1998); Edward W. Soja, "Los Angeles, 1965–1992: from crisis-generated restructuring to restructuring-generated

crisis," in Scott and Soja, *The city*, 426–462; and Edward W. Soja, *Postmetropolis: critical studies of cities and regions* (Oxford: Blackwell, 2000).

21. Wagner, *Los Angeles*, 249.

22. The willow-and-cottonwood-dominated forests that occupied much of the floodplain blended into grasslands and more sparse semiarid landscapes toward the San Fernando Valley. See Blake Gumprecht, *The Los Angeles River: its life, death and possible rebirth* (Baltimore: Johns Hopkins University Press, 1999).

23. Bernice Eastman Johnston, *California's Gabrielino Indians* (Los Angeles: Southwest Museum, 1964); A. L. Kroeber, *Handbook of the Indians of California* (Washington: Bureau of American Ethnology, 1925).

24. Gumprecht, *The Los Angeles River*.

25. Ibid.

26. See Kahrl, *Water and power*, and Walton, *Western times and water wars*.

27. Gumprecht, *The Los Angeles River*.

28. William Mulholland cited in Patt Morrison and Mark Lamonica, *Río L.A.: tales from the Los Angeles River* (Santa Monica: Angel City Press, 2001), 57.

29. Gumprecht, *The Los Angeles River*, 111. See also Blake Gumprecht, "Who killed the Los Angeles River?," in Deverell and Hise, *Land of sunshine*, 115–134. With a further extension of the aqueduct 105 miles to Mono Lake in 1940 and a second parallel aqueduct completed in 1970, "in effect," argues Gumprecht, "the Los Angeles-Owens River Aqueduct *did* become the new Los Angeles River." Gumprecht, *The Los Angeles River*, 105.

30. Olmsted Brothers and Bartholomew and Associates, *Parks, playgrounds and beaches for the Los Angeles Region. A report submitted to the Citizen's Committee on Parks, Playgrounds and Beaches* (Los Angeles: Citizens' Committee, 1930), 16. A facsimile of the original report is reproduced in Greg Hise and William Deverell, *Eden by design: the 1930 Olmsted-Bartholomew plan for the Los Angeles region* (Berkeley: University of California Press, 2000).

31. Mike Davis, "How Eden lost its garden: a political history of the Los Angeles Landscape," in Scott and Soja, *The city*, 160–185; Greg Hise and William Deverell, "Private power, public space," in *Eden by design*, 1–63.

32. Gumprecht, *The Los Angeles River*; Leonard Leader, *Los Angeles and the Great Depression* (New York: Garland Publishing, 1991).

33. Davis, "How Eden lost its garden."

34. Wagner, *Los Angeles*, 186 and 214.

35. See, for example, Dana Cuff, "Fugitive plans in the provisional city: slums and public housing in Los Angeles," in Charles G. Salas and Michael S. Roth, eds., *Looking for Los Angeles: architecture, film, photography, and the urban landscape* (Los Angeles: Getty Research Institute, 2001), 97–131; William Deverell, *Whitewashed adobe: the rise of Los Angeles and the remaking of its Mexican past* (Berkeley: University of California Press, 2004).

36. See Laura Pulido, "Rethinking environmental racism: white privilege and urban development in Southern California," *Annals of the Association of American Geographers* 90 (1) (2000): 12–40.

37. Gumprecht, *The Los Angeles River*.

38. Gumprecht, *The Los Angeles River*. See also William L. Graf, "Damage control: dams and the physical integrity of America's rivers," *Annals of the Association of American Geographers* 91 (1) (2001): 1–27. We can trace a connection between the devastating floods of the early twentieth century and the transformation of the scope of federal government in the New Deal era: there is a lineage between the Galveston flood of 1900 (in which at least 8,000 people died) and the eventual creation of the Tennessee Valley Authority in 1933, marked by modern approaches to hydrology, flood defenses, and river basin management.

39. Gumprecht, *The Los Angeles River*.

40. Ibid., 5.

41. Los Angeles County, *Drainage plan: an element of the master plan of the City of Los Angeles* (Los Angeles: Department of City Planning and the Department of Public Works Engineering Bureau, 1968).

42. See Orsi, *Hazardous metropolis*.

43. A recent estimate suggests that fully 70 percent of the land area of Los Angeles is now occupied by cars through roads, garages, and parking spaces, compared with just 5 percent devoted to parks and open space. Danielle Nierenberg, "Matters of scale—road rage," *World Watch Magazine* 13 (5) (September-October 2000).

44. See Orsi, *Hazardous metropolis*.

45. See US Army Corps of Engineers, *Los Angeles County drainage area review: final feasibility study interim report and environmental impact assessment* (Los Angeles, 1992).

46. Orsi, *Hazardous metropolis*. A sense of the changing science of river management, including the growing recognition of complexity and the need to incorporate a wider range of factors, is provided by Graf, "*Dam*age control."

47. Los Angeles County, *Los Angeles River master plan* (Los Angeles County: Departments of Public Works, Parks and Recreation, Regional Planning, 1996).

48. Orsi, *Hazardous metropolis*.

49. Southern California Institute of Architecture, *Recreating the river: a prescription for Los Angeles* (Los Angeles, 1990).

50. See Gene Desfor and Roger Keil, "Every river tells a story: the Don River (Toronto) and the Los Angeles River (Los Angeles) as articulating landscapes," *Journal of Environmental Policy and Planning* (2) (2000): 5–23.

51. See Orsi, *Hazardous metropolis*.

52. L. A. Murillo, *Verses from the poet to the river of Los Angeles porciuncula: three fragments* (Los Angeles: Patrick Reagh, 1999).

53. On aesthetics, landscape, and driving see, for example, László Moholy-Nagy, *Vision in motion* (Chicago: Paul Theobold, 1947).

54. Wagner, *Los Angeles*.

55. The history of the highway system in Los Angeles mirrors aspects of road building across the United States, and also parts of Europe, that emerged in the interwar era. See Scott Bottles, *Los Angeles and the automobile: the making of the modern city* (Berkeley: University of California Press, 1987); Edward Dimendberg, "The will to motorization: cinema, highways, and modernity," *October* 73 (1995): 90–137; and Martin Wachs, "The evolution of transportation policy in Los Angeles: images of past policies and future prospects," in Scott and Soja, *The city*, 106–159.

56. Eric Avila, "The folklore of the freeway: space, culture, and identity in postwar Los Angeles," *Aztlán: A Journal of Chicano Studies* 23(1): 15–31; Eric Avila, "All freeways lead to East Los Angeles: rethinking the L.A. Freeway and its meanings," in Wim De Wit and Christopher James Alexander, eds., *Overdrive: L.A. constructs the future, 1940–1990* (Los Angeles: Getty Research Institute, 2013), 35–48.

57. See Alexandra Schwartz, *Ed Ruscha's Los Angeles* (Cambridge, MA: MIT Press, 2010).

58. Doolin's work is used for the covers of Mike Davis's *Ecology of fear* (1998) and Scott and Soja's edited collection *The city* (1996).

59. See also Patricia Hickson, "James Doolin: illusionistic vision," in *Urban invasion* (San Jose: San Jose Museum of Art, 2001), 9–27.

60. On weather and culture see, for example, Jonathan Hill, *Weather architecture* (London: Routledge, 2012).

61. Doug Harvey, "James Doolin, 1932–2002," *LA Weekly* (22 August 2002).

62. Rodríguez, *The concrete river*. See also Stephen Flusty, "Thrashing downtown: play as resistance to the spatial and representational regulation of Los Angeles," *Cities* 17 (2000): 149–158; Julian Murphet, *Literature and race in Los Angeles* (Cambridge: Cambridge University Press, 2001).

63. See Patricia Price, "Inscribing the border: schizophrenia and the aesthetics of Aztlán," *Social and Cultural Geography* 1 (1) (2000): 101–116. See also Amalia Mesa-Bains, "El mundo femenino: chicana artists of the movement—a commentary on development and production," in Richard Griswold del Castillo, ed., *Chicano art: resistance and affirmation* (Los Angeles: Frederick S. Wight Galleries, 1991), 131–140; Sarah Schrank, *Art and the city: civic imagination and cultural authority in Los Angeles* (Philadelphia: University of Pennsylvania Press, 2009).

64. David Fletcher, "Flood control freakology: Los Angeles River watershed," in Kazys Varnelis, ed., *The infrastructural city: networked ecologies in Los Angeles* (Barcelona: Actar, 2008), 50.

65. Southern California Institute of Architecture, *Recreating the river*.

66. See Kimball L. Garrett, ed., *The biota of the Los Angeles River: an overview of the historical and present plant and animal life of the Los Angeles River drainage* (Los Angeles: Natural History Museum of Los Angeles County, 1993). In Garrett's study, completed in the early 1990s, he describes the lowland parts of the Los Angeles River as highly modified environments in which between 30 and 60 percent of vascular plant species are not native to the region.

67. Gumprecht, *The Los Angeles River*.

68. For FOVICKS (Friends of Vast Industrial Concrete Kafkaesque Structures), see http://seriss.com/people/erco/fovicks/ (accessed 1 September 2012).

69. Camilo José Vergara, "in praise of concrete: the l.a. river," *Metropolis* (April 1997), 27, 31.

70. Vergara, "in praise of concrete." See also Kim Sorvig, "Abstract and concrete: in Los Angeles, a park creates an abstraction of an abused river," *Landscape Architecture* 92 (5) (May 2002): 34–40.

71. Adrian Forty, *Concrete and culture: a material history* (London: Reaktion, 2012), 10.

72. Ludwig Trepl, professor of landscape ecology, Technical University, Munich, interview with the author (5 March 2013).

73. Forty, *Concrete and culture*, 15.

74. In some cases urban location shooting was actively encouraged as a form of culturally led urban regeneration, most notably in Lindsay-era New York. See Lawrence Meirion Webb, "Restructuring and representation: cinematic space and the built environment in the 1970s," PhD thesis, King's College London, 2010. In a more light-hearted vein the channel has also been used as a race track in a much parodied scene in Randal Kleiser's *Grease* (1978) or as a setting for dystopian future worlds in one of the foremost B movies of the postwar era, Gordon Douglas's *Them* (1954). The invasion by giant ants in *Them* might also be read as an early example of Cold War unease about the effects of radiation as well as a more generalized paranoia toward the threats of urban infestations. See Richard Jones, *Mosquito* (London: Reaktion, 2012).

75. Gumprecht, *The Los Angeles River*.

76. Marc Gregory Blake, "A study of the Los Angeles River," MA thesis, University of California, Los Angeles, 1990; Jon Regardie, "The anti-landmarks: a tour on the periphery of the city's history," *Downtown News* (31 October 1994). See also Penelope McMillan and Roxane Arnold, "Bradley proposes temporary camp for L.A.'s homeless," *Los Angeles Times* (4 June 1987); Penelope McMillan, "Homeless camp ends much like it began: a grim place of refuge closes today—and no one involved calls it a success," *Los Angeles Times* (25 September 1987).

77. Kevin Starr, foreword to Morrison and Lamonica, *Río L.A.*, 16–17.

78. See, for example, Perry Anderson, *The origins of postmodernity* (London: Verso, 1998); Michael Dear and Stephen Flusty, "Postmodern urbanism," *Annals of the Association of American Geographers* 88 (1998): 50–73; and Edward W. Soja, "Postmodern urbanization: the six restructurings of Los Angeles," in Sophie Watson and Katherine Gibson, eds., *Postmodern cities and spaces* (Oxford: Blackwell, 1995), 125–137.

79. See Eric Mann, *L.A.'s lethal air: new strategies for policy, organizing, and action* (Los Angeles: Labor/Community Strategy Center, 1991). Los Angeles County has some of the widest socioeconomic disparities of any municipality in the United States: the richest fifty individuals out of a population of nearly ten million own more wealth than the poorest two million inhabitants, and a third of those who actually have work are living below the poverty line. See also Keil, *Los Angeles: globalization, urbanization and social struggles*; Mary Pardo, "Mexican American women grassroots community activists: Mothers of East Los Angeles," in James E. Sherow, ed., *A sense of the American West: an anthology of environmental history* (Albuquerque: University of New Mexico Press, 1998), 243–260; Laura Pulido, *Environmentalism and economic justice: two Chicano struggles in the Southwest* (Tucson: University of Arizona Press, 1996); and Laura Pulido, "Multiracial organizing among environmental justice activists in Los Angeles," in Michael Dear, H. Eric Schockman, and Greg Hise, eds., *Rethinking Los Angeles* (Thousand Oaks, CA: Sage Publications, 1996), 171–188.

80. Avila, "The folklore of the freeway," 28.

81. For an overview of FoLAR see Joe Linton, *Down by the Los Angeles River: Friends of the Los Angeles River's official guide* (Berkeley: Wilderness Press, 2005). Other precedents for the emergence of an ecological perspective include Dilara El-Assaad, "Redefining the role of the L.A. River in the urban landscape of Southern California," MA thesis, Graduate School of Landscape Architecture, University of Southern California, 1988.

82. Lewis MacAdams, co-founder and president of FoLAR, interview with the author (12 April 2013). During the course of the interview MacAdams clearly indicated that the spread of the bamboo-like giant cane *Arundo donax* held ideological as well as ecological implications.

83. Orsi, *Hazardous metropolis*, 132.

84. Gumprecht, *The Los Angeles River*.

85. Wolch, "Green urban worlds."

86. Paul Stanton Kibel, "Bankside urban: an introduction," in Paul Stanton Kibel, ed., *Rivertown: rethinking urban rivers* (Cambridge, MA: MIT Press, 2007), 1–21.

87. Gumprecht, *The Los Angeles River*. See also Jessica Garrison, "River plan rolls along: efforts to revitalize LA's concrete-lined waterway will get a boost as city leaders begin a search for a design consultant," *Los Angeles Times* (4 November 2004).

88. D. J. Waldie, "Notes from Los Angeles; reclaiming a lost river, building a community," *New York Times* (10 July 2002).

89. Cara Mia Dimassa, "Living gets loftier in downtown LA," *Los Angeles Times* (22 November 2007). Similar arguments have been made by the Southern California Association of Non-Profit Housing. See http://www.scanph.org.

90. Cited in Paul Stanton Kibel, "Los Angeles' Cornfield: an old blueprint for new greenspace," *Stanford Environmental Law Review* 23 (2004): 298.

91. See also Heather Barnett, "The Chinatown Cornfields: including environmental benefits in environmental justice struggles," *Critical Planning* (Summer 2001): 50–61; Robert Gottlieb and Andrea Misako Azuma, "Bankside Los Angeles," in Kibel, *Rivertown*, 23–46; Kibel, "Los Angeles' Cornfield."

92. Orsi, *Hazardous metropolis*.

93. City of Los Angeles, *Los Angeles River revitalization master plan* (Los Angeles: City of Los Angeles, Department of Public Works, Bureau of Engineering, 2007).

94. Samuel B. Morris, "Water problems of the metropolitan area," in George W. Robbins and L. Deming Tilton, eds., *Los Angeles. Preface to a master plan* (Los Angeles: Pacific Southwest Academy, 1941), 77–89.

95. Fogelson, *The fragmented metropolis*.

96. Julie-Anne Boudreau and Roger Keil, "Seceding from responsibility? Secession movements in Los Angeles," *Urban Studies* 38 (10) (2001): 1701–1731.

97. Orsi, *Hazardous metropolis*.

98. Gumprecht, *The Los Angeles River*; Jared Orsi, "Flood control engineering in the urban ecosystem," in Deverell and Hise, *Land of sunshine*, 135–151.

99. Southern California Institute of Architecture, *Recreating the river*, 3.

100. City of Los Angeles, *Los Angeles River revitalization master plan*.

101. Ibid., 3.

102. The increasing quantity of wastewater pumped into the river since 1984 has begun to have an effect on some stretches of the riverbed, producing what the ornithologist Kimball Garrett has termed "habitat by mistake." Cited in Gumprecht, *The Los Angeles River*, 245. With increasing water flowing through the channel, the decades-long ban on boating has now been lifted, allowing recently formed groups such as LA River Expeditions to organize kayaking and canoeing. See Rory Carroll, "Mystic river," *The Guardian* (28 August 2012). See also William Deverell, "Dreams deferred: parks and open space," in De Wit and Alexander, *Overdrive*, 23–33.

103. See Matthew W. Roth, "Mullholland Highway and the engineering culture of Los Angeles in the 1920," in Tom Sitton and William Deverell, eds., *Metropolis in the making: Los Angeles in the 1920s* (Berkeley: University of California Press, 2001), 45–76.

## Chapter 6

1. J. G. Ballard, *The drowned world* (London: Gollancz, 1999 [1962]), 13.

2. See, for example, Fredric Jameson's critique of Ballard, "Progress versus utopia: or, can we imagine the future?," in *Archaeologies of the future* (London: Verso, 2005 [1982]), 288.

3. J. G. Ballard, "Gewalt ohne Ende," interview with Evelyn Finger, *Die Zeit* (8 September 2005).

4. Richard Jefferies, *After London* (London: Cassell, 1885).

5. John Wyndham, *The day of the triffids* (Harmondsworth: Penguin, 1954 [1951]).

6. See Laura Spinney, "Return to paradise," *New Scientist* 2039 (20 July 1996): 27–31. See also Alan Weisman, *The world without us* (London: Virgin, 2007). Musical evocations of London under water include the punk-era anthem "London Calling" by The Clash, dating from 1977, and Burial's eponymous first album released in 1996, which was described in its press release as set in a "near future South London under water."

7. Jonathan Glancey, "Liquid economy," *The Guardian* (1 July 2008). These comments were made in response to an exhibition of photographs of the Thames estuary by Frank Watson at the Southwark Art Gallery in 2008.

8. See, for example, James A. Galloway, ed., *Tides and floods: new research on London and the tidal Thames from the Middle Ages to the twentieth century* (London: University of London, Centre for Metropolitan History, 2010); Gustav Milne, *The port of Roman London* (London: Batsford, 1985); Gustav Milne, *The port of medieval London* (Stroud: Tempus, 2003); Stuart Oliver, "The Thames Embankment and the disciplining of nature in modernity," *Geographical Journal* 166 (2000): 227–238; Dale H. Porter, *The Thames Embankment: environment, technology, and society in Victorian London* (Akron, OH: University of Akron Press, 1998).

9. Stuart Gilbert and Ray Horner, *The Thames Barrier* (London: Thomas Telford, 1984), 4. In Roman times the Thames was smaller with a much larger floodplain, within which occasional surges could be easily contained. From the eleventh century onward, however, with rising sea levels the Thames becomes increasingly tidal. The first historical record of a serious flood affecting London is 1099, with further exceptional tides in 1236, 1663, 1791, 1834, 1852, 1874, 1875, and 1881. The London diarist Samuel Pepys, for instance, records the inundation of 7 December 1663: "I hear there was last night the greatest tide that ever was remembered in England to have been in this river: all White Hall having been drowned." Cited in Ken Wilson, *The story of the Thames Barrier* (London: Lanthorn, 1984).

10. The record tides of 1874 and 1875 led to the enactment of the Metropolis Management (Thames River Prevention of Floods) Amendment Act of 1879 which gave responsibility for flood defense to local government. See also Arthur Thomas Doodson, *Report on Thames floods* (London: HMSO, 1929).

11. Allan McRobie, Tom Spencer, and Herman Gerritsen, "The big flood: North Sea storm surge," *Philosophical Transactions of the Royal Society A* 363 (2005): 1263–1270; J. A. Steers, "The east coast floods, January 31–February 1 1953," *Geographical Journal* 119 (1953): 280–298.

12. Ministry of Housing and Local Government, "London flood barrier," report by Professor Hermann Bondi, July 1967, National Archives, London, HLG 120/1378.

13. Ibid., 2, 7, 9.

14. On the history and future prospects of flood risk in Hamburg see Iris Grossmann, Katja Woth, and Hans von Storch, "Localization of global climate change: storm surge scenarios for Hamburg in 2030 and 2085," *Die Küste* 71 (2007): 169–182.

15. Sarah Lavery and Bill Donovan, "Flood risk management in the Thames estuary looking ahead 100 years," *Philosophical Transactions of the Royal Society A* 363 (2005): 1456.

16. See, for example, Julius Henri Otto Bunge, *Tideless Thames in future London* (London: Frederick Muller, 1944); Henry Robinson, "On the past and present condition of the River Thames," *Proceedings of the Institute of Civil Engineers* 15 (1856): 195–206; and Terry Larkin, "Thames Barrier," *Architect and Builder* 31 (8) (March 1981): 19–23. The old London Bridge, demolished in 1831 after some six centuries of use, with its narrow sluicelike openings, had also acted as a form of barrage in the past, although its shape had effectively funneled water at greater velocities through the river.

17. See P. J. Clark and R. G. Tappin, "Final design of Thames Barrier gate structure," in *Thames Barrier design*, proceedings of a conference held in London on 5 October 1977 (London: Institution of Civil Engineers, 1978); Gilbert and Horner, *The Thames Barrier*; R. W. Horner, "Current proposals for the Thames Barrier and the organization of the investigations," *Philosophical Transactions of the Royal Society A* 272 (1972): 179–185; Mary P. Kendrick, "Siltation problems in relation to the Thames barrier," *Philosophical Transactions of the Royal Society A* 272 (1972): 223–243; Mary Kendrick, "The Thames Barrier," *Landscape and Urban Planning* 16 (1–2) (1988): 57–68.

18. Samantha Hardingham, *London: a guide to recent architecture* (London: Ellipsis, 1996).

19. Kendrick, "The Thames Barrier," 68.

20. A. M. Petty and F. O. Brown, "Architectural aspects to the Thames Barrier," in *Thames Barrier design*, 180.

21. See Gilbert and Horner, *The Thames Barrier*. A photograph of the new barrier adorns the cover of Bebbe Klatt's compendium of London's municipal architecture in the decade 1976–1986, which serves as a valedictory footnote to the postwar politics of local government in London. See Bebbe Klatt, ed., *GLC/ILEA architecture 1976–1986* (London: Architectural Press, 1986). See also Hardingham, *London: a guide to recent architecture*.

22. On the history of water infrastructure in London see, for example, Asok Kumar Mukhopadhyay, "Politics of London's water supply 1871–1971," PhD dissertation, London School of Economics, 1972; Oliver, "The Thames Embankment and the disciplining of nature in modernity"; and Roy Porter, "Social problems, social improvements: 1820–1890," in Porter, *London: a social history* (London: Hamish Hamilton, 1994), 257–278. The GLC, and its highly regarded Public Health Engineering Department, were disbanded by the Thatcher administration in 1986.

23. On the cultural and political limitations of "municipal managerialism" see, for example, M. Christine Boyer, *Dreaming the rational city: the myth of American city planning* (Cambridge, MA: MIT Press, 1983); John Gyford, *The politics of local socialism* (London: George Allen and Unwin, 1985); Eric Reade, *British town and country planning* (Milton Keynes: Open University Press, 1987).

24. See Wiebe E. Bijker, "The Oosterschelde storm surge barrier: a test case for Dutch water technology, management, and politics," *Technology and Culture* 43 (3) (2002): 569–584.

25. Lavery and Donovan, "Flood risk management in the Thames estuary looking ahead 100 years."

26. David King, "Climate change science: adapt, mitigate, or ignore," *Science* 303 (9 January 2004): 176–177.

27. See, for example, Jennifer A. Francis and Stephen J. Vavrus, "Evidence linking Arctic amplification to extreme weather in mid-latitudes," *Geophysical Research Letters* 39 (2012): L06801; Edward Hanna, John Cappelen, Rob Allan, Trausti Jónsson, Frank le Blancq, Tim Lillington, and Kieran Hickey, "New insights into North European and North Atlantic surface pressure variability, storminess and related climatic change since 1830," *Journal of Climate* 21 (24) (2008): 6739–6766.

28. Robert L. Wilby and George L. W. Perry, "Climate change, biodiversity and the urban environment: a critical review based on London, UK," *Progress in Physical Geography* 30 (1) (2006): 73–98. See also Raymond J. C. Cannon, "The implications of predicted climate change for insect pests in the UK, with emphasis on non-indigenous species," *Global Change Biology* 4 (1998): 785–796; Sotiris Vardoulakis and Clare Heaviside, eds., *Health effects of climate change in the UK 2012: current evidence, recommendations and research gaps* (London: UK Health Protection Agency, 2012). There has been some debate about the possibility of a future return of malaria to London, but this would depend on a critical mass of infected people rather than changes in the distribution of mosquito species: the *Anopheles* vector is already present in brackish coastland habitats. In the 1660s, for example, the London diarist Samuel Pepys makes repeated reference to an affliction known as "ague," also referred to as "marsh fever,"

which we now know to be malaria (most likely *Plasmodium vivax*), a disease that was endemic through most of Europe at the time. The troubling "gnatts" to which Pepys refers are probably mosquitoes, which would have bred very extensively in the marshlands around London. See Richard Jones, *Mosquito* (London: Reaktion, 2012); K. R. Snow and J. M. Medlock, "The potential impact of climate change on the distribution and prevalence of mosquitoes in Britain," *European Mosquito Bulletin* 21 (2006): 1–10. The new mapping project for London's flora, being coordinated by Mark Spencer of the London Natural History Society, is revealing dramatic changes in plant distribution since the last survey completed by Rodney Burton in the early 1980s.

29. Caldeira cited in King, "Climate change science."

30. R. J. Dawson, J. W. Hall, P. D. Bates, and R. J. Nicholls, "Quantified analysis of the probability of flooding in the Thames estuary under imaginable worst-case sea-level rise scenarios," *Water Resources Development* 21 (4) (2005): 577–591; K. G. Lonsdale, T. E. Downing, R. J. Nicholls, D. Parker, A. T. Vafeidis, R. Dawson, and J. Hall, "Plausible responses to rapid sea-level rise in the Thames estuary," *Climate Change* 91 (2008): 145–169; Richard S. J. Tol, Maria Bohn, Thomas E. Downing, Marie-Laure Guilleminet, Eva Hizsnyik, Roger Kasperson, Kate Lonsdale, Claire Mays, Robert J. Nicholls, Alexander A. Olsthoorn, Gabriele Pfeifle, Marc Poumadere, Ferenc L. Toth, Athanasios T. Vafeidis, Peter E. van der Werff, and I. Hakan Yetkiner, "Adaptation to five metres of sea level rise," *Journal of Risk Research* 9 (5) (2006): 467–482. The new London plan also assumes a higher incidence of fluvial flooding from all tributaries of the Thames due to climate-change-induced changes in precipitation. Greater London Authority, *The London plan* (London: GLA, 2011).

31. See, for example, Sarah L. Bradley, Glenn A. Milne, Ian Shennan, and Robin Edwards, "An improved glacial isostatic adjustment model for the British Isles," *Journal of Quaternary Science* 26 (5) (2011): 541–552.

32. See, for example, Isabel Allen and Richard Waite, "Farrell in tube warning," *Architect's Journal* 223 (8) (2006): 11; Colin Blackstock, "Flood chaos on roads and rail," *The Guardian* (8 August 2002); Greater London Authority, *London regional flood risk appraisal* (October 2009).

33. See William Stephen Eyre, "The flood hydrology of urban catchments in Greater London," PhD thesis, Department of Geography, University College London, 1979; GLA London Assembly, Environment Committee, "Crazy paving: the environmental importance of London's front gardens" (September 2005). See also Andy Karvonen, *Politics of urban runoff: nature, technology, and the sustainable city* (Cambridge, MA: MIT Press, 2011).

34. Pilita Clark, "Flood strain grows on Thames Barrier," *Financial Times* (17 February 2014).

35. Lavery and Donovan, "Flood risk management in the Thames estuary looking ahead 100 years."

36. See Environment Agency, *Review and update of the 'Tidal Thames Strategy Review'* (Bristol: Environment Agency, 2003); Environment Agency, *Thames Gateway and flood risk management—a preliminary assessment* (Bristol: Environment Agency, 2003); Environment Agency, *Planning for flood risk management in the Thames estuary* (Bristol: Environment Agency, 2003); GLA London Assembly, Environment Committee, "London under threat? Flooding risk in the Thames Gateway" (October 2005). In the preface to the GLA report the Green Party member of the London Assembly and chair of the Environment Committee, Darren Johnson, notes: "The Committee investigated flooding risk with the awareness that it is too expensive for London to be protected by simply building bigger and more extensive defences, which has been the pattern since 1879," and that "perhaps more shocking was the fact that outside of the London area, the condition of 65 per cent of the defences are not known. If a defence fails due to pressure from flood water the repercussions could be catastrophic." It is also suggested that the development of a "green grid" could help absorb future water inundation. GLA, "London under threat?," 6, 11, and 13.

37. British Antarctic Survey and Ice2Sea Consortium, *From ice to high seas* (Cambridge: British Antarctic Survey, 2013). See also Dawson et al., "Quantified analysis of the probability of flooding in the Thames estuary"; Greater London Authority, *London regional flood risk appraisal* (London: GLA, October 2009); Environment Agency, *Thames estuary 2100* (London: Environment Agency, 2009).

38. See Lavery and Donovan, "Flood risk management in the Thames estuary looking ahead 100 years." The estimate used here is based on recent census data and the distribution of schools, hospitals, and other key functions within the flood risk zone.

39. Lonsdale et al., "Plausible responses to rapid sea-level rise in the Thames estuary." See also J. A. Lowe and J. M. Gregory, "The Effects of Climate Change on Storm Surges around the United Kingdom," *Philosophical Transactions of the Royal Society A* 363 (June 2005): 1313–1328.

40. UK Office for National Statistics, 2011Census for England and Wales, http://www.ons.gov.uk/ons/guide-method/census/2011/index.html (accessed 18 September 2012).

41. See, for example, Philip Cohen and Michael Rustin, eds., *London's turning: the making of the Thames Gateway: prospects and legacy* (Aldershot: Ashgate, 2008).

42. Zoë Blackler, "Mayor's Thames Gateway plans at risk from flood threats," *Architect's Journal* 216 (15) (2002): 10; GLA, "London under threat?"

43. http://www.environment_agency.gov.uk/static/documents/Leisure/Thames_Barrier_Project_pack_2012.pdf (accessed 29 November 2013); http://www.environment-agency.gov.uk/research/library/publications/41065.aspx (accessed 29 November 2013).

44. Compare, for example, Jonathan Leake, "Ten-mile barrier to stop London flood," *Sunday Times* (9 January 2005), with Paul Ruff and Glen Morley, "Freedom from the city," in Royal Institute of British Architects, *Living with water: visions of a flooded future* (London: RIBA, 2007), 17–21. On possible "hard engineering" responses see Tim Reeder, John Wicks, Luke Lovell, and Owen Tarrant, "Protecting London from tidal flooding: limits to engineering adaptation," in W. Neil Adger, Irene Lorenzoni, and Karen L. O'Brien, eds., *Adapting to climate change: thresholds, values, governance* (Cambridge: Cambridge University Press, 2009), 54–63.

45. The key policy document signaling this shift of emphasis is UK Department for Environment, Food and Rural Affairs, *Making space for water: taking forward a new government strategy for flood and coastal erosion risk management in England* (London: Defra, 2005).

46. Excerpt from speech made by the fictional London gangster Harold Shand (played by Bob Hoskins) in the film *The Long Good Friday* (1979), directed by John MacKenzie.

47. Jonathan Glancey cited in Edward Platt, "The cockney Siberia: the Thames Gateway development is the largest urban regeneration scheme ever attempted in Britain," *New Statesman* (5 March 2010).

48. Jonathan Glancey, "The Thames Gateway: here be monsters," *The Guardian* (29 October 2003).

49. Ben Campkin, *Remaking London: decline and regeneration in urban culture* (London: I. B. Taurus, 2013).

50. Charles Dickens, *Great expectations*, ed. Charlotte Mitchell (London: Penguin, 1996 [1860–1861]), 3–4.

51. Joseph Conrad, *Heart of darkness* (Harmondsworth: Penguin, 1973 [1902]), 1. The vast scale of the estuary is also evoked by Conrad in his autobiographical collection of essays *The mirror of the sea* (1906), where he describes its "strange air of mysteriousness." Joseph Conrad, *The mirror of the sea: memories and impressions* (London: J. M. Dent, 1946 [1906]), 101. In a more uplifting vein William Morris describes the "wide green sea of the Essex marshland, with the great domed line of the sky, and the

sun shining down in one flood of peaceful light over the long distance." William Morris, *News from nowhere*, ed. Krishan Kumar (Cambridge: Cambridge University Press, 1995 [1890]), 71.

52. See, for example, the photographs by Jason Orton to accompany the essay by David Price and Reg Ward (former CEO of the London Docklands Development Corporation) where they argue for a series of prestige developments along the "banks that blight the River Thames." David Price and Reg Ward, "Where is the organic open-ended thinking needed to 'imagineer' at such a scale," *Architect's Journal* 223 (8) (2006): 26. See also Jason Orton and Ken Worpole, *350 miles: an Essex journey* (Chelmsford: Essex Development and Regeneration Agency, 2005).

53. Price and Ward, "Where is the organic open-ended thinking."

54. "Lord Foster's plans for a new London airport," *Daily Telegraph* (3 December 2011).

55. Rob Sharp, "Insurers endorse 'homes on stilts'," *Architect's Journal* 221 (14) (2005): 14.

56. See James Porter and David Demeritt, "Flood-risk management, mapping, and planning: the institutional politics of decision support in England," *Environment and Planning A* 44 (2012): 2359–2378.

57. B. Willis, "Water, water, everywhere," *Regeneration and Renewal* 18 (November 2005): 14–15.

58. Laurens M. Bouwer, Ryan P. Crompton, Eberhard Faust, Peter Höppe, and Roger A. Pielke, "Confronting disaster losses," *Science* 318 (2 November 2007): 753; Association of British Insurers, *Strategic planning for flood risk in the growth areas* (London: ABI, 2004); Association of British Insurers, *Making communities sustainable: managing flood risks in the government's growth areas* (London: ABI, 2005); James Lewis and Ilan Kelman, "Housing, flooding and risk-ecology: Thames estuary South-Shoreland and North Kent," *Journal of Architectural and Planning Research* 26 (1) (2009): 14–29. On the complexities of predicting flood risk see also James D. Brown and Sarah L. Damery, "Managing flood risk in the UK: towards an integration of social and technical perspectives," *Transactions of the Institute of British Geographers* 27 (2002): 412–426; Stuart N. Lane, Catharina Landström, and Sarah J. Whatmore, "Imagining flood futures: risk assessment and management in practice," *Philosophical Transactions of the Royal Society A* 369 (2011): 1784–1806.

59. David Crichton, "What can cities do to increase resilience?," *Philosophical Transactions of the Royal Society A* 365 (2007): 2731–2739. The UK Environment Agency was also responsible for managing flood risk in Wales until 2013 (after which this power was devolved).

60. Porter and Demeritt, "Flood-risk management, mapping, and planning," 2363.

61. Ibid., 2359.

62. Ibid. On the methodological difficulties of flood risk modeling see also Catharina Landström, Sarah J. Whatmore, and Stuart N. Lane, "Virtual engineering: computer simulation modeling for flood risk management in England," *Science Studies* 24 (2) (2011): 3–22.

63. In some cases, sites of exceptionally high biodiversity have been created accidentally: the disposal of waste such as construction rubble or pulverized fuel ash, for example, has inadvertently produced unusual ecological sites in which rare plants or invertebrates thrive. See, for example, Annie Chipchase and M. Frith, *Brownfield? Greenfield? The threat to London's unofficial countryside* (London: London Wildlife Trust, 2002); M. Eyre, M. Luff, and J. Woodward, "Beetles (coleoptera) on brownfield sites in England: an important conservation resource?," *Journal of Insect Conservation* 7 (4) (2003): 223–231; John Vidal, "A bleak corner of Essex is being hailed as England's rainforest," *The Guardian* (3 May 2003). The postindustrial Thames has itself been a focus of renewed interest, as dozens of species of fish, including salmon, have returned for the first time since the nineteenth century. Leslie B Wood, *The restoration of the tidal Thames* (Bristol: Adam Hilger, 1982).

64. Ken Worpole, "East of Eden," *Journal of Landscape Architecture* 7 (2) (Autumn 2012): 20.

65. http://www.sssi.naturalengland.org.uk/citation/citation_photo/1001732.pdf (accessed 29 November 2013). The legal protection for 479 hectares of marshland on the eastern border of London was to prove controversial, however, with an attempt to develop the site for a leisure complex in the late 1980s. See Carolyn M. Harrison and Jacquelin Burgess, "Social constructions of nature: a case study of conflicts over the development of Rainham," *Transactions of the Institute of British Geographers* 19 (3) (1994): 291–310; Nature Conservancy Council, *Inner Thames grazing marshes* (Peterborough: JNCC, 1991).

66. Citation taken from the information sheet on Ramsar wetlands (RIS) compiled by the Joint Nature Conservation Committee on behalf of the Ramsar site, "Thames estuary and marshes," designated 31 March 2000 (updated 13 June 2008). The marshes also qualify as a Special Protection Area under the European Union's Birds Directive in recognition of the internationally important population concentrations for rare species such as the avocet, *Recurvirostra avosetta*, and hen harrier, *Circus cyaneus*, along with its high concentrations of waterfowl and migratory species.

67. Royal Institute of British Architects, *Living with water.*

68. See, for example, M. Margaret Bryant, "Urban landscape conservation and the role of ecological greenways at local and metropolitan scales," *Landscape and Urban Planning* 76 (1–4) (2006): 23–44.

69. Greater London Authority, *Green infrastructure and open environments: the All London Green Grid: London Plan 2011 supplementary planning guidance* (London: Greater London Authority, 2012).

70. See, for example, Lara Almarcegui, *Guide to the wastelands of the Lea Valley: 12 empty spaces await the London Olympics* (London: Barbican Art Gallery, 2009); Mark Atkins and Iain Sinclair, *Liquid city* (London: Reaktion, 1999); Francesco Manacorda and Ariella Yedgar, eds., *Radical nature: art and architecture for a changing planet 1969–2009* (London: Koenig Books, 2009). For the nature writer Richard Mabey, the "ragged Arcadia" to be encountered in London is far more interesting than the heavily manicured landscapes of farmland and the affluent commuter belt beyond the urban fringe. See Richard Mabey, *Weeds* (London: Profile, 2010), 3. On the scientific and design implications of "wild" urban nature see, for example, James Hitchmough and Nigel Dunnett, "Introduction to naturalistic planting in urban landscapes," in Nigel Dunnett and James Hitchmough, eds., *The dynamic landscape* (London: Spon, 2004), 1–22; Ingo Kowarik, "Wild urban woodlands: towards a conceptual framework," in Ingo Kowarik and Stephan Körner, eds., *Wild urban woodlands* (Berlin: Springer, 2005), 1–32; Norbert Kühn, "Intentions for the unintentional: spontaneous vegetation as the basis for innovative planting design in urban areas," *Journal of Landscape Architecture* (Autumn 2006): 46–53.

71. See John C. Calloway, V. Thomas Parker, Michael C. Vesey, and Lisa M. Schile, "Emerging issues for the restoration of tidal marsh ecosystems in the context of predicted climate change," *Madroño* 54 (3) (2007): 234–248; John Pethick, "Estuarine and tidal wetland restoration in the United Kingdom: policy versus practice," *Restoration Ecology* 10 (3) (2002): 431–437; John M. Teal and Susan Peterson, "Restoration benefits in a watershed context," *Journal of Coastal Research* 40 (2005): 132–140; Mineke Wolters, Angus Garbutt, and Jan P. Bakker, "Salt-marsh restoration: evaluating the success of de-embankments in north-west Europe," *Biological Conservation* 123 (2005): 249–268.

72. UK Department for Environment, Food and Rural Affairs, *Making space for water*.

73. See, for example, Royal Society for the Protection of Birds, "Making space for water: the RSPB's response," www.rspb.org.uk/Images/makingspaceforwater_tcm9-132948.pdf (accessed 3 December 2013).

74. In an international context see, for example, Anuradha Mathur and Dilip da Cunha, *Mississippi floods: designing a shifting landscape* (New Haven: Yale University Press, 2001).

75. Lawrence Beale Collins, "From first to second nature: a study of the river Ravensbourne in South East London," MSc thesis, University College London, 2010; Kevin Patrick, "The battle for the Quaggy," *Landscape Design* 274 (October 1998): 21–24.

76. See, for example, Andrew Brookes and F. Douglas Shields, eds., *River channel restoration: guiding principles for sustainable projects* (Chichester: John Wiley, 1996); John Cairns, "The status of the theoretical and applied science of restoration ecology," *Environmental Professional* 13 (3) (1991): 186–194; Nigel Dudley, *Authenticity in nature: making choices about the naturalness of ecosystems* (London: Earthscan, 2011); and Eric Higgs, *Nature by design: people, natural process, and ecological restoration* (Cambridge, MA: MIT Press, 2003).

77. Some of the most bitter recent conflicts in the UK between agricultural and environmental interests have been in relation to the protection of wetlands such as the Somerset Levels in the West of England. See Jeremy Purseglove, *Taming the flood: a history and natural history of rivers and wetlands* (Oxford: Oxford University Press, 1988). On the elimination of wetlands see also Rodney James Giblett, *Postmodern wetlands: culture, history, ecology* (Edinburgh: Edinburgh University Press, 1996).

78. See R. J. N. Devoy, "Flandrian sea level changes and vegetational history of the lower Thames estuary," *Philosophical Transactions of the Royal Society B* 285 (1979): 355–407; H. Godwin, *History of the British flora*, 2nd ed. (Cambridge: Cambridge University Press, 1975); F. H. Perring, "Changes in our native vascular plant flora," in D. L. Hawksworth, ed., *The changing flora and fauna of Britain* (London: Academic Press, 1974), 7–26.

79. Sophie Penelope Stephenson Seel, "Later prehistoric woodlands and wood use on the lower Thames floodplain," PhD thesis, Institute of Archaeology, University College London, 2001.

80. See Christopher Tilley, *A phenomenology of landscape: places, paths and monuments* (Oxford: Berg, 1994).

81. See Matthew Gandy, "Entropy by design: Gilles Clément, Parc Henri Matisse and the limits to avant-garde urbanism," *International Journal of Urban and Regional Research* 37 (1) (2013): 259–278.

82. G. H. Willcox, "Exotic plants on Roman waterlogged sites in London," *Journal of Archaeological Science* 4 (1977): 269–282.

83. Initial elements of an emerging postwar critique of modernity include Theodor W. Adorno and Max Horkheimer, *Dialectic of enlightenment*, trans. John Cumming (London: Verso, 1979 [1944]).

84. Morris, *News from nowhere*. The utopian vision for London offered by Morris has some similarities with the modern concept of the bioregion. See Martin Delveaux, "'O me! O me! How I love the earth': William Morris's News from nowhere and the birth of sustainable society," *Contemporary Justice Review* 8 (2) (2005): 131–146.

85. Morris, *News from nowhere*, 44.

86. Ibid.

87. Ibid.

88. The exchange is cited in Mark Hewitt and Susannah Hagan, eds., *City fights: debates on urban sustainability* (London: James and James, 2001), 21–28.

89. David Gissen, *Subnature: architecture's other environments* (New York: Princeton Architectural Press, 2009), 104.

90. Rory Olcayto, "Flooded London exhibition imagines the capital in 2090," *Building Design* (19 June 2008).

91. Marcus Fairs, "Flooded London by Squint/Opera," *Dezeen* (18 June 2008).

92. Ivan Yefremov, *Andromeda: a space-age tale*, trans. George Hanna (Moscow: Foreign Languages Publishing House, 1959).

93. Jameson, "Progress versus utopia," 292.

94. See, for example, Kristina Hill, "Climate-resilient urban waterfronts," in Jeroen Aerts, Wouter Botzen, Malcolm J. Bowman, Philip J. Ward, and Piet Dircke, eds., *Climate adaptation and flood risk in coastal cities* (London: Earthscan, 2012), 123–144; S. Pickett, M. Cadenasso, and J. Grove, "Resilient cities: meaning, models, and metaphor for integrating the ecological, socio-economic, and planning realms," *Landscape and Urban Planning* 69 (2004): 369–384; N. Walker et al., "Resilience, adaptability and transformability in social-ecological systems," *Ecology and Society* 9 (2) (2004); Chris Zevenbergen et al., *Urban flood management* (Leiden: CRC Press/Balkena, 2011); Chris Zevenbergen et al., "Challenges in urban flood management: travelling across spatial and temporal scales," *Journal of Flood Risk Management* 1 (2) (2008): 81–88.

95. The concept of "resilience" is very ambiguous: it can also mean resistance to change as well as disruption. The writer Peter Ackroyd, for example, presents London's history as one of resilience in the face of the recurring threat of violence, the city having repeatedly withstood the threat of social or political disorder. See Peter Ackroyd, *London: the biography* (London: Chatto and Windus, 2000).

96. See David Harvey, *A brief history of neo-liberalism* (Oxford: Oxford University Press, 2005).

97. See, for example, James Evans, "Resilience, ecology and adaptation in the experimental city," *Transactions of the Institute of British Geographers* 36 (2011): 223–237.

98. Porter, *London: a social history*, 389.

---

## Epilogue

1. The little egret was pointed out to me by the journalist and environmental activist Lawrence Beale Collins on a guided walk along the Ravensbourne on 3 January 2013. On London's lesser-known rivers see Nicholas J. Barton, *The lost rivers of London: a study of their effects upon London and Londoners and the effects of London and Londoners upon them* (London: Phoenix House, 1965 [1962]); Tom Bolton, *London's lost rivers: a walker's guide* (London: Strange Attractor, 2011); Paul Talling, *London's lost rivers* (London: Random House, 2011).

2. The redesign of Ladywell Fields was funded by a partnership between municipal government and the European Union's LIFE Environment Programme entitled QUERCUS (Quality Urban Environments for River Corridor Users and Stakeholders). The initial pretext for the QUERCUS project was to improve social well-being rather than concerns with flood control or urban biodiversity, but these aspects have gained significance through the involvement of groups such as Thames21 and the coordinating role of Lewisham Biodiversity Partnership. For further details see London Borough of Lewisham/Environment Agency, "Ravensbourne River Corridor Improvement Plan" (September 2010).

3. Jennifer Wolch, "Green urban worlds," *Annals of the Association of American Geographers* 97 (2) (2007): 373–384. On the ethical dimensions to urban nature see also Jennifer Wolch, "Zoöpolis," *Capitalism, Nature, Socialism* 7 (2) (1996): 21–47.

4. http://www.indesignlive.sg/articles/BishanAng-Mo-Kio-Park-by-Atelier-Dreiseitl-Asia (accessed 30 November 2013). See, for example, Herbert Dreiseitl and Dieter Grau, *Recent waterscapes: planning, building and designing with water* (Basel: Birkhäuser, 2009).

5. Vijay Prashad, "The technology of sanitation in colonial Delhi," *Modern Asian Studies* 35 (1) (2001): 120.

6. On comparative aspects of water culture see, for example, Veronica Strang, "Common senses: water, sensory experience and the generation of meaning," *Journal of Material Culture* 10 (1) (2005): 92–120; and James L. Wescoat Jr., "The 'right of thirst' for animals in Islamic law: a comparative approach," *Environment and Planning D: Society and Space* 13 (1995): 637–654.

7. Bruno Rudolf and Joerg Rapp, "The century flood of the River Elbe in August 2002: synoptic weather development and climatological aspects," *Quarterly Report of the German NWP-System of the Deutscher Wetterdienst* 2 (1) (2003): 8–23.

8. See, for example, Reimund Schwarze and Gert G. Wagner, "The political economy of natural disaster insurance: lessons from the failure of a proposed compulsory insurance scheme in Germany," discussion paper, Technische Universität Berlin, Fakultät Wirtschaft und Management, 2006.

9. See W. J. Wouter Botzen and Jeroen C. J. M. van den Berg, "Insurance against climate change and flooding in the Netherlands: present, future, and comparison with other countries," *Risk Analysis* 28 (2008): 413–426; John O'Neil and Martin O'Neil, *Social justice and the future of flood insurance* (York: Joseph Rowntree Foundation, 2012).

10. Kevin Grove, "Pre-empting the next disaster: catastrophe insurance and the financialization of disaster management," *Security Dialogue* 43(2) (2013): 139–155.

11. Ulrich Beck, *Risikogesellschaft. Auf dem Weg in eine andere Moderne* (Frankfurt: Suhrkamp, 1986), 63.

12. See Roberto Esposito, *Bíos: biopolitics and philosophy*, trans. Timothy Campbell (Minneapolis: University of Minnesota Press, 2008). On the biopolitics of environmental risk see also Austin Zeiderman, "Living dangerously: biopolitics and urban citizenship in Bogotá, Colombia," *American Ethnologist* 40 (1) (2013): 71–87.

13. See Mick Hulme, "Abrupt climate change: can society cope?," *Philosophical Transactions A* 36 (2003): 2001–2021.

14. Walter Gropius and Martin Wagner, "The new city pattern for the people and by the people," *Andean Quarterly* (Winter 1943): 4–12.

15. See, for example, Patrick Carroll, *Science, culture, and modern state formation* (Berkeley: University of California Press, 2006).

16. Arjun Appadurai, interview with the author, in the documentary *Liquid city: water, landscape and social formation in 21st century Mumbai* (UK/India, 2007; director Matthew Gandy; available from the UCL Urban Laboratory).

17. Nikhil Anand, "Pressure: the polytechnics of water supply in Mumbai," *Cultural Anthropology* 26 (4) (2011): 542–564. See also Nikhil Anand, "Municipal disconnect: on abject water and its urban infrastructures," *Ethnography* 13 (4) (2012): 487–509; Michelle Kooy and Karen Bakker, "Technologies of government: constituting subjectivities, spaces, and infrastructures in colonial and contemporary Jakarta," *International Journal of Urban and Regional Research* 32(2) (2008): 375–391; Katharine Meehan, "Disciplining de facto development: water theft and hydrosocial order in Tijuana,"

*Environment and Planning D: Society and Space* 31 (2) (2013): 319–336; Dennis Rodgers, "Haussmannization in the tropics: abject urbanism and infrastructural violence in Nicaragua," *Ethnography* 13 (4) (2012): 413–438; and Antina Von Schnitzler, "Traveling technologies: infrastructure, ethical regimes and the materiality of politics in South Africa," *Cultural Anthropology* 28 (4) (2013): 670–693.

18. Lisa Björkman, "Becoming a slum: from municipal colony to illegal settlement in liberalization-era Mumbai," *International Journal of Urban and Regional Research* 38 (1) (2014): 36–59.

19. See Timothy Mitchell, *Carbon democracy: political power in the age of oil* (London: Verso, 2013 [2011]).

20. See, for example, Erik Swyngedouw, "Apocalypse forever? Post-political populism and the spectre of climate change," *Theory, Culture and Society* 27 (2–3) (2009): 213–232. See also Mustafa Dikeç, *Badlands of the republic: space, politics and urban policy* (Oxford: Blackwell, 2007).

21. Examples of systems-based terminology can be found in, for example, J. Arjan Wardekker, Arie de Jong, Joost M. Knoop, and Jeroen P. van der Sluijs, "Operationalising a resilience approach to adapting an urban delta to uncertain climate changes," *Technological Forecasting and Social Change* 77 (2010): 987–998; and Chris Zevenbergen, Adrian Cashman, Niki Evelpidou, and Erik Pasche, *Urban flood management* (Leiden: CRC Press/Balkena, 2011). See also Marina Alberti and John H. Marzluff, "Ecological resilience in urban ecosystems: linking urban patterns to human and ecological functions," *Urban Ecosystems* 7 (2004): 241–265. For a range of more nuanced or critically engaged perspectives on urban resilience see Lawrence J. Vale and Thomas J. Campanella, eds., *The resilient city: how modern cities recover from disaster* (Oxford: Oxford University Press, 2005).

22. Antoine Picon, "Anxious landscapes: from the ruin to rust," *Grey Room* 1 (2000): 64–83.

23. James Ferguson, "Decomposing modernity: history and hierarchy after development," in *Global shadows: Africa in the neoliberal world order* (Durham: Duke University Press, 2006), 176–193. See also Gurminder K. Bhambra, *Rethinking modernity: postcolonialism and the sociological imagination* (Basingstoke: Palgrave Macmillan, 2007); Dilip Parameshwar Gaonkar, "On alternative modernities," *Public Culture* 11 (1) (1999): 1–18; Dipesh Chakrabarty, *Provincializing Europe: postcolonial thought and historical difference* (Princeton: Princeton University Press, 2000).

24. David Harvey, *The limits to capital* (Oxford: Blackwell, 1982), 226. See also Fredric Jameson, "The brick and the balloon," *New Left Review* 228 (1998): 25–46.

## Selected Bibliography

Abani, Chris. *Graceland*. New York: Farrar, Straus and Giroux, 2004.

Abriani, Alberto. "Dal sifone alla città / From the siphon to the city." *Casabella* 542/543 (1988): 24–29, 117.

Abu-Lughod, Janet. *Rabat: urban apartheid in Morocco*. Princeton: Princeton University Press, 1980.

Achebe, Chinua. *No longer at ease*. London: Heinemann, 1960.

Adorno, Theodor W., and Max Horkheimer. *Dialectic of enlightenment*. Trans. John Cumming. London: Verso, 1979 [1944].

Aerts, Jeroen, Wouter Botzen, Malcolm J. Bowman, Philip J. Ward, and Piet Dircke, eds. *Climate adaptation and flood risk in coastal cities*. London: Earthscan, 2012.

Agde, Günter, and Karin Herbst-Meßlinger, eds. *Die rote Traumfabrik. Meschrabpom-Film und Prometheus (1921–1936)*. Berlin: Bertz and Fischer, 2012.

Ahiamba, J. E., K. O. Dimuna, and G. R. A. Okogun. "Built environment decay and urban health in Nigeria." *Journal of Human Ecology* (Delhi, India) 23 (3) (2008): 259–265.

Alberti, Marina, and John H. Marzluff. "Ecological resilience in urban ecosystems: linking urban patterns to human and ecological functions." *Urban Ecosystems* 7 (2004): 241–265.

Anand, Nikhil. "Municipal disconnect: on abject water and its urban infrastructures." *Ethnography* 13 (4) (2012): 487–509.

Anand, Nikhil. "Pressure: the polytechnics of water supply in Mumbai." *Cultural Anthropology* 26 (4) (2011): 542–564.

Anderson, Bonnie S., and Judith P. Zinsser. *A history of their own: women in Europe from prehistory to the present.* New York: Harper and Row, 1988.

Anderson, Perry. *The origins of postmodernity.* London: Verso, 1998.

Anderson, Susan C., and Bruce H. Tabb, eds. *Water, leisure and culture: European historical perspectives.* Oxford: Berg, 2003.

Antonelli, Carlo F. "Acque sporche: Londra e il 'Metropolitan Board of Works', 1855–1865." *Storia Urbana* 16 (61) (1992): 61–81.

Appadurai, Arjun. "Deep democracy: urban governmentality and the horizon of politics." *Public Culture* 14 (1) (2002): 21–47.

Appadurai, Arjun. "Spectral housing and urban cleansing: notes on millennial Mumbai." *Public Culture* 12 (3) (2000): 627–651.

Apter, Andrew. *The pan-African nation: oil and the spectacle of culture in Nigeria.* Chicago: University of Chicago Press, 2005.

Arabindoo, Pushpa. "Mobilising for water: hydro-politics of rainwater harvesting in Chennai." *International Journal of Urban Sustainable Development* 3 (1) (2011): 106–126.

Avila, Eric. "The folklore of the freeway: space, culture, and identity in postwar Los Angeles." *Aztlán: A Journal of Chicano Studies* 23 (1) (1998): 15–31.

Bachelard, Gaston. *L'eau et les rêves: essai sur l'imagination de la matière.* Paris: José Corti, 1947 [1942].

Bacon, Edmund N. *Design of cities.* London: Thames and Hudson, 1967.

Baker, Pauline H. *Urbanization and political change: the politics of Lagos 1917–1967.* Berkeley: University of California Press, 1974.

Bakker, Karen. "Archipelagos and networks: urbanisation and water privatisation in the South." *Geographical Journal* 169 (4) (2003): 328–341.

Bakker, Karen. "Neoliberal versus postneoliberal water: geographies of privatization and resistance." *Annals of the Association of American Geographers* 103 (2) (2013): 253–260.

Bakker, Karen. *Privatizing water: governance failure and the world's urban water crisis.* Ithaca: Cornell University Press, 2010.

Ballard, J. G. *The drowned world.* London: Gollancz, 1999 [1962].

Banerjee, Soumyajit, Gautam Aditya, and Goutam K. Saha. "Household disposables as breeding habitats of dengue vectors: linking wastes and public health." *Waste Management* (New York) 33 (2013): 233–239.

Banham, Reyner. *Los Angeles: the architecture of four ecologies*. London: Penguin Allen Lane, 1971.

Banham, Reyner. "Tennessee Valley Authority: l'ingegneria dell'utopia / Tennesse Valley Authority: the engineering of utopia." *Casabella* 542/543 (1988): 74–81, 125.

Banham, Reyner. *Theory and design in the first machine age*. Oxford: Butterworth-Heinemann, 1960.

Barles, Sabine, *La ville délétère. Médecins et ingénieurs dans l'espace urbain XVIIIe-XIXe siècle*. Seyssel: Champ Vallon, 1999.

Barry, Andrew. "Technological zones." *European Journal of Social Theory* 9 (2) (2006): 239–253.

Barry, Andrew, and Georgina Born, eds. *Interdisciplinarity: reconfigurations of the social and natural sciences*. London: Routledge, 2013.

Bärthel, Hilmar. *Wasser für Berlin*. Berlin: Verlag für Bauwesn, 1997.

Barton, Nicholas J. *The lost rivers of London: a study of their effects upon London and Londoners and the effects of London and Londoners upon them*. London: Phoenix House, 1965 [1962].

Baviskar, Amita. "Between micro-politics and administrative imperatives: decentralization and the watershed mission in Madhya Pradesh, India." *European Journal of Development Research* 16 (1) (2004): 26–40.

Baviskar, Amita. "Between violence and desire: space, power, and identity in the making of metropolitan Delhi." *International Social Science Journal* 55 (175) (2003): 89–98.

Baviskar, Amita. *In the belly of the river: tribal conflicts over development in the Narmada Valley*. 2nd ed. Delhi: Oxford University Press, 2005.

Baviskar, Amita. "The politics of the city." *Seminar* 516 (2002). http://www.india-seminar.com.

Bayly, Susan. *Caste, society and politics in India from the eighteenth century to the modern age*. Cambridge: Cambridge University Press, 1999.

Beaumont-Maillet, Laure. *L'eau à Paris*. Paris: Hazan, 1991.

Beck, Ulrich. *Risikogesellschaft. Auf dem Weg in eine andere Moderne*. Frankfurt: Suhrkamp, 1986.

Bélanger, Pierre. "Landscape as infrastructure." *Landscape Journal* 28 (1) (2009): 79–95.

Benevolo, Leonardo. *The European city*. Trans. Carl Ipsen. Oxford: Blackwell, 1993.

Benjamin, Walter. *Einbahnstrasse*. Berlin: Rowohlt, 1928.

Benjamin, Walter. "Paris, the capital of the nineteenth century." Trans. Howard Eiland. In Walter Benjamin, *The writer of modern life: essays on Charles Baudelaire*, ed. Michael W. Jennings, 30–45. Cambridge, MA: Belknap Press of Harvard University Press, 2006 [1935].

Bennett, Jane. "The agency of assemblages and the North American blackout." *Public Culture* 17 (3) (2005): 445–465.

Benson, Timothy O., ed. *Expressionist utopias: paradise, metropolis, architectural fantasy*. Berkeley: University of California Press, 2001.

Benvenuto, Edoardo. "Architetti, ingegneri e matematici / Architects, engineers and mathematicians." *Casabella* 542/543 (1988): 16–19, 115–116.

Berman, Marshall. *All that is solid melts into air: the experience of modernity*. London: Verso, 1982.

Bernet, Claus. "The 'Hobrecht Plan' (1862) and Berlin's urban structure." *Urban History* 31 (3) (2004): 400–419.

Bernheimer, Charles. "Of whores and sewers: Parent-Duchâtelet, engineer of abjection." *Raritan Quarterly Review* 6 (3) (1987): 72–90.

Beveridge, Ross. *A politics of inevitability: the privatisation of the Berlin Water Company, the global city discourse, and governance in 1990s Berlin*. Wiesbaden: VS Research, 2012.

Beveridge, Ross, and Matthias Naumann. "The Berlin Water Company: from 'inevitable' privatization to 'impossible' remunicipalization." In Matthias Bernt, Britta Grell, and Andrej Holm, eds., *The Berlin Reader*. Bielefeld: transcript, 2013.

Bhambra, Gurminder K. *Rethinking modernity: postcolonialism and the sociological imagination*. Basingstoke: Palgrave Macmillan, 2007.

Bijker, Wiebe E. "Dikes and dams, thick with politics." *Isis* 98 (1) (2007): 109–123.

Bijker, Wiebe E. "The Oosterschelde storm surge barrier: a test case for Dutch water technology, management, and politics." *Technology and Culture* 43 (3) (2002): 569–584.

Biro, Matthew. *The dada cyborg: visions of the new human in Weimar Berlin*. Minneapolis: University of Minnesota Press, 2009.

Biro, Matthew. "From analogue to digital photography: Bernd and Hilla Becher and Andreas Gursky." *History of Photography* 36 (3) (2012): 353–366.

Björkman, Lisa. "Becoming a slum: from municipal colony to illegal settlement in liberalization-era Mumbai." *International Journal of Urban and Regional Research* 38 (1) (2014): 36–59.

Blackbourn, David. *The conquest of nature: water, landscape and the making of modern Germany*. London: Jonathan Cape, 2006.

Blau, Eve. *The architecture of Red Vienna, 1919–1934*. Cambridge, MA: MIT Press, 1999.

Bloch, Ernst. *The principle of hope*. Trans. Neville Plaice, Stephen Plaice, and Paul Knight. Cambridge, MA: MIT Press, 1986 [1954].

Bocquet, Denis, Konstantinos Chatzis, and Agnès Sander. "From free good to commodity: universalizing the provision of water in Paris (1830–1930)." *Geoforum* 39 (2008): 1821–1832.

Böhme, Hartmut. *Kulturgeschichte des Wassers*. Frankfurt am Main: Suhrkamp, 1988.

Bottles, Scott. *Los Angeles and the automobile: the making of the modern city*. Berkeley: University of California Press, 1987.

Botzen, W. J. Wouter, and Jeroen C. J. M. van den Berg. "Insurance against climate change and flooding in the Netherlands: present, future, and comparison with other countries." *Risk Analysis* 28 (2008): 413–426.

Boudreau, Julie-Anne, and Roger Keil. "Seceding from responsibility? Secession movements in Los Angeles." *Urban Studies* 38 (10) (2001): 1701–1731.

Boudriot, Pierre-Denis. "Les égouts de Paris aux XVIIe et XVIIIe siècles: les humeurs de la ville préindustrielle." *Histoire, Economie et Société* 9 (2) (1990): 197–211.

Boyer, M. Christine. *Dreaming the rational city: the myth of American city planning*. Cambridge, MA: MIT Press, 1983.

Bradley, Sarah L., Glenn A. Milne, Ian Shennan, and Robin Edwards. "An improved glacial isostatic adjustment model for the British Isles." *Journal of Quaternary Science* 26 (5) (2011): 541–552.

Brash, Alexander, Jamie Hand, and Kate Orff, eds. *Gateway: visions for an urban national park*. New York: Princeton Architectural Press, 2011.

Breman, Jan. *Footloose labour: working conditions in India's informal sector*. Cambridge: Cambridge University Press, 1996.

Brenner, Neil, ed. *Implosions/explosions: towards a theory of planetary urbanization*. Berlin: jovis, 2014.

Bresnick, Adam. "Prosopoetic compulsion: reading the uncanny in Freud and Hoffmann." *Germanic Review* 71 (1996): 114–132.

Broich, John. *London: water and the making of the modern city*. Pittsburgh: University of Pittsburgh Press, 2013.

Brookes, Andrew, and F. Douglas Shields, eds. *River channel restoration: guiding principles for sustainable projects*. Chichester: John Wiley, 1996.

Brown, James D., and Sarah L. Damery. "Managing flood risk in the UK: towards an integration of social and technical perspectives." *Transactions of the Institute of British Geographers* 27 (2002): 412–426.

Bryant, M. Margaret. "Urban landscape conservation and the role of ecological greenways at local and metropolitan scales." *Landscape and Urban Planning* 76 (1–4) (2006): 23–44.

Budds, Jessica. "Water, power, and the production of neoliberalism in Chile, 1973–2005." *Environment and Planning D: Society and Space* 31 (2) (2013): 301–318.

Burchell, Graham, Colin Gordon, and Peter Miller, eds. *The Foucault effect: studies in governmentality*. Chicago: University of Chicago Press, 1991.

Busch, Bernd, and Larissa Förster, eds. *Wasser*. Cologne: Wienand, 2000.

Butler, Judith. *Gender trouble: feminism and the subversion of identity*. London: Routledge, 1990.

Cairns, John. "The status of the theoretical and applied science of restoration ecology." *Environment and Progress* 13 (3) (1991): 186–194.

Callaghan, Helen, and Martin Höpner. "Changing ideas: organized capitalism and the German Left." *West European Politics* 35 (3) (2012): 551–573.

Campkin, Ben. *Remaking London: decline and regeneration in urban culture*. London: I. B. Taurus, 2013.

Caprotti, Federico. "Malaria and technological networks: medical geography in the Pontine Marshes, Italy, in the 1930s." *Geographical Journal* 172 (2) (2006): 145–155.

Caprotti, Federico, and Joanna Romanowicz. "Thermal eco-cities: green building and thermal urban metabolism." *International Journal of Urban and Regional Research* 37 (6) (2013): 1949–1967.

Carroll, Patrick. *Science, culture, and modern state formation*. Berkeley: University of California Press, 2006.

Castle, Terry. *The female thermometer: eighteenth-century culture and the invention of the uncanny*. New York: Oxford University Press, 1995.

Chadwick, George F. *The park and the town: public landscape in the nineteenth and twentieth centuries*. London: Architectural Press, 1966.

Chakrabarty, Dipesh. *Habitations of modernity: essays in the wake of subaltern studies*. Chicago: University of Chicago Press, 2002.

Chakrabarty, Dipesh. *Provincializing Europe: postcolonial thought and historical difference*. Princeton: Princeton University Press, 2000.

Chakrabarty, Dipesh. *Rethinking working-class history: Bengal 1890–1940*. Delhi: Oxford University Press, 1989.

Chandavarkar, Rajnarayan. *The origins of industrial capitalism in India: business strategies and the working class in Bombay 1900–1940*. Cambridge: Cambridge University Press, 1994.

Chandrasekhar, Indira, and Peter C. Seel, eds. *body.city: siting contemporary culture in India*. Berlin: Haus der Kulturen der Welt; Delhi: Tulika Books, 2003.

Chapman, J. M., and Brian Chapman. *The life and times of Baron Haussmann: Paris in the Second Empire*. 1957. London: Weidenfeld and Nicholson.

Chatterjee, Partha. *The politics of the governed*. New York: Columbia University Press, 2004.

Chattopadhyay, Swati. *Unlearning the city: infrastructure in a new optical field*. Minneapolis: University of Minnesota Press, 2012.

Chatzis, Konstantinos, and Olivier Coutard. "Water and gas: early developments in the utility networks of Paris." *Journal of Urban Technology* 12 (3) (2005): 1–17.

Chevalier, Louis. *Laboring classes and dangerous classes in Paris during the first half of the nineteenth century*. Trans. Frank Jellinck. Princeton: Princeton University Press, 1973 [1958].

Chevallier, Fabienne. *Le Paris moderne: histoire des politiques d'hygiène (1855–1898)*. Rennes: Presses Universitaires de Rennes, 2010.

Cho, Myung-Rae. "The politics of urban nature restoration: the case of Cheonggyecheon restoration in Seoul, Korea." *International Development Planning Review* 32 (2) (2010): 146–165.

Clark, Peter. *European cities and towns: 400–2000*. Oxford: Oxford University Press, 2009.

Clark, T. J. *The painting of modern life: Paris in the art of Manet and his followers*. London: Thames and Hudson, 1985.

Classen, Constance, David Howes, and Anthony Synnott. *Aroma: the cultural history of smell*. London: Routledge, 1994.

Cobban, James. "Public housing in colonial Indonesia 1900–1940." *Modern Asian Studies* 27 (4) (1993): 871–896.

Cohen, Philip, and Michael Rustin, eds. *London's turning: the making of the Thames Gateway: prospects and legacy*. Aldershot: Ashgate, 2008.

Coleman, William. *Death is a social disease: public health and political economy in early industrial France*. Madison: University of Wisconsin Press, 1982.

Collier, Peter, and Robert Lethbridge, eds. *Artistic relations: literature and the visual arts in nineteenth-century France*. New Haven: Yale University Press, 1994.

Conrad, Joseph. *Heart of darkness*. Harmondsworth: Penguin, 1973 [1902].

Conrad, Joseph. *The mirror of the sea: memories and impressions*. London: J. M. Dent, 1946 [1906].

Conrad, Sebastian. *German colonialism: a short history*. Trans. Sorcha O'Hagan. Cambridge: Cambridge University Press.

Corbin, Alain. *The foul and the fragrant: odor and the French social imagination*. Trans. Miriam L. Kochan, Roy Porter, and Christopher Prendergast. Cambridge, MA: Harvard University Press, 1988 [1986]

Corbin, Alain. *Time, desire and horror: towards a history of the senses*. Trans. Jean Birrell. Cambridge: Polity, 1995 [1991].

Corbin, Alain. *Women for hire: prostitution and sexuality in France after 1850*. Trans. Alan Sheridan. Cambridge, MA: Harvard University Press, 1990 [1978].

Corbridge, Stuart, Glyn Williams, Manoj Srivastava, and René Véron, eds. *Seeing the state: governance and governmentality in India*. Cambridge: Cambridge University Press, 2005.

Cosgrove, Denis, and Stephen Daniels, eds. *The iconography of landscape: essays on the symbolic representation, design and use of past environments*. Cambridge: Cambridge University Press, 1988.

Crichton, David. "What can cities do to increase resilience?" *Philosophical Transactions of the Royal Society A* 365 (2007): 2731–2739.

Cuff, Dana. "Fugitive plans in the provisional city: slums and public housing in Los Angeles." In *Looking for Los Angeles: architecture, film, photography, and the urban landscape*, ed. Charles G. Salas and Michael S. Roth, 97–131. Los Angeles: Getty Research Institute, 2001.

Daunton, Martin, and Bernhard Rieger, eds. *Meanings of modernity: Britain from the late-Victorian era to World War II*. Oxford: Berg, 2001.

Davis, Jennifer. "Corruption in public service delivery: experience from South Asia's water and sanitation sector." *World Development* 32 (1) (2004): 53–71.

Davis, Mike. "Cannibal city: Los Angeles and the destruction of nature." In *Urban revisions: current projects for the public realm*, ed. Russell Ferguson, 39–57. Cambridge, MA: MIT Press, 1994.

Davis, Mike. *Ecology of fear: Los Angeles and the imagination of disaster*. New York: Metropolitan Books, 1998.

Dawson, Richard J., J. W. Hall, P. D. Bates, and Robert J. Nicholls. "Quantified analysis of the probability of flooding in the Thames estuary under imaginable worst-case sea-level rise scenarios." *Water Resources Development* 21 (4) (2005): 577–591.

Dear, Michael. *The postmodern urban condition*. Oxford: Blackwell, 2000.

Dear, Michael, and Stephen Flusty. "Postmodern urbanism." *Annals of the Association of American Geographers* 88 (1998): 50–73.

de Bonneville, Françoise. *Le livre du bain*. Paris: Flammarion, 1997.

Delaporte, François. *Disease and civilization: the cholera in Paris, 1832*. Trans. Arthur Goldhammer. Cambridge, MA: MIT Press, 1986.

de Lisle, Philippe Cebron. "L'eau a Paris au dix-neuvieme siècle." PhD dissertation, Université de Lille III, 1991.

Delveaux, Martin. "'O me! O me! How I love the earth': William Morris's News from nowhere and the birth of sustainable society." *Contemporary Justice Review* 8 (2) (2005): 131–146.

Dennis, Richard. *Cities in modernity*. Cambridge: Cambridge University Press, 2008.

Desfor, Gene, and Roger Keil. "Every river tells a story: the Don River (Toronto) and the Los Angeles River (Los Angeles) as articulating landscapes." *Journal of Environmental Policy and Planning* 2 (1) (2000): 5–23.

d'Eugny, Anne. *Au temps de Baudelaire, Guys et Nadar*. Paris: Éditions du Chêne, 1945.

Deutsche, Rosalyn. *Evictions: art and spatial politics*. Cambridge, MA: MIT Press, 1996.

Deverell, William. *Whitewashed adobe: the rise of Los Angeles and the remaking of its Mexican past*. Berkeley: University of California Press, 2004.

Deverell, William, and Greg Hise, eds. *Land of sunshine: an environmental history of metropolitan Los Angeles*. Pittsburgh: University of Pittsburgh Press, 2005.

Devoy, R. J. N. "Flandrian sea level changes and vegetational history of the lower Thames estuary." *Philosophical Transactions of the Royal Society B* 285 (1979): 355–407.

De Wit, Wim, and Christopher James Alexander, eds. *Overdrive: L.A. constructs the future, 1940–1990*. Los Angeles: Getty Research Institute, 2013.

Dickens, Charles. *Great expectations*. Ed. Charlotte Mitchell. London: Penguin, 1996 [1860–1861].

Dikeç, Mustafa. *Badlands of the republic: space, politics and urban policy*. Oxford: Blackwell, 2007.

Dimendberg, Edward. "The kinetic icon: Reyner Banham on Los Angeles as mobile metropolis." *Urban History* 33 (1) (2006): 106–125.

Dimendberg, Edward. "The will to motorization: cinema, highways, and modernity." *October* 73 (1995): 90–137.

Dirks, Nicholas B. *Castes of mind: colonialism and the making of modern India*. Princeton: Princeton University Press, 2001.

Dobraszczyk, Paul. *Into the belly of the beast: exploring London's Victorian sewers*. Reading, UK: Spire, 2009.

Dossal, Mariam. "Henry Conybeare and the politics of centralised water supply in mid-nineteenth century Bombay." *Indian Economic and Social History Review* 25 (1988): 79–96.

Dossal, Mariam. *Imperial designs and Indian realities: the planning of Bombay City, 1845–1875*. Bombay: Oxford University Press, 1991.

Douglas, Mary. *Purity and danger: an analysis of the concepts of pollution and taboo*. London: Routledge and Kegan Paul, 1978.

Dreiseitl, Herbert, and Dieter Grau. *Recent waterscapes: planning, building and designing with water*. Basel: Birkhäuser, 2009.

Dudley, Nigel. *Authenticity in nature: making choices about the naturalness of ecosystems*. London: Earthscan, 2011.

Duncan, James. "The superorganic in American cultural geography." *Annals of the Association of American Geographers* 70 (1980): 181–198.

Dunnett, Nigel, and James Hitchmough, eds. *The dynamic landscape*. London: Spon, 2004.

Echenberg, Myron J. *Black death, white medicine: bubonic plague and the politics of public health in colonial Senegal, 1914–1945*. Portsmouth, NH: Heinemann, 2002.

Eder, Ernst Gerhard. "Sonnenanbeter und Wasserratten: Körperkultur und Freiluftbadebewegung in Wiens Donaulandschaft." *Archiv für Sozialgeschichte* 33 (1993): 245–274.

Edholm, Felicity. "The view from below: Paris in the 1880s." In *Landscape: politics and perspectives*, ed. Barbara Bender, 139–169. Oxford: Berg, 1993.

Edwards, Paul N. "Infrastructure and modernity: force, time, and social organization in the history of sociotechnical systems." In *Modernity and technology*, ed. Thomas J. Misa, Philip Brey, and Andrew Feenberg, 185–225. Cambridge, MA: MIT Press, 2003.

Eisner, Lotte. *The haunted screen: expressionism in the German cinema and the influence of Max Reinhardt*. Trans. Roger Greaves. Berkeley: University of California Press, 1969 [1965/1952].

Ekwensi, Cyprian. *Jagua Nana*. London: Hutchinson, 1961.

Ekwensi, Cyprian. *People of the city*. London: Heinemann, 1963 [1954].

Elger, Dietmar. *Expressionism: a revolution in German art*. Cologne: Benedikte Taschen, 1989.

Elias, Norbert. *The civilizing process*. Vol. 1, *The history of manners*. Trans. Edmund Jephcott. Oxford: Basil Blackwell, 1978 [1939].

Elimelech, Menachem, and William A. Phillip. "The future of seawater desalination: energy, technology, and the environment." *Science* 333 (2011): 712–717.

Enwezor, Okwui, Carlos Basualdo, Ute Meta Bauer, Susanne Ghez, Sarat Maharaj, Mark Nash, and Octavio Zaya, eds. *Créolité and creolization. Documenta11_Platform3*. Ostfildern-Ruit: Hatje Cantz, 2003.

Enwezor, Okwui, Carlos Basualdo, Ute Meta Bauer, Susanne Ghez, Sarat Maharaj, Mark Nash, and Octavio Zaya, eds. *Under siege: four African cities. Freetown, Johannesburg, Kinshasa, Lagos. Documenta11 Platform4*. Ostfildern-Ruit: Hatje Cantz, 2003.

Esposito, Roberto. *Bíos: biopolitics and philosophy*. Trans. Timothy Campbell. Minneapolis: University of Minnesota Press, 2008.

Evans, James. "Resilience, ecology and adaptation in the experimental city." *Transactions of the Institute of British Geographers* 36 (2011): 223–237.

Evans, Richard J. *Death in Hamburg: society and politics in the cholera years, 1830–1910*. Oxford: Clarendon Press, 1987.

Feenberg, Andrew. *Between reason and experience: essays in technology and modernity*. Cambridge, MA: MIT Press, 2010.

Ferguson, James. *Global shadows: Africa in the neoliberal world order*. Durham: Duke University Press, 2006.

Ferguson, James, and Akhil Gupta. "Spatializing states: towards an ethnography of neoliberal governmentality." *American Ethnologist* 29 (4) (2002): 981–1002.

Ferrari, Laura. *L'acqua nel paesaggio urbano: letture esplorazioni ricerche scenari*. Florence: Florence University Press, 2005.

Florman, Samuel C. *The existential pleasures of engineering.* New York: St. Martin's Press, 1976.

Flusty, Stephen. "Thrashing downtown: play as resistance to the spatial and representational regulation of Los Angeles." *Cities* 17 (2000): 149–158.

Fogelson, Robert M. *The fragmented metropolis: Los Angeles 1850–1930.* Berkeley: University of California Press, 1993 [1967].

Fontane, Theodor. *Irrungen, Wirrungen.* Stuttgart: Reclam, 1994 [1891].

Fontane, Theodor. *Wanderungen durch die Mark Brandenburg.* Wiesbaden: Emil Vollmer, 1980 [1880].

Forty, Adrian. *Concrete and culture: a material history.* London: Reaktion, 2012.

Foster, Hal. *Prosthetic gods.* Cambridge, MA: MIT Press, 2004.

Fourchard, Laurent. "Lagos, Koolhaas and partisan politics in Nigeria." *International Journal of Urban and Regional Research* 35 (1) (2011): 40–56.

Frank, Susanne. *Stadtplanung im Geschlechterkampf: Stadt und Geschlecht in der Großstadtentwicklung des 19. und 20. Jahrhunderts.* Opladen: Leske und Budrich, 2003.

Frascina, Francis, Nigel Blake, Briony Fer, Tamar Garb, and Charles Harrison. *Modernity and modernism: French painting in the nineteenth century.* New Haven: Yale University Press, 1993.

Frenkel, Stephen, and John Western. "Pretext or prophylaxis? Racial segregation and malarial mosquitoes in a British tropical colony: Sierra Leone." *Annals of the Association of American Geographers* 78 (2) (1988): 211–228.

Freud, Sigmund. "The 'uncanny.'" In *The standard edition of the complete psychological works of Sigmund Freud.* Trans. James Strachey. Vol. 17, 217–256. London: Hogarth Press, 1955 [1919].

Fried, Michael. *Menzel's realism: art and embodiment in nineteenth-century Berlin.* New Haven: Yale University Press, 2002.

Friedländer, Saul. *Nazi Germany and the Jews: the years of persecution, 1933–1939.* New York: Harper Collins, 1997.

Frisby, David. *Cityscapes of modernity: critical explorations.* Cambridge: Polity, 2001.

Frisby, David. *Fragments of modernity: theories of modernity in the work of Simmel, Kracauer and Benjamin.* Cambridge, MA: MIT Press, 1986.

Fry, Maxwell, and Jane Drew. *Tropical architecture in the humid zone.* New York: Reinhold, 1956.

Furlong, Kathryn. "The dialectics of equity: consumer citizenship and the extension of water supply in Medellín, Colombia." *Annals of the Association of American Geographers* 103 (5) (2013): 1176–1192.

Gabriel, Nate. "Urban political ecology: environmental imaginary, governance, and the non-human." *Geography Compass* 8 (1) (2014): 38–48.

Gaillard, Jeanne. *Paris la Ville, 1852–1870.* Paris: Champion, 1977.

Galloway, James A., ed. *Tides and floods: new research on London and the tidal Thames from the Middle Ages to the twentieth century.* London: University of London, Centre for Metropolitan History, 2010.

Gandy, Matthew. "The bacteriological city and its discontents." *Historical Geography* 34 (2006): 14–25.

Gandy, Matthew. *Concrete and clay: reworking nature in New York City.* Cambridge, MA: MIT Press, 2002.

Gandy, Matthew. "Cyborg urbanization: complexity and monstrosity in the contemporary city." *International Journal of Urban and Regional Research* 29 (1) (2005): 26–49.

Gandy, Matthew. "Learning from Lagos." *New Left Review* 33 (2005): 37–52.

Gandy, Matthew. "Rethinking urban metabolism: water, space and the modern city." *City* 8 (3) (2004): 371–387.

Gaonkar, Dilip Parameshwar. "On alternative modernities." *Public Culture* 11 (1) (1999): 1–18.

Garrett, Bradley. *Explore everything: place-hacking and the city.* London: Verso, 2013.

Garrioch, David. *The making of revolutionary Paris.* Berkeley: University of California Press, 2002.

Gatens, Moira. *Imaginary bodies: ethics, power and corporeality.* London: Routledge, 1996.

Gay, Peter. *Weimar culture: the outsider as insider.* New York: W. W. Norton, 1968.

Gernsheim, Alison, and Helmut Gernsheim. *The history of photography from the camera obscura to the beginning of the modern era.* New York: McGraw-Hill, 1969.

Gernsheim, Alison, and Helmut Gernsheim. *L. J. M. Daguerre: the history of the diorama and the daguerreotype.* London: Secker and Warburg, 1965.

Ghertner, D. Asher. "Calculating without numbers: aesthetic governmentality in Delhi's slums." *Economy and Society* 39 (2) (2010): 185–217.

Ghertner, D. Asher. "Nuisance talk and the propriety of property: middle class discourse of a slum free Delhi." *Antipode* 44 (4) (2012): 1161–1187.

Giblett, Rodney James. *Postmodern wetlands: culture, history, ecology*. Edinburgh: Edinburgh University Press, 1996.

Giedion, Sigfried. *Mechanization takes command: a contribution to anonymous history*. Oxford: Oxford University Press, 1948.

Gilbert, Stuart, and Ray Horner. *The Thames Barrier*. London: Thomas Telford, 1984.

Gille, Bertrand. *Les ingénieurs de la Renaissance*. Paris: Hermann, 1967.

Gillet, Marcel, ed. *L'homme, la vie et la mort dans le nord au XIXe siècle*. Lille: Presses Universitaires, 1972.

Gissen, David. *Subnature: architecture's other environments*. New York: Princeton Architectural Press, 2009.

Gleber, Anke. *The art of taking a walk: flanerie, literature, and film in Weimar culture*. Princeton: Princeton University Press, 1999.

Gleber, Anke. "Female flanerie and the *Symphony of the city*." In *Women in the metropolis: gender and modernity in Weimar culture*, ed. Katharina von Ankum, 67–88. Berkeley: University of California Press, 1997.

Gordon, Peter E., and John P. McCormick, eds. *Weimar thought: a contested legacy*. Princeton: Princeton University Press, 2013.

Gosling, Nigel. *Nadar*. London: Secker and Warburg, 1976.

Gottlieb, Robert, and Margaret FitzSimmons. *Thirst for growth: water agencies as hidden government in California*. Tucson: University of Arizona Press, 1991.

Goubert, Jean-Pierre. *The conquest of water: the advent of health in the industrial age*. Trans. Andrew Wilson. Cambridge: Polity, 1989 [1986].

Gough, Jerry B. "Lavoisier's memoirs on the nature of water and their place in the chemical revolution." *Ambix* 30 (1983): 89–106.

Graf, William L. "Damage control: dams and the physical integrity of America's rivers." *Annals of the Association of American Geographers* 91 (1) (2001): 1–27.

Graham, Stephen, ed. *Disrupted cities: when infrastructure fails*. London: Routledge, 2010.

Graham, Stephen, Renu Desai, and Colin McFarlane. "Water wars in Mumbai." *Public Culture* 25 (1) (2013): 115–141.

Graham, Stephen, and Simon Marvin. *Splintering urbanism: networked infrastructures, technological mobilities and the urban condition*. London: Routledge, 2001.

Graham, Stephen, and Nigel Thrift. "Out of order: understanding repair and maintenance." *Theory, Culture and Society* 24 (1) (2007): 1–25.

Green, Nicholas. *The spectacle of nature: landscape and bourgeois culture in nineteenth-century France*. Manchester: Manchester University Press, 1990.

Gröning, Gert, and Joachim Wolschke-Bulmahn, eds. *Naturschutz und Demokratie!?* Munich: Martin Meidenbauer, 2006.

Gropius, Walter. *The new architecture and the Bauhaus*. Trans. P. Morton Shand. London: Faber, 1935.

Gropius, Walter, and Martin Wagner. "The new city pattern for the people and by the people." *Andean Quarterly* (Winter 1943): 4–12.

Grossmann, Iris, Katja Woth, and Hans von Storch. "Localization of global climate change: storm surge scenarios for Hamburg in 2030 and 2085." *Die Küste* 71 (2007): 169–182.

Grove, Kevin. "Pre-empting the next disaster: catastrophe insurance and the financialization of disaster management." *Security Dialogue* 43 (2) (2013): 139–155.

Grundmann, Reiner. "On control and shifting boundaries: modern society in the web of systems and networks." In *The governance of large technical systems*, ed. Olivier Coutard, 239–257. London: Routledge, 1999.

Gubler, Duane J., and Gary G. Clark. "Dengue/dengue hemorrhagic fever: the emergence of a global health problem." *Emerging Infectious Diseases* 1 (2) (1995): 55–57.

Guerrand, Roger-Henri. "La bataille du tout-a-l'égout." *Histoire* 53 (1983): 66–74.

Guillerme, André E., *The age of water: the urban environment of the north of France AD 300–1800*. College Station: Texas A&M University Press, 1988 [1983].

Gumprecht, Blake. *The Los Angeles River: its life, death and possible rebirth*. Baltimore: Johns Hopkins University Press, 1999.

Gupta, Kapil. "Urban flood resilience planning and management and lessons for the future: a case study of Mumbai, India." *Urban Water Journal* 4 (3) (2007): 183–194.

Gyford, John. *The politics of local socialism*. London: George Allen and Unwin, 1985.

Habermas, Jürgen. *The philosophical discourse of modernity*. Trans. Frederick Lawrence. Cambridge: Polity, 1987 [1985].

Hake, Sabine. *The cinema's third machine: writing on film in Germany, 1907–1933*. Lincoln: University of Nebraska Press, 1993.

Hake, Sabine. *Topographies of class: modern architecture and mass society in Weimar Berlin*. Ann Arbor: University of Michigan Press, 2008.

Halperin, Joan Ungersma. *Félix Fénéon: aesthete and anarchist in fin-de-siècle Paris*. New Haven: Yale University Press, 1988.

Hambourg, Maria Morris, Françoise Heilbrun, and Phillipe Néagu, eds. *Nadar*. New York: Metropolitan Museum of Modern Art, 1995.

Hamlin, Christopher. *Public health and social justice in the age of Chadwick: Britain, 1800–1854*. Cambridge: Cambridge University Press, 1988.

Hamlin, Christopher. *A science of impurity: water analysis in nineteenth-century Britain*. Berkeley: University of California Press, 1990.

Haney, David H. *When modern was green: life and work of landscape architect Leberecht Migge*. London: Routledge, 2010.

Hanna, Edward, John Cappelen, Rob Allan, Trausti Jónsson, Frank le Blancq, Tim Lillington, and Kieran Hickey. "New insights into North European and North Atlantic surface pressure variability, storminess and related climatic change since 1830." *Journal of Climate* 21 (24) (2008): 6739–6766.

Hansen, Miriam. "Early silent cinema: whose public sphere?" *New German Critique, NGC* 29 (Spring/Summer 1983): 147–184.

Hansen, Thomas Blom. *The saffron wave: democracy and Hindu nationalism in modern India*. Princeton: Princeton University Press, 1999.

Hansen, Thomas Blom. *Wages of violence: naming and identity in postcolonial Bombay*. Princeton: Princeton University Press, 2001.

Hansen, Thomas Blom, and Finn Stepputat, eds. *Sovereign bodies: citizens, migrants and state in the postcolonial world*. Princeton: Princeton University Press, 2005.

Hård, Mikael, and Andrew Jamison, eds. *The intellectual appropriation of technology: discourses on modernity, 1900–1939*. Cambridge, MA: MIT Press, 1998.

Harris, Andrew. "Concrete geographies: assembling global Mumbai through transport infrastructure." *City* 17 (3) (2013): 343–360.

Harris, Andrew. "Vertical urbanisms: opening up the geographies of the three-dimensional city." *Progress in Human Geography* (in press).

Harriss-White, Barbara. *India working: essays on society and economy*. Cambridge: Cambridge University Press, 2003.

Harvey, David. *A brief history of neo-liberalism*. Oxford: Oxford University Press, 2005.

Harvey, David. *Consciousness and the urban experience: studies in the history and theory of capitalist urbanization 1*. Oxford: Blackwell, 1985.

Harvey, David. *The limits to capital*. Oxford: Blackwell, 1982.

Hasan, Zoya, S. N. Jha, and Rasheeduddin Khan, eds. *The state, political processes and identity: reflections on modern India*. New Delhi: Sage, 1999.

Hau, Michael. *The cult of health and beauty in Germany: a social history, 1890–1930.* Chicago: University of Chicago Press, 2003.

Hegemann, Werner. *Das steinerne Berlin.* Braunschweig: Vieweg, 1988 [1930].

Heidenreich, Elisabeth. *Fliessräume. Die Vernetzung von Natur, Raum und Gesellschaft seit dem 19. Jahrhundert.* Frankfurt: Campus, 2004.

Hellweg, Uli, and Jörn Oltmans, eds. *Wasser in der Stadt: Perspektiven einer neuen Urbanität.* Berlin: Transit, 2000.

Herzog, Dagmar. *Sexuality in Europe: a twentieth-century history.* Cambridge: Cambridge University Press, 2011.

Hewitt, Mark, and Susannah Hagan, eds. *City fights: debates on urban sustainability.* London: James and James, 2001.

Higgs, Eric. *Nature by design: people, natural process, and ecological restoration.* Cambridge, MA: MIT Press, 2003.

Hilberseimer, Ludwig. *Berliner Architektur der 20er Jahre.* Berlin: Gebr. Mann Verlag, 1967.

Hildreth, Martha L. *Doctors, bureaucrats and public health in France, 1888–1902.* New York: Garland, 1987.

Hill, Donald. *A history of engineering in classical and mediaeval times.* London: Routledge, 1996.

Hill, Jonathan. *Weather architecture.* London: Routledge, 2012.

Hinchliffe, Steve, and Nick Bingham. "People, animals, and biosecurity in and through cities." In S. Harris Ali and Roger Keil, eds., *Networked disease: emerging infections in the global city*, 214–227. Oxford: Wiley-Blackwell, 2008.

Hiorns, Frederick. *Town-building in history: an outline review of conditions, influences, ideas, and methods affecting "planned" towns through five thousand years.* London: George G. Harrap, 1956.

Hise, Greg. *Magnetic Los Angeles.* Baltimore: Johns Hopkins University Press, 1997.

Hise, Greg, and William Deverell. *Eden by design: the 1930 Olmsted-Bartholomew plan for the Los Angeles region.* Berkeley: University of California Press, 2000.

Hitchcock, Henry-Russell. *Architecture: nineteenth and twentieth centuries.* New Haven: Yale University Press, 1977.

Hoag, Heather, and May-Brit Ohman. "Turning water into power: debates over the development of Tanzania's Rufiji basin, 1945–1985." *Technology and Culture* 49 (2008): 624–651.

Holston, John, ed. *Cities and citizenship*. Durham: Duke University Press, 1999.

Homann, Klaus, Martin Kieren, and Ludovica Scarpa, eds. *Martin Wagner 1885–1957. Wohnungsbau und Weltstadtplanung. Die Rationalisierung des Glücks*. Berlin: Akademie der Künste, 1987.

Home, Robert. *Of planting and planning: the making of British colonial cities*. London: E. and F. N. Spon, 1997.

Honneth, Axel. "The other of justice: Habermas and the ethical challenge of postmodernism." In *The Cambridge companion to Habermas*, ed. Stephen K. White, 289–323. Cambridge: Cambridge University Press, 1995.

Horwood, Catherine. "'Girls who arouse dangerous passions': women and bathing, 1900–39." *Women's History Review* 9 (4) (2000): 653–673.

Hughes, Thomas P. "The evolution of large technical systems." In Wiebe E. Bijker, Thomas P. Hughes, and Trevor J. Pinch, eds., *The social construction of technological systems: new directions in the sociology and history of technology*, 51–82. Cambridge, MA: MIT Press, 1987.

Hugo, Victor. *Les misérables*. Trans. Norman Denny. Harmondsworth: Penguin, 1982 [1862].

Hulme, Mick. "Abrupt climate change: can society cope?" *Philosophical Transactions of the Royal Society A* 36 (2003): 2001–2021.

Illich, Ivan. *H2O and the waters of forgetfulness: reflections on the historicity of "stuff."* Berkeley: Heyday Books, 1985.

Immerwahr, Daniel. "The politics of architecture and urbanism in postcolonial Lagos." *Journal of African Cultural Studies* 19 (2) (2007): 165–186.

Jacquemet, Gérard. "Urbanisme parisien: la bataille du tout-à-l'égout à la fin du XIXe siècle." *Revue d'Histoire Moderne et Contemporaine* 26 (October-December 1979): 505–548.

Jameson, Fredric. *Archaeologies of the future*. London: Verso, 2005.

Jameson, Fredric. "The brick and the balloon." *New Left Review* 228 (1998): 25–46.

Jameson, Fredric. *A singular modernity: essay on the ontology of the present*. London: Verso, 2002.

Jay, Martin. "The uncanny nineties." *Salmagundi* 108 (1995): 20–29.

Jefferies, Richard. *After London*. London: Cassell, 1885.

Jenkins, William. *New topographics: photographs of a man-altered landscape*. Rochester, NY: International Museum of Photography, 1975.

Jones, Richard. *Mosquito*. London: Reaktion, 2012.

Jordan, David P. *Transforming Paris: the life and labors of Baron Haussmann*. New York: Free Press, 1995.

Jørgensen, Dolly, Finn Arne Jørgensen, and Sara B. Pritchard, eds. *New natures: joining environmental history with science and technology studies*. Pittsburgh: University of Pittsburgh Press, 2013.

Joyce, James. *Ulysses*. London: Penguin, 1992 [1922].

Joyce, Peter. *The rule of freedom: liberalism and the modern city*. London: Verso, 2003.

Kaes, Anton, Martin Jay, and Edward Dimendberg, eds. *The Weimar Republic sourcebook*. Berkeley: University of California Press, 1994.

Kahrl, William L. *Water and power: the controversy over Los Angeles' water supply in the Owens Valley*. Berkeley: University of California Press, 1982.

Kaïka, Maria. *City of flows: modernity, nature, and the city*. New York: Routledge, 2005.

Kaïka, Maria. "Dams as symbols of modernization: the urbanization of nature between geographical imagination and materiality." *Annals of the Association of American Geographers* 96 (2) (2006): 276–301.

Kaïka, Maria, and Erik Swyngedouw. "Fetishising the modern city: the phantasmagoria of urban technological networks." *International Journal of Urban and Regional Research* 24 (1) (2000): 120–138.

Kane, Stephanie C. *Where rivers meet the sea: the political ecology of water*. Philadelphia: Temple University Press, 2012.

Kaplan, Marion A. *Between dignity and despair: Jewish life in Nazi Germany*. Oxford: Oxford University Press, 1998.

Karvonen, Andy. *Politics of urban runoff: nature, technology, and the sustainable city*. Cambridge, MA: MIT Press, 2011.

Kaviraj, Sudipta. "Filth and the public sphere: concepts and practices about space in Calcutta." *Public Culture* 10 (1) (1997): 83–113.

Kearns, Gerry. "Biology, class and the urban penalty." In *Urbanising Britain: essays on class and community in the nineteenth century*, ed. G. Kearns and Charles W. J. Withers, 12–30. Cambridge: Cambridge University Press, 1991.

Kearns, Gerry. "Zivilis or Hygaeia: urban public health and the epidemiological transition." In *The rise and fall of great cities: aspects of urbanization in the western world*, ed. Richard Lawton, 96–124. London: Belhaven, 1989.

Keil, Roger. *Los Angeles: globalization, urbanization and social struggles*. London: Wiley, 1998.

Kelly, Ann H., and Uli Beisel. "Neglected malarias: the frontlines and back alleys of global health." *Biosocieties* 6 (1) (2011): 71–87.

Kelman, Ari. *A river and its city: the nature of landscape in New Orleans*. Berkeley: University of California Press, 2003.

Kendrick, Mary. "The Thames Barrier." *Landscape and Urban Planning* 16 (1–2) (1988): 57–68.

Khilnani, Sunil. *The idea of India*. London: Penguin, 1997.

Kibel, Paul Stanton, ed. *Rivertown: rethinking urban rivers*. Cambridge, MA: MIT Press, 2007.

Killen, Andreas. *Berlin electropolis: shock, nerves, and German modernity*. Berkeley: University of California Press, 2006.

King, Anthony D. *Colonial urban development: culture, social power and environment*. London: Routledge and Kegan Paul, 1976.

King, David. "Climate change science: adapt, mitigate, or ignore." *Science* 303 (2004): 176–177.

Klein, Ira. "Urban development and death: Bombay City, 1870–1914." *Modern Asian Studies* 20 (1986): 725–754.

Kolsky, Pete. "Engineers and urban malaria: part of the solution, or part of the problem?" *Environment and Urbanization* 11 (1) (1999): 159–164.

Konvitz, Josef W., Mark H. Rose, and Joel A. Tarr. "Technology and the city." *Technology and Culture* 31 (1990): 284–294.

Kooy, Michelle, and Karen Bakker. "Technologies of government: constituting subjectivities, spaces, and infrastructures in colonial and contemporary Jakarta." *International Journal of Urban and Regional Research* 32 (2) (2008): 375–391.

Kowarik, Ingo, and Stephan Körner, eds. 2005. *Wild urban woodlands*. Berlin: Springer.

Kracauer, Siegfried. *From Caligari to Hitler: a psychological history of German film*. Princeton: Princeton University Press, 1947.

Kracauer, Siegfried. *Straßen in Berlin und anderswo*. Berlin: Das Arsenal, 1987 [1964].

Krauss, Rosalind. "Tracing Nadar." *October* 5 (1978): 29–47.

Kristeva, Julia. *Powers of horror: an essay on abjection*. Trans. L. S. Roudiez. New York: Columbia University Press, 1982 [1980].

Kristeva, Julia. "Psychoanalysis and the polis." Trans. M. Waller. In *The Kristeva Reader*, ed. Toril Moi, 301–320. Oxford: Blackwell, 1986 [1982].

Kühn, Norbert. "Intentions for the unintentional: spontaneous vegetation as the basis for innovative planting design in urban areas." *Journal of Landscape Architecture* (Autumn 2006): 46–53.

Kunstmann, Luzar. *La culture en clandestins. L'UX*. Paris: Hazan, 2009.

La Berge, Ann F. "Edwin Chadwick and the French connection." *Bulletin of the History of Medicine* 62 (1988): 23–42.

Ladd, Brian K. "Public baths and civic improvement in nineteenth-century German cities." *Journal of Urban History* 14 (3) (1988): 372–393.

Lahiji, Nadir, and D. S. Friedman, eds. *Plumbing: sounding modern architecture*. New York: Princeton Architectural Press, 1997.

Landström, Catharina, Sarah J. Whatmore, and Stuart N. Lane. "Virtual engineering: computer simulation modelling for flood risk management in England." *Science Studies* 24 (2) (2011): 3–22.

Lane, Barbara Miller. *Architecture and politics in Germany, 1918–1945*. Cambridge, MA: Harvard University Press, 1985 [1968].

Lane, Stuart N., Catharina Landström, and Sarah J. Whatmore. "Imagining flood futures: risk assessment and management in practice." *Philosophical Transactions of the Royal Society A* 369 (2011): 1784–1806.

Lange, Susanne. *Bernd and Hilla Becher: life and work*. Cambridge, MA: MIT Press, 2002.

Laporte, Dominique. *History of shit*. Trans. Nadia Benabid and Rudolphe el-Khoury. Cambridge, MA: MIT Press, 2000 [1978].

Lavedan, Pierre. "L'influence de Haussmann: l'haussmannisation." *Urbanisme et habitation* 34 (1953): 302–317.

Lavery, Sarah, and Bill Donovan. "Flood risk management in the Thames Estuary looking ahead 100 years." *Philosophical Transactions of the Royal Society A* 363 (2005): 1455–1474.

Lavin, Maud. *Cut with the kitchen knife: the Weimar photomontages of Hannah Höch*. New Haven: Yale University Press, 1993.

Leca, Ange-Pierre. *Et le choléra s'abattit sur Paris 1832*. Paris: Albin Michel, 1982.

Leguay, Jean-Pierre. *L'eau dans la ville au moyen âge*. Rennes: Presses Universitaires de Rennes, 2002.

Leighton, John, and R. Thomson. *Seurat and the bathers*. London: National Gallery, 1997.

Leonard, Charlene Marie. *Lyon transformed: public works of the Second Empire 1853–1864*. Berkeley: University of California Press, 1961.

Le Roux, Hannah. "Modern architecture in post-colonial Ghana and Nigeria." *Architectural History* 47 (2004): 361–392.

Lesser, Wendy. *The life below the ground: a study of the subterranean in literature*. Boston: Faber and Faber, 1987.

Lethen, Helmut. *Cool conduct: the culture of distance in Weimar Germany*. Trans. Don Reneau. Berkeley: University of California Press, 2002.

Lethen, Helmut. *Neue Sachlichkeit 1924–1932. Studien zur Literatur des "Weissen Sozialismus."* Stuttgart: J. B. Metzlersche Verlagsbuchhandlung, 1970.

Lewis, Beth Irwin. *George Grosz: art and politics in the Weimar Republic*. Princeton: Princeton University Press, 1991.

Leyden, Friedrich. *Groß-Berlin. Geographie der Weltstadt*. Berlin: Mann Verlag, 1933.

Loftus, Alex. *Everyday environmentalism: creating an urban political ecology*. Minneapolis: University of Minnesota Press, 2012.

Loftus, Alex. "Rethinking political ecologies of water." *Third World Quarterly* 30 (5) (2009): 953–968.

Lonsdale, K. G., T. E. Downing, R. J. Nicholls, D. Parker, Athanasios T. Vafeidis, Richard J. Dawson, and Jim Hall. "Plausible responses to rapid sea-level rise in the Thames estuary." *Climatic Change* 91 (2008): 145–169.

Lowe, J. A., and Jonathan M. Gregory. "The effects of climate change on storm surges around the United Kingdom." *Philosophical Transactions of the Royal Society A* 363 (2005): 1313–1328.

Loyer, François. *Paris nineteenth century: architecture and urbanism*. Trans. C. L. Clark. New York: Abbeville Press, 1988.

Mabey, Richard. *Weeds*. London: Profile, 2010.

Magnusson, Roberta J. *Water technology in the Middle Ages: cities, monasteries, and waterworks after the Roman Empire*. Baltimore: Johns Hopkins University Press, 2001.

Mahadevia, Darshini. "State supported segmentation of Mumbai: policy options in the global economy." *Review of Development and Change* 3 (1) (1998): 12–41.

Malet, Henri. *Le baron Haussmann et la rénovation de Paris*. Paris: Éditions municipales, 1973.

Mamdani, Mahmood. *Citizen and subject: contemporary Africa and the legacy of late colonialism.* Princeton: Princeton University Press, 1996.

Manacorda, Francesco, and Ariella Yedgar, eds. *Radical nature: art and architecture for a changing planet 1969–2009.* London: Koenig Books, 2009.

Masselos, Jim. "Postmodern Bombay: fractured discourses." In *Postmodern cities and spaces,* ed. Sophie Watson and Katherine Gibson, 199–215. Oxford: Blackwell, 1995.

Masson, Françoise, and Marc Gaillard. *Les fontaines de Paris.* Amiens: Martelle, 1995.

Mathur, Anuradha, and Dilip da Cunha. *Mississippi floods: designing a shifting landscape.* New Haven: Yale University Press, 2001.

Mathur, Anuradha, and Dilip da Cunha. *Soak: Mumbai in an estuary.* New Delhi: Rupa, 2009.

Mauguen, Pierre-Yves. "Les galeries souterraines d'Haussmann. Le système des égouts parisiens, prototype ou exception?" *Annales de la recherche urbaine* 44–45 (1989): 163–175.

Mazower, Mark. *Dark continent: Europe's twentieth century.* London: Allen Lane, 1998.

Mazower, Mark. *Governing the world: the history of an idea.* London: Penguin, 2012.

McCormick, Richard W. *Gender and sexuality in Weimar modernity: film, literature, and the "new objectivity".* Basingstoke: Palgrave, 2001.

McFarlane, Colin. "Geographical imaginations and spaces of political engagement: examples from the Indian Alliance." *Antipode* 36 (2004): 890–916.

McFarlane, Colin, and Jonathan Rutherford. "Political infrastructures: governing and experiencing the fabric of the city." *International Journal of Urban and Regional Research* 32 (2) (2008): 363–374.

McRobie, Allan, Tom Spencer, and Herman Gerritsen. "The big flood: North Sea storm surge." *Philosophical Transactions of the Royal Society A* 363 (2005): 1263–1270.

McWilliams, Carey. *California: the great exception.* Berkeley: University of California Press, 1999 [1949].

Meehan, Katharine. "Disciplining de facto development: water theft and hydrosocial order in Tijuana." *Environment and Planning D: Society and Space* 31 (2) (2013): 319–336.

Mehta, Suketu. *Maximum city: Bombay lost and found.* New York: Alfred A. Knopf, 2004.

Melosi, Martin. *The sanitary city: urban infrastructure in America from colonial times to the present.* Baltimore: Johns Hopkins University Press, 2000.

Merriman, John M., *The margins of city life: explorations on the French urban frontier, 1815–1851*. Oxford: Oxford University Press, 1991.

Millington, Nate. "Post-industrial imaginaries: nature, representation and ruin in Detroit, Michigan." *International Journal of Urban and Regional Research* 37 (1) (2013): 279–296.

Mitchell, Timothy. *Carbon democracy: political power in the age of oil*. London: Verso, 2013 [2011].

Mitchell, Timothy. *Rule of experts: Egypt, techno-politics, modernity*. Berkeley: University of California Press, 2002.

Moholy-Nagy, László. *Vision in motion*. Chicago: Paul Theobold, 1947.

Monstadt, Jochen. "Conceptualizing the political ecology of urban infrastructures: insights from technology and urban studies." *Environment and Planning A* 41 (2009): 1924–1942.

Moore, Charles W., and Jane Lidz. *Water and architecture*. London: Thames and Hudson, 1994.

Morizet, André. *Du vieux Paris au Paris moderne, Haussmann et ses prédécesseurs*. Paris: Hachette, 1932.

Morris, William. *News from nowhere*. Ed. Krishan Kumar. Cambridge: Cambridge University Press, 1995 [1890].

Mukerji, Chandra. *Impossible engineering: technology and territoriality on the Canal du Midi*. Princeton: Princeton University Press, 2009.

Mukta, Parita. "Wresting riches, marginalizing the poor, criminalizing dissent: the building of the Narmada Dam in Western India." *South Asia Bulletin: Comparative Studies of South Asia, Africa and the Middle East* 15 (2) (1995): 99–109.

Murard, Lion, and Patrick Zylberman. *L'hygiène dans la République: la santé publique en France, ou l'utopie contrariée 1870–1918*. Paris: Fayard, 1996.

Murphet, Julian. *Literature and race in Los Angeles*. Cambridge: Cambridge University Press, 2001.

Murray, Bruce Arthur. *Film and the German left in the Weimar Republic: from Caligari to Kuhle Wampe*. Austin: University of Texas Press, 1990.

Nading, Alex M. "Dengue mosquitoes are single mothers: biopolitics meets ecological aesthetics in Nicaraguan community health work." *Cultural Anthropology* 27 (4) (2012): 572–596.

Nagarkar, Kiran. *Ravan & Eddie*. New Delhi: Penguin, 1995.

Newhall, Beaumont. *The history of photography*. New York: Museum of Modern Art, 1982.

Nijman, Jan. "The effects of economic globalization on land use and land values in Mumbai." In Richard Grant and John R. Short, eds., *Globalization and the margins*, 150–169. Basingstoke: Palgrave Macmillan, 2002.

Nochlin, Linda. *Bathers, bodies, beauty: the visceral eye*. Cambridge, MA: Harvard University Press, 2006.

Nochlin, Linda. *The politics of vision: essays on nineteenth-century art and society*. London: Thames and Hudson, 1989.

Nussbaum, Martha C. *The clash within: democracy, religious violence, and India's future*. Cambridge, MA: Belknap Press, 2007.

Nwosu, Maik. *Invisible chapters*. Lagos: Malaika and Hyburn, 2001.

Obiechina, Emmanuel. *Culture, tradition and society in the West African novel*. Cambridge: Cambridge University Press, 1975.

Okome, Onookome. "Nollywood: spectatorship, audience and sites of consumption." *Postcolonial Text* 3 (2) (2007): 1–21.

Oliver, Stuart. "The Thames Embankment and the disciplining of nature in modernity." *Geographical Journal* 166 (2000): 227–238.

Olsen, Donald J. *The city as a work of art: London, Paris, Vienna*. New Haven: Yale University Press, 1986.

Olukoju, Ayodeji. *Infrastructure development and urban facilities in Lagos, 1861 2000*. Ibadan: Institut Français de Recherche en Afrique, University of Ibadan, 2003.

O'Malley, Mark Anthony. "Plumbing the body politic: a political ecology of water and waste in Berlin, 1850–1880." PhD thesis, Department of Geography, University of California, Berkeley, 1997.

Orsi, Jared. *Hazardous metropolis: flooding and urban ecology in Los Angeles*. Berkeley: University of California Press, 2004.

O'Shea-Meddour, Wendy. "Gaston Bachelard's *L'eau et les rêves*: conquering the feminine elements." *French Cultural Studies* 14 (1) (2003): 81–99.

Packard, Randall M. *The making of a tropical disease: a short history of malaria*. Baltimore: Johns Hopkins University Press, 2007.

Pardo, Mary. "Mexican American women grassroots community activists: Mothers of East Los Angeles." In *A sense of the American West: an anthology of environmental history*, ed. James E. Sherow, 243–260. Albuquerque: University of New Mexico Press, 1998.

Patel, Sujata, and Jim Masselos, eds. *Bombay and Mumbai: the city in transition*. New Delhi: Oxford University Press, 2003.

Pendse, Sandeep. "Toil, sweat and the city." In *Bombay: metaphor for modern India*, ed. Sujata Patel and Alice Thorner, 3–25. New Delhi: Oxford University Press, 1995.

Penner, Barbara. *Bathroom*. London: Reaktion, 2014.

Penzo, Paola. "L'urbanistica e l'amministrazione socialista a Bologna, 1914–1920." *Storia Urbana* 18 (66) (1994): 109–143.

Pethick, John. "Estuarine and tidal wetland restoration in the United Kingdom: policy versus practice." *Restoration Ecology* 10 (3) (2002): 431–437.

Peukert, Detlev J. K. *The Weimar Republic: the crisis of classical modernity*. New York: Hill and Wang, 1987.

Picon, Antoine. "Anxious landscapes: from the ruin to rust." *Grey Room* 1 (2000): 64–83.

Picon, Antoine. *Les saint-simoniens: raison, imaginaire et utopie*. Paris: Belin, 2002.

Pike, David L. *Subterranean cities: the world beneath Paris and London, 1800–1945*. Ithaca: Cornell University Press, 2005.

Pinkney, David. *Napoleon III and the rebuilding of Paris*. Princeton: Princeton University Press, 1958.

Pinol, Jean-Luc, ed. *Atlas historiques de villes de France*. Barcelona: Centre de Cultura Contemporània de Barcelona, 1996.

Pollock, Griselda. *Vision and difference: femininity, feminism and the histories of art*. London: Routledge, 1988.

Porter, Dale H. *The Thames Embankment: environment, technology, and society in Victorian London*. Akron, OH: University of Akron Press, 1998.

Porter, James, and David Demeritt. "Flood-risk management, mapping, and planning: the institutional politics of decision support in England." *Environment and Planning A* 44 (2012): 2359–2378.

Porter, Roy. *London: a social history*. London: Hamish Hamilton, 1994.

Prakash, Gyan. *Another reason: science and the imagination of modern India*. Princeton: Princeton University Press, 1999.

Prakash, Gyan. *Mumbai fables*. Princeton: Princeton University Press, 2010.

Prakash, Gyan. "Subaltern studies as postcolonial criticism." *American Historical Review* 99 (5) (December 1994): 1475–1490.

Prashad, Vijay. "The technology of sanitation in colonial Delhi." *Modern Asian Studies* 35 (2001): 113–155.

Pred, Allan, and Michael Watts, eds. *Reworking modernity: capitalism and symbolic discontent.* New Brunswick: Rutgers University Press, 1992.

Price, Patricia. "Inscribing the border: schizophrenia and the aesthetics of Aztlán." *Social and Cultural Geography* 1 (1) (2000): 101–116.

Prinet, Jean, and Antoinette Dilasser. *Nadar.* Paris: Colin, 1966.

Pritchard, Sara B. *Confluence: the nature of technology and the remaking of the Rhône.* Cambridge, MA: Harvard University Press, 2011.

Pulido, Laura. *Environmentalism and economic justice: two Chicano struggles in the Southwest.* Tucson: University of Arizona Press, 1996.

Pulido, Laura. "Multiracial organizing among environmental justice activists in Los Angeles." In Michael Dear, H. Eric Schockman, and Greg Hise, eds., *Rethinking Los Angeles*, 171–188. Thousand Oaks, CA: Sage Publications, 1996.

Pulido, Laura. "Rethinking environmental racism: white privilege and urban development in Southern California." *Annals of the Association of American Geographers* 90 (1) (2000): 12–40.

Purseglove, Jeremy. *Taming the flood: a history and natural history of rivers and wetlands.* Oxford: Oxford University Press, 1988.

Rabinbach, Anson. *The human motor: energy, fatigue, and the origins of modernity.* Berkeley: University of California Press, 1992 [1990].

Rabinow, Paul. *French modern: norms and forms of the social environment.* Cambridge, MA: MIT Press, 1989.

Ramanathan, Usha. "Illegality and the urban poor." *Economic and Political Weekly* 41 (29) (2006): 3193–3197.

Rao, Vyjayanthi. "Proximate distances: the phenomenology of density in Mumbai." *Built Environment* 33 (2) (2007): 227–248.

Rao, Vyjayanthi. "Slum as theory: the South/Asian city and globalisation." *International Journal of Urban and Regional Research* 30 (2006): 225–232.

Ratcliffe, Barrie M. "Cities and environmental decline: elites and the sewage problem in Paris from the mid 18th to the mid 19th century." *Planning Perspectives* 5 (1990): 189–222.

Reade, Eric. *British town and country planning.* Milton Keynes: Open University Press, 1987.

Reeder, Tim, John Wicks, Luke Lovell, and Owen Tarrant. "Protecting London from tidal flooding: limits to engineering adaptation." In W. Neil Adger, Irene Lorenzoni, and Karen L. O'Brien, eds., *Adapting to climate change: thresholds, values, governance*, 54–63. Cambridge: Cambridge University Press, 2009.

Reid, Donald. *Paris sewers and sewermen: realities and representations*. Cambridge, MA: Harvard University Press, 1991.

Reisen, William K., Richard M. Takahashi, Brian D. Carroll, and Rob Quiring. "Delinquent mortgages, neglected swimming pools, and West Nile Virus, California." *Emerging Infectious Diseases* 14 (11) (2008): 1747–1749.

Rey, Paul. "Du ballon de Nadar à Apollo XIII: la détection à distance des ressources naturelles." *Mémoires de l'Académie des Sciences, Inscriptions et Belles-Lettres de Toulouse* 132 (1970): 97–103.

Rice, Shelley. *Parisian views*. Cambridge, MA: MIT Press, 1997.

Rickels, Laurence A. *The case of California*. Minneapolis: University of Minnesota Press, 2001.

Rigby, Brian, ed. *French literature, thought and culture in the nineteenth century: a material world*. London: Macmillan, 1993.

Robinson, Jennifer. "Cities in a world of cities: the comparative gesture." *International Journal of Urban and Regional Research* 35 (1) (2011): 1–23.

Roche, Daniel. "Le temps de l'eau rare, du moyen âge à l'époque modern." *Annales: économies, sociétés, civilisations* 39 (1984): 383–399.

Rodgers, Dennis. "Haussmannization in the tropics: abject urbanism and infrastructural violence in Nicaragua." *Ethnography* 13 (4) (2012): 413–438.

Rodríguez, Luis J. *The concrete river*. Willimantic, CT: Curbstone Books, 1991 [1990].

Roncayolo, Marcel, and Louis Bergeron. "D'Haussmann à nos jours." In *Paris: genèse d'un paysage*, ed. Louis Bergeron, 217–298. Paris: Picard, 1989.

Roobeek, Annemieke J. M. "The crisis in Fordism and the rise of a new technological paradigm." *Futures* 19 (1987): 129–154.

Rorty, Richard. *Contingency, irony, and solidarity*. Cambridge: Cambridge University Press, 1989.

Rorty, Richard. *Philosophy and the mirror of nature*. Princeton: Princeton University Press, 1979.

Rossi, Aldo. *The architecture of the city*. Trans. Diane Ghirardo and Joan Ockman. Cambridge, MA: MIT Press, 1982 [1966].

Roth, Matthew W. "Mullholland Highway and the engineering culture of Los Angeles in the 1920." In *Metropolis in the making: Los Angeles in the 1920s*, ed. Tom Sitton and William Deverell, 45–76. Berkeley: University of California Press, 2001.

Rowe, Dorothy. *Representing Berlin: sexuality and the city in imperial and Weimar Germany*. Aldershot: Ashgate, 2003.

Rudolf, Bruno, and Joerg Rapp. "The century flood of the River Elbe in August 2002: synoptic weather development and climatological aspects." *Quarterly Report of the German NWP-System of the Deutscher Wetterdienst* 2 (1) (2003): 8–23.

Saalman, Howard. *Haussmann: Paris transformed*. New York: George Braziller, 1971.

Salomon-Bayet, Claire, ed. *Pasteur et la révolution pastorienne*. Paris: Payot, 1986.

Scarpa, Ludovica. *Martin Wagner und Berlin. Archiketur und städtebau in der Weimarer Republik*. Trans. Heinz-Dieter Held. Braunschweig: Vieweg, 1986 [1983].

Schafran, Alex. "Origins of an urban crisis: the restructuring of the San Francisco Bay area and the geography of foreclosure." *International Journal of Urban and Regional Research* 37 (2) (2013): 663–688.

Schama, Simon. *Landscape and memory*. New York: Alfred A. Knopf, 1995.

Scharf, Aaron. *Art and photography*. London: Allen Lane, 1968.

Schivelbusch, Wolfgang. *Disenchanted night: the industrialization of light in the nineteenth century*. Trans. Angela Davies. Oxford: Berg, 1988 [1983].

Schorske, Carl E. *German social democracy, 1905–1917: the development of the great schism*. Cambridge, MA: Harvard University Press, 1955.

Schrader, Bärbel. *The golden twenties: art and literature in the Weimar Republic*. New Haven: Yale University Press, 1988.

Schrank, Sarah. *Art and the city: civic imagination and cultural authority in Los Angeles*. Philadelphia: University of Pennsylvania Press, 2009.

Schwartz, Alexandra. *Ed Ruscha's Los Angeles*. Cambridge, MA: MIT Press, 2010.

Schwarze, Reimund, and Gert G. Wagner. "The political economy of natural disaster insurance: lessons from the failure of a proposed compulsory insurance scheme in Germany." Discussion paper, Technische Universität Berlin, Fakultät Wirtschaft und Management, 2006.

Scott, Allen J., and Edward W. Soja, eds. *The city: Los Angeles and urban theory at the end of the twentieth century*. Berkeley: University of California Press, 1996.

Seel, Sophie Penelope Stephenson. "Later prehistoric woodlands and wood use on the lower Thames floodplain." PhD thesis, Institute of Archaeology, University College London, 2001.

Sennett, Richard. *Flesh and stone: the body and the city in Western civilization*. London: Faber and Faber, 1994.

Shaw, Ian Graham Ronald, Paul F. Robbins, and John Paul Jones III. "A bug's life: the spatial ontologies of mosquito management." *Annals of the Association of American Geographers* 100 (2) (2010): 373–392.

Shipway, Martin. *Decolonization and its impact: a comparative approach to the end of the colonial empires*. Malden, MA: Blackwell, 2008.

Silverman, Kaja. *The acoustic mirror: the female voice in psychoanalysis and cinema*. Bloomington: Indiana University Press, 1988.

Simone, AbdouMaliq. "People as infrastructure: intersecting fragments in Johannesburg." *Public Culture* 16 (3) (2009): 407–429.

Smith, Carl. *City water, city life: water and the infrastructure of ideas in urbanizing Philadelphia, Boston, and Chicago*. Chicago: University of Chicago Press, 2013.

Smith, Paul. *Seurat and the avant-garde*. New Haven: Yale University Press, 1996.

Snowden, Frank M. *Naples in the time of cholera, 1884–1911*. Cambridge: Cambridge University Press, 1995.

Soja, Edward W. *Postmetropolis: critical studies of cities and regions*. Oxford: Blackwell, 2000.

Soll, David. *Empire of water: an environmental and political history of New York City*. Ithaca: Cornell University Press, 2013.

Speziale, Salvatore. "Società e malattia: Tunisi di fronte al colera del 1885." *Africa* 49 (2) (1994): 275–298.

Spinney, Laura. "Return to paradise." *New Scientist* 2039 (20 July 1996): 27–31.

Stallabrass, Julian. "Sebastião Salgado and fine art journalism." *New Left Review* 157 (1997): 131–162.

Stallybrass, Peter, and Allon White. *The politics and poetics of transgression*. London: Methuen, 1986.

Star, Susan Leigh. "The ethnography of infrastructure." *American Behavioral Scientist* 43 (3) (1999): 377–391.

Steers, James Alfred. "The east coast floods, January 31–February 1 1953." *Geographical Journal* 119 (1953): 280–298.

Stepan, Nancy Leys. *Picturing tropical nature*. London: Reaktion, 2001.

Stoler, Ann Laura. "Imperial debris: reflections on ruins and ruination." *Cultural Anthropology* 33 (2) (2008): 191–219.

Stolz, Robert. *Bad water: nature, pollution, and politics in Japan, 1870–1950*. Durham: Duke University Press, 2014.

Strang, Veronica. "Common senses: water, sensory experience and the generation of meaning." *Journal of Material Culture* 10 (1) (2005): 92–120.

Strumingher, Laura S. "L'ange de la maison: mothers and daughters in nineteenth-century France." *International Journal of Women's Studies* 2 (1979): 51–61.

Stürmer, Rainer. *Freiflächenpolitik in Berlin in der Weimarer Republik: ein Beitrag zur Sozial- und Umweltschutzpolitik einer modernen Industriestadt*. Berlin: Arno Spitz, 1991.

Sutcliffe, Anthony. *The autumn of central Paris*. London: Edward Arnold, 1970.

Sutcliffe, Anthony. "The impressionists and Haussmann's Paris." *French Cultural Studies* 6 (2) (1995): 197–219.

Sutcliffe, Anthony. *Paris: an architectural history*. New Haven: Yale University Press, 1993.

Sutcliffe, Anthony. *Towards the planned city: Germany, Britain, the United States and France 1780–1914*. Oxford: Basil Blackwell, 1981.

Sutter, Paul S. "Nature's agents or agents of empire? Entomological workers and environmental change during the construction of the Panama Canal." *Isis* 98 (2007): 724–754.

Swanson, Maynard W. "The sanitation syndrome: bubonic plague and urban native policy in the Cape Colony." *Journal of African History* 18 (3) (1977): 387–410.

Swyngedouw, Erik. "Apocalypse forever? Post-political populism and the spectre of climate change." *Theory, Culture and Society* 27 (2–3) (2009): 213–232.

Swyngedouw, Erik. "The city as a hybrid: on nature, society and cyborg urbanization." *Capitalism, Nature, Socialism* 7 (1996): 65–80.

Swyngedouw, Erik. "Into the sea: desalination and the hydro-fix in Spain." *Annals of the Association of American Geographers* 103 (2) (2013): 261–270.

Swyngedouw, Erik. "Power, nature, and the city: the conquest of water and the political ecology of urbanization in Guayaquil, Ecuador: 1880–1990." *Environment and Planning A* 29 (2) (1997): 311–332.

Swyngedouw, Erik. *Social power and the urbanization of water: flows of power*. Oxford: Oxford University Press, 2004.

Swyngedouw, Erik, and Nik Heynen. "Urban political ecology, justice, and the politics of scale." *Antipode* 35 (5) (2003): 898–918.

Szarkowski, John, *Photography until now*. New York: Museum of Modern Art, 1989.

Szreter, Simon. "Economic growth, disruption, deprivation, disease and death: on the importance of the politics of public health for development." *Population and Development Review* 23 (4) (1997): 693–728.

Tafuri, Manfredo. *The sphere and the labyrinth: avant-gardes and architecture from Piranesi to the 1970s*. Trans. Pellegrino d'Acierno and Robert Connolly. Cambridge, MA: MIT Press, 1987 [1980].

Tarr, Joel A. "The metabolism of the industrial city: the case of Pittsburgh." *Journal of Urban History* 28 (2002): 511–545.

Tarr, Joel A., and Gabriel Dupuy, eds. *Technology and the rise of the networked city in Europe and America*. Philadelphia: Temple University Press, 1988.

Tatar, Maria M. "The houses of fiction: toward a definition of the uncanny." *Comparative Literature* 33 (1981): 167–182.

Tatar, Maria M. *Lustmord: sexual murder in Weimar Germany*. Princeton: Princeton University Press, 1995.

Taylor, Vanessa, and Frank Trentmann. "Liquid politics: water and the politics of everyday life in the modern city." *Past and Present* 211 (2011): 199–241.

Teich, Mikulas. "Circulation, transformation, conservation of matter and the balancing of the biological world in the eighteenth century." *Ambix* 29 (1982): 17–28.

Tepasse, Heinrich. *Stadttechnik im Städtebau Berlins*, vol. 1, *19. Jahrhundert*. Berlin: Gebr. Mann, 2001.

Tilley, Christopher. *A phenomenology of landscape: places, paths and monuments*. Oxford: Berg, 1994.

Toepfer, Karl. *Empire of ecstasy: nudity and movement in German body culture*. Berkeley: University of California Press, 1997.

Tol, Richard S. J., Maria Bohn, Thomas E. Downing, Marie-Laure Guillerminet, Eva Hizsnyik, Roger Kasperson, Kate Lonsdale, et al. "Adaptation to five metres of sea level rise." *Journal of Risk Research* 9 (5) (2006): 467–482.

Trupat, Christina. "'Bade zu Hause!' Zur Geschichte des Badezimmers in Deutschland seit der Mitte des 19. Jahrhunderts." *Technikgeschichte* 63 (3) (1996): 219–236.

Tucker, Paul Hayes. *Monet at Argenteuil*. New Haven: Yale University Press, 1982.

Tvedt, Terje. *The river Nile in the age of the British: political ecology and the quest for economic power*. London: I. B. Taurus, 2004.

Uduku, Ola. "Modernist architecture and the 'tropical' in West Africa: the tropical architecture movement in West Africa, 1948–1970." *Habitat International* 30 (3) (2006): 396–411.

Vale, Lawrence J., and Thomas J. Campanella, eds. *The resilient city: how modern cities recover from disaster*. Oxford: Oxford University Press, 2005.

van der Brug, P. H. "Malaria in Batavia in the 18th century." *Tropical Medicine and International Health* 2 (9) (1997): 892–902.

van Laak, Dirk. "Der Begriff 'Infrastruktur' und was er vor seiner Erfindung besagte." *Archiv für Begriffsgeschichte* 41 (1999): 280–299.

van Leeuwen, Thomas A. P. *The springboard in the pond: an intimate history of the swimming pool*. Cambridge, MA: MIT Press, 1998.

Van Zanten, David. *Building Paris: architectural institutions and the transformation of the French capital 1830–1870*. Cambridge: Cambridge University Press, 1994.

Varnelis, Kazys, ed. *The infrastructural city: networked ecologies in Los Angeles*. Barcelona: Actar, 2008.

Vidler, Anthony. *The architectural uncanny: essays in the modern unhomely*. Cambridge, MA: MIT Press, 1992.

Vidler, Anthony. *Histories of the immediate present: inventing architectural modernism*. Cambridge, MA: MIT Press, 2008.

Vigarello, Georges. *Concepts of cleanliness: changing attitudes in France since the Middle Ages*. Trans. Jean Birrell. Cambridge: Cambridge University Press, 1988 [1985].

von Ankum, Katharina, ed. *Women in the metropolis: gender and modernity in Weimar culture*. Berkeley: University of California Press, 1992.

Von Schnitzler, Antina. "Traveling technologies: infrastructure, ethical regimes and the materiality of politics in South Africa." *Cultural Anthropology* 28 (4) (2013): 670–693.

Wagner, Anton. *Los Angeles: Werden, Leben und Gestalt der Zweimillionenstadt in Südkalifornien*. Leipzig: Bibliographisches Institut, 1935.

Walton, John. *Western times and water wars: state, culture, and rebellion in California*. Berkeley: University of California Press, 1992.

Ward, Janet. *Weimar surfaces: urban visual culture in 1920s Germany*. Berkeley: University of California Press, 2001.

Wardekker, J. Arjan, Arie de Jong, Joost M. Knoop, and Jeroen P. van der Sluijs. "Operationalising a resilience approach to adapting an urban delta to uncertain climate changes." *Technological Forecasting and Social Change* 77 (2010): 987–998.

Watkins, David. *Morality and architecture: the development of a theme in architectural history and theory from the Gothic revival to the modern movement*. Oxford: Clarendon Press, 1977.

Wedermeyer-Kolwe, Bernd. *"Der neue Mensch": Körperkultur im Kaiserreich und in der Weimarer Republik*. Würzburg: Königshausen and Neumann, 2004.

Weinstein, Liza. "'One man-handled': fragmented power and political entrepreneurship in globalizing Mumbai." *International Journal of Urban and Regional Research* 38 (1) (2014): 14–35.

Weisman, Alan. *The world without us*. London: Virgin, 2007.

Weitz, Eric D. *Weimar Germany: promise and tragedy*. Princeton: Princeton University Press, 2007.

Weizman, Eyal. *Hollow land: Israel's architecture of occupation*. London: Verso, 2007.

Welton, Donn, ed. *Body and flesh: a philosophical reader*. Oxford: Blackwell, 1998.

Werner, Dietrich. *Wasser für antike Rom*. Berlin: VEB Verlag für Bauwesen, 1986.

Werner, Frank. *Stadtplanung, Berlin 1900–1950*. Berlin: Kiepert, 1969.

Wescoat, James L., Jr. "The 'right of thirst' for animals in Islamic law: a comparative approach." *Environment and Planning D: Society and Space* 13 (1995): 637–654.

White, Richard. *The organic machine: the remaking of the Columbia River*. New York: Hill and Wang, 1995.

Whitely, Nigel. *Reyner Banham: historian of the immediate future*. Cambridge, MA: MIT Press, 2002.

Wilby, Robert L., and George L. W. Perry. "Climate change, biodiversity and the urban environment: a critical review based on London, UK." *Progress in Physical Geography* 30 (1) (2006): 73–98.

Willcox, G. H. "Exotic plants on Roman waterlogged sites in London." *Journal of Archaeological Science* 4 (1977): 269–282.

Williams, Raymond. *The country and the city*. Oxford: Oxford University Press, 1973.

Williams, Rosalind H. *Notes on the underground: an essay on technology, society and the imagination*. 2nd ed. Cambridge, MA: MIT Press, 2008.

Wilson, Elizabeth. *The sphinx in the city: urban life, the control of disorder, and women*. London: Virago, 1991.

Wolch, Jennifer. "Green urban worlds." *Annals of the Association of American Geographers* 97 (2) (2007): 373–384.

Wolch, Jennifer. "Zoöpolis." *Capitalism, Nature, Socialism* 7 (2) (1996): 21–47.

Wolff, Janet. "The invisible *flâneuse*: women and the literature of modernity." *Theory, Culture and Society* 2 (1985): 37–46.

Wolff, Kurt H., ed. *The sociology of Georg Simmel*. New York: Free Press, 1950.

Wolters, Mineke, Angus Garbutt, and Jan P. Bakker. "Salt-marsh restoration: evaluating the success of de-embankments in north-west Europe." *Biological Conservation* 123 (2005): 249–268.

Worster, Donald. *Rivers of empire: water, aridity, and the growth of the American West*. New York: Pantheon Books, 1985.

Wright, Gwendolyn. *The politics of design in French colonial urbanism*. Chicago: University of Chicago Press, 1991.

Wright, Lawrence. *Clean and decent: the history of the bathroom and the W.C.* London: Routledge and Kegan Paul, 1960.

Wulf, Meike. "Towards an urban epidemiology." In Matthew Gandy, ed., *Urban constellations*, 75–79. Berlin: jovis, 2011.

Wyndham, John. *The day of the triffids*. Harmondsworth: Penguin, 1954 [1951].

Yack, Bernard. *The fetishism of modernities: epochal self-consciousness in contemporary social and political thought*. Notre Dame, IN: University of Notre Dame Press, 1997.

Yefremov, Ivan. *Andromeda: a space-age tale*. Trans. George Hanna. Moscow: Foreign Languages Publishing House, 1959.

Yeoh, Brenda. "The global cultural city? Spatial imagineering and politics in the (multi) cultural marketplaces of south-east Asia." *Urban Studies* 42 (2005): 945–958.

Zeiderman, Austin. "Living dangerously: biopolitics and urban citizenship in Bogotá, Colombia." *American Ethnologist* 40 (1) (2013): 71–87.

Zérah, Marie-Hélène. "Splintering urbanism in Mumbai: contrasting trends in a multi-layered society." *Geoforum* 39 (6) (2008): 1922–1932.

Zola, Émile. *L'assommoir*. Paris: Flammarion, 1997 [1876].

# Index

*Note: Page numbers in italics indicate illustrations.*

Abani, Chris, 108
Achebe, Chinua, 99
Algeria, 82
Amsterdam, 42
Anand, Nikhil, 221
Andersen, Thom, 174
*Andromeda* (Yefremov), 211
Antonioni, Michelangelo, 174
Appadurai, Arjun, 6, 109, 129, 130
Archigram, 148
Architectonic palimpsest, 223
Architects and Engineers for Social Responsibility, 209
Architecture, 2, 56–57, 101, 202, 204
Arnheim, Rudolf, 77
*Assommoir, L'* (Zola), 18
*Ausflug* culture, 69–72
"Autopia" (Banham), 148, 158, 178
Avant-garde urbanism, 148, 171, 209–210
Avila, Eric, 283 (n. 56)

Baca, Judith, 169
Bachelard, Gaston, 1
Bacon, Edmund, 28
"Bacteriological city," 9, 81, 142, 221. *See also* Disease
Ballard, J. G., 20, 169, 185–186, 211
Bangkok, 5, 187
Banham, Reyner, 145, 147, 148, 158
Barry, Andrew, 12
Bartholomew, Harlan, 152
Bartning, Otto, 61
Batavia (Jakarta), 89, 91
*Bathers at Asnières* (Seurat), *46*
Bathhouses, 13, 43, 64, 66
Bathing, 13, 43, 64, 71, 244 (n. 55)
Bathroom, 13, 43–44, 73
Baudelaire, Charles, 32, 54
Bauhaus, 59
Baviskar, Amita, 17
Becher, Bernd and Hilla, 20
Beck, Ulrich, 220
Beijing, 187
Belgrand, Eugène, 27, 32, 38, *39*
Bellamy, Edward, 61, 209

Belle Époque, 48
Benegal, Dev, 125
Bengaluru (Bangalore), 111, 141
Benjamin, Walter, 32, 58
Berkeley School of Geography, 148
Berlin, 4, 9, 24, 55–79
    Berliner Freibäder-Verein (Berlin Open Air Bathing Club), 70
    *Berliner Wassertisch*, 16
    boundary extension, 60
    city building exhibition of 1910, 62
    Friedrichshain, 57
    Grünewald, 65, 69–70
    Hobrecht plan, 60
    Köpenick, 60
    Kreuzberg, 9, 57
    late urbanization, 251 (n. 10)
    as modernist utopia, 56
    "Das neue Berlin" (Wagner), 61
    Schlachtensee, 66
    Spandau, 60
    Steglitz, 57, 60
    Wannsee, 59, 66, *67*, *68*, 255 (n. 49)
    Wedding, 57
    Zehlendorf, 57
Berlioz, Hector, 32
Bertsch, Adolphe, 31
Beveridge, Ross, 233 (n. 54)
Biodiversity, 17
    urban, 161, 217
Biopolitics, 66, 220
Bishan-Ang Mo Kio Park, 218
Blum, Albrecht Viktor, 58
Bombay. *See* Mumbai
Bondi, Hermann, 189
Boorman, John, 174, *175*
Böß, Gustav, 60
"Boundary aesthetic," 49
"Bourgeois environmentalism," 9, 124, 143
Boyle, Danny, 187
Bradley, Tom, 178
Brenner, Neil, 232 (n. 41)
Brisbane, 187
Bruce-Chwatt, Leonard, 96, *97*, 102
Buenos Aires, 134
Buhari, Muhammadu, 108
Burnham, Daniel, 180
Burtynsky, Edward, 21
Busquet, Raoul, 53

Caillebotte, Gustave, 18
California Native Plants Society, 177
Campkin, Ben, 200
Capital, 24, 37, 215, 221, 223, 224
    "capital connections" (Wagner), 61
    circulation of capital, 136
    "fictitious capital," 224
    fixed forms of capital, 224
    global banking and finance, 214
Capitalist urbanization, 5, 29, 37, 49, 50, 79, 124, 143
Carné, Marcel, 76
"Caste-based apartheid," 114
"Catastrophe bonds," 219
Chamberlain, Joseph, 87
Chatterjee, Partha, 132, 138
Chennai, 111
*Children of Men* (Cuarón), 187
China, 116
*Chinatown* (Polanski), 182
"City of buckets," 6
Civil engineering, 16, 95, 110, 182. *See also* Engineered metropolis
Clajus, Hermann, 78
Clark, T. J., 51
Climate change, 110, 188, 213, 219, 220, 224
Cochabamba, 134
Cold War, 10
Collective memory, 25, 169

Colonialism, 4, 9, 82, 86–96, 104, 114–121, 138, 219
Comte, Auguste, 49
Concrete, 173
"Concrete jungle," 173
Condé, Maryse, 87
Conrad, Joseph, 201
Corbin, Alain, 13, 44
Corporeal metaphors, 10, 48–49
Corporeal politics, 12, 135, 219. *See also* Miasmic politics
Corruption, 79, 100, 127
Counterdystopian architectonic projections, 211
Counter-Reformation, 43
*Créolité*, 87
Crespi, Juan, 147
Cuarón, Alfonso, 187
Cuba, 108
Cyborg urbanization, 11

Da Cunha, Dilip, 142
Dakar, 134
Dams
    Aswan (Egypt), 11
    Garrison (US), 10
    Kakhovka (Ukraine), 10
    Modak Sagar (India), 119, *120*
    Narmada (India), 122
D'Arnaud, Camille, 31
Davis, Mike, 149
*Day of the Triffids, The* (Wyndham), 186
DDT, 84, 104
De Bellaigue, Christopher, 137
Debussy, Claude, 32
*Déjeuner sur l'herbe, Le* (Manet), 73
De Kaufmann, Richard, 53
Delhi, 111, 134, 219
Demeritt, David, 203
Democratic deliberation, 16, 179, 192
De Sequeira, Ruy, 83

Detroit, 21
Dhaka, 114
Dichotomous metaphors, 49
Dichotomous olfactory universe, 48
Dickens, Charles, 18, 54, 201
Dikeç, Mustafa, 222
Disease
    bubonic plague, 13, 94, 118
    cholera, 13, 36, 42, 84
    dengue fever, 5, 84
    dysentery, 84
    encephalitis, 5, 84
    lymphatic filariasis, 102
    malaria, 25, 81–108, 135
    tuberculosis, 62, 85
    typhoid, 42
    yellow fever, 102
Döblin, Alfred, 57
Doolin, James, 168, *170*
Doré, Gustave, 32
D'Ormesson, Wladimir, 53
Doyle, Richard, 187
Dreiseitl, Herbert, 218
Dresden, 219
Drew, Jane, 101
"Drilling tower forests" (*Bohrturmwälder*) (Wagner), 148–149
*Drowned World, The* (Ballard), 185
Dudow, Slatan, 58, 259 (n. 84)
Dürer, Albrecht, 13, *15*

East Asian economic crisis (1998), 5
Ecology, 5, 222
    "cosmopolitan ecologies," 17
    ecological critique of modernity, 208
    ecological design, 207
    ecological footprint, 223
    "ecological frontier," 11, 122, 222
    "ecological gentrification," 17, 178, 183
    ecological imaginaries, 169, 177, 218

"ecological irrigation," 164
ecological restoration, 17, 180, 181, 182, 207
"ecological simulacrum," 182
"ecological urban citizenship" (Wolch), 149, 218
"ecological zone" (in malaria control), 90
"human ecology" (Banham), 148
"microecologies" (Stoler), 107
Eisner, Lotte, 259 (n. 84)
Ekwensi, Cyprian, 99
Elias, Norbert, 12
El Niño, 182
Engels, Friedrich, 54
Engineered metropolis, 213. See also Civil engineering
"Enterprise state" (Mahadevia), 131
Environmental risk, 220, 222. See also "Catastrophe bonds"
Environmental security, 207
Epidemiology, 2, 5, 81, 99, 104, 107, 218. See also Disease
Ercolano, Greg, 171
Ermisch, Richard, 65
Essex, 189, 200
Expressionism, 56

Fairs, Marcus, 211
Fascism, 96
Feenberg, Andrew, 23
Feld, Hans, 75, 77
Feminist theory, 50
Fénéon, Félix, 45
Ferguson, James, 223
Fiscal conservatism, 9
Fletcher, David, 171
*Fließräume* (Heidenreich), 13
*Flood* (Mitchell), 187
Flood insurance, 160, 219

Floodplain restoration, 17, 25, 199, 217. See also Ecology: ecological restoration
Floods, 17, 83, 110, 118, 177, 181–182, 187, 219
Fogelson, Robert, 147
Fontane, Theodor, 69
Forbát, Fred, 61
Fordism, 20
Forty, Adrian, 173
Foster, Norman, 202
Foucault, Michel, 14
"Four ecologies" (Banham), 148
FOVICKS (Friends of Vast Industrial Concrete Kafkaesque Structures), 171
Franco-Prussian War, 53
Frankfurt, 72
*Frauenbad, Das* (Dürer), 13, *15*
Freetown, 87, 88, 264 (n. 39)
Freud, Sigmund, 49
Fry, Maxwell, 101

Gandhi, Indira, 139
Gandhi, Rajiv, 141
"Geometrization of the masses," 72
German People's Park Association, 62
Giedion, Sigfried, 8
Gilroy, Alan, 91
Gissen, David, 210
Glancey, Jonathan, 188, 200
Glasgow, 9
Godwin, William, 53
Gourguet, Jean, 76
Governmentality, 82, 96, 142, 222
*Graceland* (Abani), 108
Graham, Stephen, 8
Gramsci, Antonio, 14
Graves, Robert, 210
Great Depression, 155
*Great Expectations* (Dickens), 201

Gropius, Walter, 61, 78
Gumprecht, Blake, 157

Habermas, Jürgen, 23
Hambourg, Maria, 33
Hamburg, 189
Häring, Hugo, 61
Harris, Andrew, 8
Harriss-White, Barbara, 276 (n. 77)
Hartlaub, Gustav Friedrich, 58
Haussmann, Georges-Eugène, 27, 36–37, 52–53, 221
Havana, 91
*Heart of Darkness* (Conrad), 201
Hegemann, Werner, 58
Hegemony (Gramsci), 14
Heidenreich, Elisabeth, 13
Heliotherapy, 62
Henning, Paul Rudolf, 61
Hessel, Franz, 72
Hilferding, Rudolf, 79
Hindu mythology, 114
Hiorns, Frederick, 27
"Historiographies of absence," 24, 96–100
Höch, Hannah, 57
Hoffmann, Ludwig, 64
Honduras, 83
Hopper, Edward, 169
"Horizontal cities," 6
Howard, Ebenezer, 62
Hugo, Victor, 18, 28, 33
Hunter, William, 118
Hybridization, 4
Hyderabad, 111, 141
Hygiene, 10, 43
    hygienism, 47, 64

Impressionism, late, 69, 168
India, 82, 86, 182

"Indirect rule," 96
Infant mortality, 85
Infrastructure
    bonds, 133, 214, 275 (n. 62)
    contamination, 6, 129
    failure, 6, 129
    "green," 204
    maintenance, 6, 129
    modernization, 43, 110
    origins of term, 2–3, 225–226 (n. 5)
    unseen networks, 14, 28, 49, 110, 210
Interdisciplinarity, 2, 224
Interstitial spaces, 18
*Invisible Chapters* (Nwosu), 108
*Irrungen, Wirrungen* (Fontane), 69
Italian futurism, 148
Italy, 82, 96, 108, 111

Jakarta, 5, 134, 187
James, P. D., 187
Jameson, Fredric, 213
Jansen, Hermann, 65
Java, 89
Jefferies, Richard, 186
Jelliffe, Derrick B., 85
Jhering, Herbert, 77
Jos, 87
Joyce, James, 20, 57

Kaduna, 87
Kahn, Fritz, 10
Kaviraj, Sudipta, 121
Kazan, Elia, 3
Keményi, Alfréd, 77
Kendrick, Mary, 191
Kent, 189, 200–201, 207
Kirchner, Ludwig, 75
Koch, Robert, 66, 89
Kolkata (Calcutta), 5, 111
Kracauer, Siegfried, 58, 72, 77
Kraszna-Krausz, Andor, 55, 76

Lagos, 4, 24, 25, 81–108, 114, 207
    Ajegunle, *7*, 84
    Amukoko, *106*
    Apapa, 84, 90, 94
    colonial origins, 83
    "disease topography," 89
    draining swamps, 88–96
    Isale-Eko, 87
    Kokomaiko swamp, 89
    Lagos Executive Development Board, 94
    Lagos Island, 84
    Lagos Ladies' League, 89
    Lekki Peninsula, 84
    as "Liverpool of West Africa," 83
    Oshodi, *105*
    resurgence of malaria in 1990s, 268 (n. 79)
    Victoria Island, 84
Landscape
    "anxious landscapes" (Picon), 223
    concrete, 168, 173, 181
    cultural, 70, 204
    engineered, 176
    of epidemiological uncertainty, 218
    "facade landscapes" (Wagner), 164
    landscape design, 62, 206, 217–218
    "landscapes of disaster" (Benegal), 125
    landscape restoration, 182 (*see also* Ecology: ecological restoration)
    marginalized, 174
    origins of term, 3
    quotidian, 171
    riparian, 173, 178, 180, 217–218
    *Stadtlandschaft* (Wagner), 167
    swale-type, 161, 164
    technological, 149
    utilitarian, 164, 168
Lang, Fritz, 76
Lefebvre, Henri, 11
Legouvé, Ernest, 49

Leistikow, Walter, 69
Lenné, Peter Josef, 62
Lesser, Ludwig, 62
"Leveling pressure" (Beck), 220
Leyden, Friedrich, 57, *63*
London, 25, 57, 117, 185–215
    *After London* (Jefferies), 186
    All London Green Grid, 204
    Charlton, 191
    Deptford, 200
    *Flooded London* (Squint/Opera 2), 210–211, *212*
    flood protection system, 195, 213, 292 (n. 36)
    floods and flood risk, 189, *190*, 195, 207
    future threat of malaria, 290 (n. 28)
    Greater London Authority, 192, 196
    Greater London Council, 191
    Heathrow airport, 202
    Herbert Commission, 192
    and history of Thames, 188–189, 288 (n. 9)
    Honor Oak, 211, *212*
    Lea Valley, 200
    Lewisham, 217
    London County Council, 192
    *London Plan*, 204
    Metropolitan Board of Works, 192
    *Postcards from the Future* (Graves and Madoc-Jones), 210
    premodern floodplain, 208
    rising groundwater, 195
    Silvertown, 191
    Teddington, 189, 196
    Thames Barrier, 191–192, *193*, *194*, 196, *197*, 199
    Thames estuary, 199–200, 204, 207
    Thames Gateway, 198
    Woolwich, 200

*Longue durée*, 8
Loos, Adolf, 59
Los Angeles, 23, 145–183
    Burbank, 146
    Canoga Park, 146
    Carson, 161, *162*, *165*, *166*
    cinematic representations, 285 (n. 74)
    Cornfield, 178
    Culver City, 148
    Dominguez Gap Wetlands Project, 161, *163*
    East Los Angeles, 167
    flash floods, 146, 155
    freeway system, 147, 148, 167, 176
    Friends of the Los Angeles River (FoLAR), 177, 179
    Glendale, 146
    Glendale Narrows, *172*
    *Great Wall of Los Angeles, The* (Baca), 169
    Griffith Park, 174
    Hollywood, 78, 148
    industrial-riparian zone, 156
    *LA Plays Itself* (Andersen), 174
    Long Beach, 146
    Los Angeles and San Gabriel Rivers Conservancy, 179
    Los Angeles Aqueduct, 151
    Los Angeles Flood Control District, 155, 160, 179
    Los Angeles River Master Plan, 158, 164
    Los Angeles River Revitalization Master Plan, 180
    Los Angeles River Task Force, 178
    Macy Street Bridge, *154*
    Mothers of East Los Angeles (MELA), 176, 179
    Olmsted-Bartholomew Plan, 152
    Ord Plan, *153*
    and "postmodern urbanism," 23
    and Proposition 13, 179
    Pueblo de los Angeles, 150
    San Bernardino, 151
    Santa Monica, 151
    Tujunga Wash, 169
    Tujunga Wash Greenway, 164
Lovelock, James, 210
Lubitsch, Ernst, 76
Lugard, Frederick, 87
Lynch, Kevin, 223

MacGregor, William, 86, 101
Mächler, Martin, 62
Macintosh, Kate, 209
Madoc-Jones, Didier, 210
Mahadevia, Darshini, 131
Manet, Edouard, 73
Mangrove swamps, 83, 111
Manila, 134, 187
Marvin, Simon, 8
Marx, Karl, 209
Mathur, Anuradha, 142
May, Ernst, 72
Mehta, Suketu, 114
Meissen, 219
Mendelsohn, Erich, 56
*Menschen am Sonntag* (Siodmak and Ulmer), 72–78, *74*
Menzel, Adolph, 18, *19*, 69
Metaphors of spatial disorder, 47, 248 (n. 74)
Meunier, Constantin, 168
Mexican-American War, 151
Miasmic politics, 9, 108
Michelet, Jules, 49
Middle class, 9, 33, 37, 49, 51, 53, 77, 99, 108, 111, 134–135, 141, 143, 186, 222
Mies van der Rohe, Ludwig, 4, 101
*Mietskaserne* (rental barracks), 58, 60, 64

Migge, Leberecht, 61, 62
Mille, A. A., 42
Mitchell, Timothy, 11, 102
Mitchell, Tony, 187
"Modern infrastructural ideal" (Graham and Marvin), 8
Modernism, 4, 78
    architectural, 57, 59, 142, 173
    technological, 192, 221
    tropical modernism, 4, 101
    in urban design, 64, 70
Modernity, 2–4, 11, 117
    as contradictory amalgam, 82, 138
    "countervalent forces of modernity" (Avila), 177
    "decomposing modernity" (Ferguson), 223
    defined against nature, 14
    and drainage of wetlands, 207
    fractured modernities, 101
    functional, 124
    "imitative modernity" (Kaviraj), 121
    late modernity, 223
    literary representations, 20
    multiple modernities, 223
    premodern conditions, 52
    "reflexive modernity" (Beck), 220
    science-led, 121
    sensory aspects, 44
    technology and, 147
    "uneven modernity," 52
    utopian visions, 109, 211
Modernization, 3–4, 24, 43
Moholy-Nagy, László, 58, 75
Möller, Otto, 57
Monet, Claude, 169
"Money economy" (Simmel), 58
Monzet, André, 53
Morley, 9
Morris, William, 61, 209

Moses, Robert, 66
"Mosquito coast," 83
Mosquitoes, 5, 81–108, 218
Mouffe, Chantal, 222
Mulholland, William, 152, 181
Mumbai (Bombay), 10, 25, 109–143, 187
    Andheri, 125, 127, 133
    Bhandup water treatment works, 121, *123*
    Bhayandar, 125
    Bombay First, 140
    Bombay Improvement Trust, 119
    *Bombay Our City* (Patwardhan), 139
    Borivali, 125
    Chembur, 125, 127
    colonial, 24, 114–119, 271 (nn. 16, 21)
    Dharavi, 140
    Dhobi Ghat, 18
    extent of slums, 110, 270 (n. 4)
    Ghatkopar, 127
    Goregaon, 127
    Juhu, 110
    Kandivali, 127
    Mahim Creek, *126*, 140
    Malabar Hill, 110, 111, *115*
    manufacturing decline, 136–137
    Mira Road, 125
    monsoon of 2005, 110
    Mumbai Metropolitan Region Development Authority, 138–139
    Shiv Sena, 127, 137
    Society for the Promotion of Area Resource Centres (SPARC), 130
    Worli, 110
Municipal managerialism, 221
Munkacsy, Martin, 58
Murillo, L. A., 160
Murnau, F. W., 56, 76

Nabokov, Vladimir, 57
Nadar, Félix, vii, 24, 27, 30–36, *34*, *35*
Nagarkar, Kiran, 109
Napoléon III, 27, 31, 36
National Slum Dwellers Federation (India), 130–131
Nature
    agency of, 5, 84, 102, 107
    conservation of, 65
    cultures of, 56, 218
    essentialist conceptions, 1
    gender and, 1, 75
    ideological aspects, 48, 138
    metaphors drawn from, 149
    metropolitan, 23, 25, 30, 56, 59, 218
    premodern, 114, 207–208
    "re-wilding," 17, 207
    "second nature," 124
    "transitory nature" (Vergara), 171
    urbanization of, 138
    "wild urban nature," 17
Nazism, 55, 78
Nehru, Jawaharlal, 121
Neocolonialism, 104
Neo-Haussmannite planning, 108, 140–141, 142, 221
Neo-Hippocratic doctrines, 52
Neo-Hobbesianism, 185, 211
Neoliberalism, 9, 16, 213, 214
Neo-organicism, 11
Neorealism, 75
Neoromanticism, 69, 208
Neo-Weberianism, 61
Netherlands, 187, 189, 219
New Deal, 157, 182, 282 (n. 38)
New Guinea, 89
"New narcissism" (Corbin), 44
"New objectivity" (*Neue Sachlichkeit*), 58, 75
New Orleans, 186, 187

*News from Nowhere* (Morris), 209
New York, 17, 57, 58, 66, 111, 187
Nicaragua, 83
Niépce, Nicéphore, 31
Nigeria, 87, 90, 91, 96, 99, 102, 104
Njoku, Amby, 81, 108
*No Longer at Ease* (Achebe), 99
Nwosu, Maik, 108

Olfactory realm, 13
"Olfactory revolution" (Corbin), 44, 50
Olfactory topography, 47
Olmsted, Frederick Law, Jr., 152
Oosterschelderkering, 195
"Open space politics" (*Freiflächenpolitik*), 59, 62
"Operational landscapes" (Brenner), 11
"Orchestra of specialists" (Wagner and Gropius), 78, 221
Ord, Edward, 152
"Organic economy," 45
Organic metaphors, 10, 52, 223
"Organized capitalism" (Hilferding), 79

Pabst, Georg Wilhelm, 56
Panama Canal, 83
Parent-Duchâtelet, Alexandre, 47, 246 (n. 66)
Paris, 18, 37–54, 109, 117
    Asnières, 45
    *banlieue*, 51
    Boncy, 41
    "capital of the nineteenth century," 23
    catacombs, 32
    Clichy, 42
    Exposition Universelle (1867), 33
    Gennevilliers, 42, 45
    Montfaucon, 41
    Saint-Denis, 45

    Second Empire reconstruction, 27, 36–43
    sewers, 27–54, 240 (n. 36)
Pasteur, Louis, 42
Patwardhan, Anand, 139
Pawar, Sharad, 141
Pechstein, Max, 75
*People of the City* (Ekwensi), 99
Pettenkofer, Max von, 66
Philipon, Charles, 31
Phoenix, 21
Photography, 20–21, 30–36, 58, 168, 171
Picon, Antoine, 223
Pinkney, David, 36
Piranesi, Giovanni Battista, 33
"Planetary urbanization" (Brenner and Schmid), 11
Pliny the Elder (Gaius Plinius Secundus), 53
Poelzig, Hans, 255 (n. 49)
*Point Blank* (Boorman), 174, *175*
Polanski, Roman, 182
Political economy, 5
"Politics of inevitability" (Beveridge), 16
"Politics of light and air" (Wagner), 62
Politics of "spatial elimination," 140
Pollution control, 45
Porter, James, 203
Porter, Roy, 215
Postcolonial cities, 6, 9, 10, 221–222
Postcolonial state, 102, 121, 136
Postcolonial theory, 99, 223
Postimpressionism, 45
Postmodernism, 4
"Postpolitical condition," 222
Power, 14, 125, 126, 136, 182, 221
Prague, 187
Prakash, Gyan, 99
Prashad, Vijay, 219
Privatization, 16, 133–134, 151, 192

Public health, 6, 8, 25, 62, 66, 82, 83, 86, 95, 108, 135
Public realm, 16, 25, 132, 213, 220, 221
Public space, 135, 217, 218

Racial segregation, 86–87, 94–95, 156
Racism, 78, 87, 96, 118, 167
    "scientific," 9, 88
Rainham Marshes, 203–204, *205*
Rancière, Jacques, 222
"Rationalization of happiness" (Wagner), 61
Rawlinson, Robert, 117
*Rear Courtyard and House* (Menzel), 18, *19*
Renaissance, 48
Residential segregation, 9, 51, 156
Reuter, Ernst, 66
Riis, Jacob, 58
Rivers
    Charles, 181
    Cheonggyecheon, 17
    Dhuys, 53
    Gargai, 122
    Hudson, 181
    Kallang, 218
    Los Angeles, 24, 25, 145–183
    Marne, 53
    Mississippi, 177
    Mithi, 125
    Pinjal, 122
    Ravensbourne, 206, 217
    Sacramento, 181
    Seine, 38
    Spree, 62
    Tennessee, 3
    Thames, 187
    Ulhas, 122
    Vaitarna, 122
    Vanne, 53

Rodríguez, Luis J., 146, 169
Rome, 37
Roraback, Dick, 145
Rorty, Richard, 23
Ross, Roland, 86
Rossellini, Roberto, 76, 259 (n. 82)
Rossi, Aldo, 79
Roth, Joseph, 72
Ruscha, Ed, 168
Ruskin, John, 209
Ruttman, Walter, 76

Saalman, Anthony, 37
"Saffron capitalism," 137
Saint-Simon, Henri de, 37, 239 (n.31)
Salgado, Sebastião, 21
Sand, George, 31
Sauer, Carl, 148
Scarpa, Ludovica, 65
Schama, Simon, 3
Schapiro, Meyer, 45
Scharoun, Hans, 57, 61, 192, 260 (n. 101)
Schmieder, Oskar, 148
Schüfftan, Eugen, 75, 76
Schütte-Lihotzky, Margarete, 254 (n. 33)
Science fiction, 220
"Scientific politics," 10
"Scientific realism," 45
"Scientific urbanism," 11, 221
Seeler, Moritz, 75
Semiarid urbanization, 222
Sengupta, Somini, 110
Seoul, 17
Seurat, Georges, 30, 45, *46*, 75
*Sewers, The* (Nadar), *34, 35*
Sewers
    colonial disruption of existing system, 219
    Lieurnur system, 42
    open, 5, 101, *106*, 107, 219
    as tourist attraction, 33
    *tout à l'égout*, 43, 242 (n. 45), 243 (nn. 49, 51)
Sexuality, 43, 48, 73
"Shadow state" (Harriss-White), 136
Shanghai, 141
Sierra Club, 177
Sierra Leone, 87
Sieweke, Jorg, 81
Simmel, Georg, 58
Singapore, 86, 141, 218
Siodmak, Kurt, 75
Siodmak, Robert, 72, *74*, 75, 78, 259 (n. 82)
Slave trade, 83
Slum clearance, 140, 167
Slums, 100, 110, 119, 135, 139
*Soak* (Mathur and da Cunha), 142
*South Corridor, Hoover Dam* (Jane and Louise Wilson), 21, *22*
Southern California Institute of Architecture, 160, 171
Spatial fetishism, 50
Spatial legibility, 223
"Spectacle of enlightenment," 33
Spinney, Laura, 187
Squint/Opera 2, 211, *212*
"Stage set cities" (*Kulissenstädte*) (Wagner), 148
Stallabrass, Julian, 21
State formation, 3, 8, 138, 214
*Steinerne Berlin, Das* (Hegemann), 58
Stoler, Ann Laura, 269 (n. 83)
Strategic planning. *See* Urban planning
"Structural adjustment programs," 102
*Study #4 for Connections* (Doolin), *170*
"Subalternity" (Prakash), 99
"Subpolitical realm" (Beck), 220

Subterranean realm, 210. *See also* Infrastructure; Sewers
Surat, 135
Swimming pools, 20, 168, 181
Swyngedouw, Erik, 222
Symbolism (in European art), 69

Tafuri, Manfredo, 79
Taut, Bruno, 56, 61
Taylorism, 79
"Technical ideology of metropolis" (Vidler), 29
Technological fixes, 12
Technological networks, 2, 11, 13–14, 23
Technological politics, 16
"Technological zone" (Barry), 12
Technomanagerialism, 130, 192, 217
"Technoscience," 3
Teige, Karel, 79
Tennessee Valley Authority, 3
Texier, Edmond, 51
Thane, 125, 128
Toepfer, Karl, 71
Tokyo, 57
"Tranquil utopia" (Fairs), 211
Tulloch, Hector, 117
*28 Days Later* (Boyle), 187

Ulmer, Edgar J., 72, *74*, 76
*Ulysses* (Joyce), 20, 57
"Unconventional hydrocarbons," 12
United Kingdom
    Environment Agency, 203
    Royal Commission on Population, 85
United States
    Army Corps of Engineers, 156–157, 177, 182
    Federal Emergency Management Agency (FEMA), 160
    Works Progress Administration (WPA), 156
Urban adaptation, 17, 25, 188
Urban agriculture, 17, 104
"Urban alimentary system," 54
Urban fragmentation, 147, 148, 215
Urban fringe, 10, 125, 127
Urban Gothicism, 50
"Urban imaginary," 141
Urban metabolism, 10–11
"Urban pastoral," 47
Urban periphery, 51
Urban planning, 8, 27, 36–43, 60–69, 78–79, 86–87, 94–95, 110, 118, 121, 138–139, 142, 147, 152, 158, 164, 179, 180, 192, 199, 203, 213, 221
"Urban political ecology," 5
Urban resilience, 17, 25, 214, 222, 298 (n. 95)
Urban sustainability, 209–210, 214, 222
"Urban uncanny," 49–50, 247–248 (nn. 72, 74)
Utopian socialism, 79
Utzon, Jørn, 192

Vergara, Camilo José, 171
"Vertical urbanism," 6
Vertov, Dziga, 75
Victorian liberalism, 116
Vidler, Anthony, 28–29
Vidor, King, 77
Viertel, Berthold, 76
Vigarello, Georges, 14
Virchow, Rudolf, 66
Visconti, Luchino, 76
Vitalism, 6

Wagner, Anton, 148–149, 150, 155, 164, 167

Wagner, Martin, 55, 60–69, 78–79, 260 (n. 98)
Waldie, D. J., 178
Waste
    cesspits and cesspools, 42, 44, 244 (n. 56)
    "cesspool city," 53
    human waste, 38, 44, 50
    night soil collection, 41, 44, 241 (n. 40), 243 (n. 47)
Water. *See also* Bathhouses; Bathing; Bathrooms; Dams; Floodplain restoration; Floods; Infrastructure; Landscape; Rivers; Sewers; Swimming pools
    conservation of, 130
    desalination, 12
    effects of drought, 118, 129
    elemental purity, 52
    gendered associations, 1, 75
    groundwater extraction, 127, 195
    "hydro-social cycle," 5
    "hydro subject," 221
    hydrotherapy, 71
    land drainage, 88–96, 188, 207
    leakage rates, 128
    luxury use, 43, 71, 130, 222
    as metabolic necessity, 224
    metaphorical aspects, 1, 48
    mythological dimensions, 75
    ornamental features, 48, 247 (n. 69)
    pilferage, 128
    "plumbed city," 6
    plumbing, 73, 245 (n. 58)
    premodern circulation, 52, 114, 219
    *qanat* technologies, 219
    rainwater harvesting, 130, 219
    role in landscape design, 217–218 (*see also* Landscape)
    rural water crisis, 10, 128
    as speculative chimera, 224
    tankers, 6, 122
    vendors, 6
    wastewater treatment, 17, 121, 287 (n. 102)
    "water mafia," 125
    "water revolution," 14
    watershed protection, 17
Weibel, Peter, 209
Weimar Republic, 55
Whistler, James, 169
Wiene, Robert, 56
Wilder, Billy, 72, 78
*Wild River* (Kazan), 3
Williams, Raymond, 36, 96
Wilson, Harold, 189
Wilson, Jane and Louise, 21, *22*
Wolch, Jennifer, 149, 218
Wolff-Plottegg, Manfred, 209
Women
    dichotomous cultural representations, 48
    relegation to realm of nature, 48–49, 75
Working class, 18, 51, 58, 64, 66, 136, 156, 161, 167, 176–177
World Bank, 133
World Health Organization, 102, 104
Worpole, Ken, 204
Wyndham, John, 186, 187

Yefremov, Ivan, 211

*Zabriskie Point* (Antonioni), 174
*Zanjas*, 150–151, 178
Zero carbon cities, 215
Žižek, Slavoj, 222
Zola, Émile, 18
*Zopadpatti*, 110
Zuckerman, Solly, 189